THE GEOLOGY OF CORNWALL

Cornwall is renowned for its spectacular rocky coastline and the diversity and complexity of its geology. This geology, and its particular relation to the mineral wealth of the county, has been the subject of continuing investigation since the end of the seventeenth century.

This book reviews the importance of Cornwall in the development of geological thought in Britain, and highlights the impact of new concepts and methodologies that have recently revolutionised earth studies. Leading academic and commercial geologists present a wide-ranging review of the current position and assess the environmental consequences of rock and mineral exploitation.

The Geology of Cornwall will be of interest to all those, whether amateur or professional, who wish to explore this fascinating subject. Fieldworkers will find the book particularly helpful: the book includes a listing of sites of notable scientific interest.

The Geology of Cornwall is a companion volume to *The Geology of Devon*, also published by University of Exeter Press.

The Editors: E.B. SELWOOD is Senior Lecturer in Earth Studies, University of Exeter. E.M. DURRANCE is Emeritus Professor of Geology, University of Nebraska. C.M. BRISTOW is Visiting Professor in Industrial Geology, Camborne School of Mines

THE GEOLOGY OF CORNWALL

AND THE ISLES OF SCILLY

Edited by
E.B. Selwood
E.M. Durrance
and
C.M. Bristow

UNIVERSITY
of
EXETER
PRESS

First published in 1998 by
University of Exeter Press
Reed Hall, Streatham Drive
Exeter, EX4 4QR
UK
www.exeterpress.co.uk

Hardback ISBN 978 0 85989 529 3
Paperback ISBN 978 0 85989 432 6

Reprinted 1999, 2007, 2013
Printed digitally since 2022

British Library Cataloguing in Publication Data
A catalogue record of this book is available
from the British Library

The University of Exeter gratefully acknowledges the financial
support of the following organizations:

ECC International Limited
Riofinex North Limited
Goonvean and Rostowrack China Clay Company Limited
Redland Minerals Limited

Typeset in 10 on 12 Times New Roman

Dedicated to the memory of

Dr E.M.L. HENDRIKS

A pioneer in the modern interpretation
of the Geology of Cornwall

Contents

Plates

(Located between xvi and xvii)

The cover shows Crowns Engine Houses, Botallack [SW 3623 3352]

Crowns Engine (*front*): a 36-inch diameter engine pumped water from the Crowns section of Botallack Mine. Pearce's Whim (*back*): a 24-inch diameter engine hauled men and ore from the Boscawen Diagonal Shaft. The buildings were renovated by the Carn Brea Mining Society in 1984–85. (*Photograph © A.J.J. Goode*).

Figures

Tables

Contributors

J.R. ANDREWS
Department of Geology, Southampton Oceanography Centre, University of Southampton

K. ATKINSON
Camborne School of Mines, University of Exeter

C.M. BRISTOW
Camborne School of Mines, University of Exeter

R. BURT
Department of Economic and Social History, University of Exeter

R.A. CULLINGFORD
Department of Geography, University of Exeter

E.M. DURRANCE
Earth Resources Centre, University of Exeter

R.P. EDWARDS
Camborne School of Mines, University of Exeter

P. GRAINGER
School of Engineering, University of Exeter

M.B. HART
Department of Geological Sciences, University of Plymouth

M.J. HEATH
Earth Resources Centre, University of Exeter

K.P. ISAAC
Enterprise Oil plc.

F.W.A.A. LUCAS
Loeb Aron & Co Ltd

D.A.C. MANNING
Department of Geology, University of Manchester

D. ROBINSON
Department of Geology, University of Bristol

R.C. SCRIVENER
British Geological Survey

E.B. SELWOOD
Earth Resources Centre, University of Exeter

R.K. SHAIL
Camborne School of Mines, University of Exeter

T.J. SHEPHERD
British Geological Survey

J.M. THOMAS
Department of Geography, University of Exeter

J. WILLIS-RICHARDS
Loeb Aron & Co Ltd

Abbreviations

Elements

Ag	silver	Fe	iron	Po	polonium
Al	aluminium	Ga	gallium	Ra	radium
Ar	argon	H	hydrogen	Rb	rubidium
As	arsenic	K	potassium	Rn	radon
B	boron	Li	lithium	Si	silicon
Ba	barium	Mg	magnesium	Sm	samarium
Bi	bismuth	Mn	manganese	Sn	tin
Be	beryllium	N	nitrogen	Sr	strontium
C	carbon	Na	sodium	Th	thorium
Ca	calcium	Nb	niobium	Ti	titanium
Cd	cadmium	Nd	neodymium	U	uranium
Cl	chlorine	Ni	nickel	V	vanadium
Cr	chromium	O	oxygen	W	tungsten
Cs	caesium	P	phosphorus	Zn	zinc
Cu	copper	Pb	lead	Zr	zirconium
F	fluorine				

Chemical compounds

B_2O_3	boric oxide	H_2O	water
$CaCl_2$	calcium chloride	KCl	potassium chloride
$CaCO_3$	calcium carbonate	K_2O	potassium monoxide
CaO	calcium oxide	Li_2O	lithium monoxide
CH_4	methane	NaCl	sodium chloride (common salt)
CO_2	carbon dioxide	Na_2O	sodium monoxide
HCl	hydrochloric acid	SiO_2	silicon dioxide (silica)

Miscellaneous

BGS	British Geological Survey	LREE	Light rare earth elements
BP	Before present	μm	micron, one millionth of a metre
BIRPS	British Institutions Reflection Profiling Syndicate	Ma	Million years
D1, D2...	Successive phases of deformation	Moho	Mohorovicic discontinuity
DOE	Department of the Environment	MORB	Mid ocean ridge basalt
DSDP	Deep Sea Drilling Project	NRA	National Rivers Authority (now part of the Environment Agency)
ECORS	Etude de la Croûte Continentale et Océanique par Réflexion et Réfraction Sismique	NRPB	National Radiological Protection Board
F1, F2...	Successive phases of folding	OD	Ordnance datum
HREE	Heavy rare earth elements	REE	Rare earth elements
IC	Illite crystallinity	S1, S2 ...	Successive phases of slaty cleavage formation
IGS	Institute of Geological Sciences	sd	Standard deviation
IPOD	International Programme of Ocean Drilling	SPL	Start – Perranporth Line
ka	Thousand years	TDS	Total dissolved solids
kbar	Kilobar	WR	Whole rock
		XRD	X-ray diffraction

Plate 1. Pillow lavas at Nare Head in Roseland [SW 918 370] (*Photograph © C.M. Bristow*)

Plate 2. Small duplex structure developed on the south limb of an antiform at Sandymouth [SS 201 100] (*Photograph © J.R. Andrews*)

Plate 3. Lizard Complex (mainly serpentinite), Kynance Cove [SW 685 133] (*Photograph © C.M. Bristow*)

Plate 4. Granite cupola with associated veining, Porthmeor Cove [SW 4254 3764] (*Photograph © A.J.J. Goode*)

Plate 5. Granite seascape. Armed Knight and Longships, Land's End [SW 345 246] (*Photograph © A.J.J. Goode*)

Plate 6. Steep bedding in Bude Sandstone Formation, Compass Point [SS 2003 0635], Bude (*Photograph © J. Saunders*)

Plate 7. Secondary copper mineralization in a stope, Kenneggy Mine (part of Wheal Speed) [SW 565 285] (*Photograph © A.J.J. Goode*)

Plate 8a. Arsenic calciner, Botallack Mine [SW 3641 3322] (*Photograph © A.J.J. Goode*)
Arsenic ores were roasted in the Brunton calciner and the resultant arsenical fumes were condensed in a 'labyrinth'

Plate 8b. Dressing floors, Wheal Basset, Vanner House [SW 6910 3990] (*Photograph © A.J.J. Goode*).
Lighter fractions of the crushed ores were washed away here, leaving heavy concentrate

Preface

Geological Field Work

It has often been remarked that geology is a subject best studied by actually looking at the rocks, minerals and fossils, and their structures and relationships, in the field. Therefore, although this book mainly deals with descriptions from an interpretative viewpoint, an Appendix has been added listing geological and geomorphological conservation sites where field relationships may be observed.

Geological field work is an interesting and rewarding endeavour, but it should be borne in mind that in many cases rock exposures are situated on private land, and may be in hazardous localities or be hazardous themselves. The inclusion of a locality in this book, even where specific mention of the need for access permission or the presence of hazardous conditions is not made, does not imply that access is freely available and the localities are safe: it is the responsibility of the user of this book to obtain permission to enter private land and to avoid hazards of all kinds.

To help establish an acceptable framework of conduct, the Geologists' Association has published a Code for Geological Field Work which is reproduced (with permission) overleaf. It is hoped that everyone will follow all recommendations given in the Code.

January 1998

E.B. Selwood
E.M. Durrance
C.M. Bristow

A Code for Geological Field Work

A geological code of conduct has become essential if opportunities for field work in the future are to be preserved. The rapid increase in field studies in recent years has tended to concentrate attention upon a limited number of localities, so that sheer collecting pressure is destroying the scientific value of irreplaceable sites. At the same time, the volume of field work is causing concern to many site owners. Geologists must be seen to use the countryside with responsibility; to achieve this, the following general points should be observed.

1. Obey the Country Code, and observe local byelaws. Remember to shut gates and leave no litter.
2. Always seek prior permission before entering private land.
3. Don't interfere with machinery.
4. Don't litter fields or roads with rock fragments which might cause injury to livestock, or be a hazard to pedestrians or vehicles.
5. Avoid undue disturbance to wildlife. Plants and animals may inadvertently be displaced or destroyed by careless actions.
6. On coastal sections, consult the local Coastguard Service whenever possible, to learn of local hazards such as unstable cliffs, or tides which might jeopardise excursions possible at other times.
7. When working in mountainous or remote areas, follow the advice given in the pamphlet 'Mountain Safety', issued by the Central Council for Physical Education, and, in particular, inform someone of your intended route.
8. When exploring underground, be sure you have the proper equipment, and the necessary experience. Never go alone. Report to someone your departure, location, estimated time underground, and your actual return.
9. Don't take risks on insecure cliffs or rock faces. Take care not to dislodge rock, since other people may be below.
10. Be considerate. By your actions in collecting, do not render an exposure untidy or dangerous for those who follow you.

Collecting and Field Parties

1. Students should be encouraged to observe and record but not hammer indiscriminately.
2. Keep collecting to a minimum. Avoid removing *in situ* fossils, rocks or minerals unless they are genuinely needed for serious study.
3. For teaching, the use of replicas is commended. The collecting of actual specimens should be restricted to those localities where there is a plentiful supply, or to scree, fallen blocks and waste tips.

4. Never collect from walls or buildings. Take care not to undermine fences, walls bridges or other structures.
5. The leader of a field party is asked to ensure that the spirit of this Code is fulfilled, and to remind his party of the need for care and consideration at all times. He should remember that his supervisory role is of prime importance. He must be supported by adequate assistance in the field. This is particularly important on coastal sections, or over difficult terrain, where there might be a tendency for parties to become dispersed.

Health and Safety at Work Act

Since the introduction of this Act, safety measures are more strictly enforced on sites including quarries. Protective clothing, particularly safety helmets must be worn by employees, so visitors are expected to observe the same precaution, often as a condition of entry. Suitable helmets are readily available, cheap to purchase, and should be part of the necessary equipment of all geologists. They must be worn at all times in quarries.

Visiting Quarries

1. An individual, or the leader of the party, should have obtained prior permission to visit.
2. The leader of a party should have made himself familiar with the current state of a quarry. He should have consulted with the Manager as to where visitors may go, and what local hazards should be avoided.
3. On each visit, both arrival and departure must be reported.
4. In the quarry, the wearing of safety hats and stout boots is recommended.
5. Keep clear of vehicles and machinery.
6. Be sure that blast warning procedures are understood.
7. Beware of rock falls. Quarry faces may be highly dangerous and liable to collapse without warning.
8. Beware of sludge lagoons.

Research Workers

1. No research worker has the special right to 'dig out' any site.
2. Excavations should be back-filled where necessary to avoid hazard to men and animals and to protect vulnerable outcrops from casual collecting.
3. Don't disfigure rock surfaces with numbers or symbols in brightly coloured paint.
4. Ensure that your research material and notebook eventually become available to others by depositing them with an appropriate institution.
5. Take care that publication of details does not lead to the destruction of vulnerable exposures. In these cases, do not give the precise location of such sites, unless this is essential to scientific argument. The details of such localities could be deposited in a national data centre for Geology.

Societies, Schools and Universities

1. Foster an interest in geological sites and their wise conservation. Remember that much may be done by collective effort to help clean up overgrown sites (with permission of the owner), and in consultation with English Nature.
2. Create working groups for those amateurs who wish to do field work and collect, providing leadership to direct their studies.
3. Make contact with your local County Naturalists' Trust, Field Studies Centre or Natural History Society, to ensure that there is co-ordination in attempts to conserve geological sites and retain access to them.

Reprinted by permission of the Geologists' Association. Further copies may be obtained from the Geologists' Association, Burlington House, Piccadilly, London W1V 9AG.

Acknowledgements

The editors wish to thank Mr A.J.J. Goode (British Geological Survey), Dr J.R. Andrews (University of Southampton) and Mr J. Saunders (University of Exeter) for colour plates, and Mr T. Bacon (University of Exeter) for preparing the text figures. Assistance by the staff of the English Nature Cornwall Office in the preparation of the Appendix is gratefully acknowledged.

The editors and publisher have made every effort to trace original copyright holders of materials used in this book in order to obtain their permission. We would like to take this opportunity of making acknowledgement to any copyright holder that we have failed to contact. We are particularly grateful to the following for permission to reproduce and modify text figures from their publications:

British Geological Survey. Figures are reproduced by permission of the Director, British Geological Survey, © NERC. All rights reserved. Catalogues of BGS's maps, books and other publications are available on request from: Sales Desk, British Geological Survey, Kingsley Dunham Centre, Keyworth, Nottingham, NG12 5GG.

Elsevier Science-NL

The Geological Society of London

The Geologists' Association

Graham & Trotman Ltd (Kluwer Academic Publishers)

The Royal Society

Science Reviews (Science & Technology Letters)

The Ussher Society

Chapter One

Introduction

The spectacular coastline and the high moors of Cornwall feature some of the most attractive scenery in Britain. Although the areas of Cornwall and the Isles of Scilly only reach 3549.2 km^2 and 16.35 km^2 respectively, Cornwall alone has a coastline of 525 km. This affords magnificent exposures of the rocks and their geological structures. Together with data obtained from mines and quarries, these sections have enabled researchers to gain detailed knowledge of a remarkable geology (*Figure 1.1*). Inevitably the offshore geology (*Figure 1.2*) is less well known, but it contrasts markedly with that onshore. It should be remembered that the geological lines represented on these maps are based on lithology and only approximate to chronostratigraphic boundaries.

The present book reviews the rapid, recent developments in the geology of Cornwall, and attempts to explain the controversial interpretations that have arisen, in terms that can be understood by scientists not actively engaged in the region. Beginners in geology will find *Cornwall's Geology and Scenery* (Bristow, 1996) a useful introduction to this volume.

Geologically, Cornwall forms the western part of the Cornubian Massif that is overstepped in east Devon and parts of west Somerset by terrigenous sediments of Permo–Triassic age. The massif is composed of an unexposed Proterozoic basement, overlain by Palaeozoic sediments (killas), contemporaneous volcanic rocks, and minor igneous intrusions. All were deformed into a mountain chain during the Variscan orogeny between 350 and 290 Ma (million years ago), and intruded by granite plutons between 290 and 270 Ma. The buoyancy conferred by the rising granites contributed significantly to the overall uplift of the area.

Today, erosion has reduced the Cornish landscape to a gently sloping plateau, from which the granite masses rise as conspicuous uplands. The granite moors of St Austell, Carnmenellis and Land's End, reach 240–300 m above sea-level, while the larger mass of Bodmin Moor has a maximum height of 420 m. Notable granite tors, such as the Cheesewring, are found along the northern and eastern margins of Bodmin Moor, but elsewhere such majesty is rarely observed.

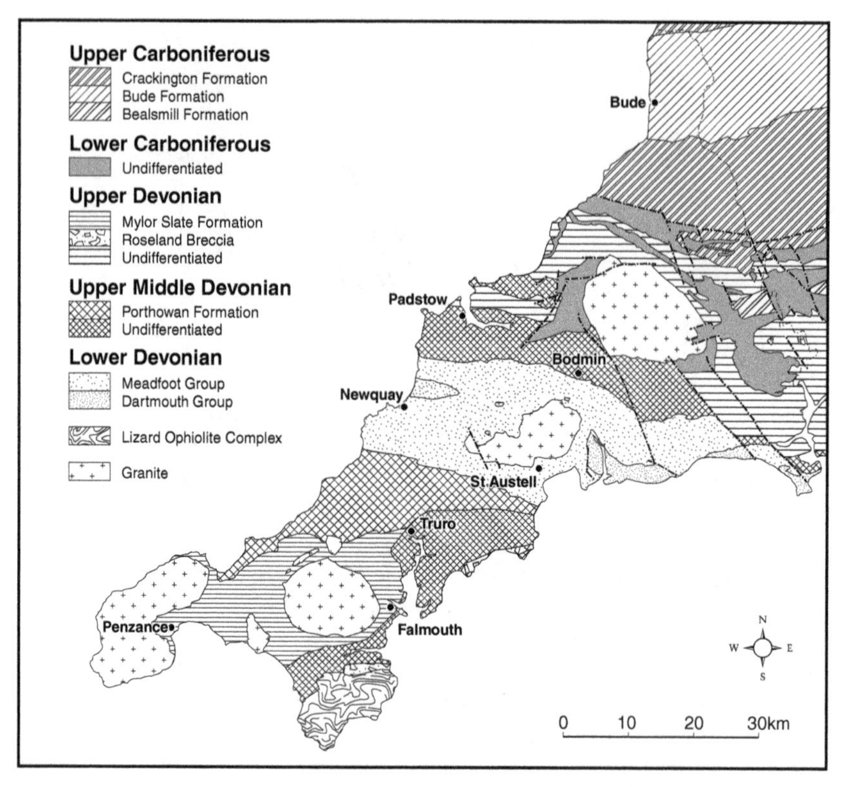

Figure 1.1. A simplified geological map of Cornwall. Mainly after BGS 1:250 000 Sheets 50°N–06°W (Land's End) and 49°N–06°W (Lizard)

Below the granite moors, the killas platform, which also includes the Lizard peninsula, stands lowest in the west at about 60 m above sea-level, and rises gently to around 180 m in the east. However, a significant topographical change occurs at the structural lineament known as the Start–Perranporth Line (*Figure 6.1*). Small patches of Tertiary sediments resting on the Palaeozoic rocks of the platform southwest of the lineament, suggest that the platform in that area had reached approximately its present form by the early Palaeogene. Since no such sediments occur to the northeast, it is possible that movement along the Start–Perranporth Line has down-faulted the country to the south and west by 50–100 m.

The killas platform is deeply dissected by river valleys, mainly oriented to the south. These were deeply incised during glacial low stands of sea-level in the Pleistocene. Later, when sea-levels rose, offshore islands and large rias such as the Fal and Fowey estuaries were formed. Farther inland the overdeepened valleys were filled with Holocene sediments.

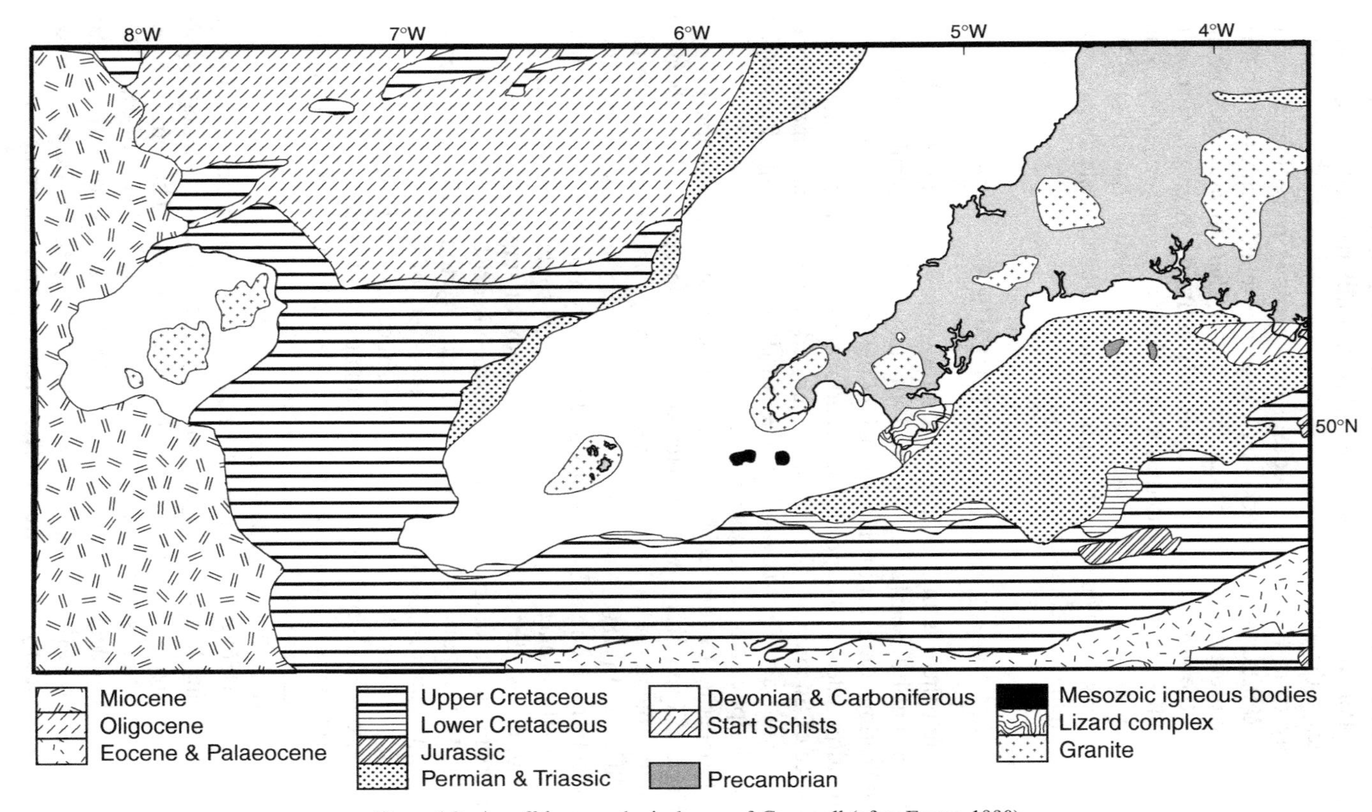

Figure 1.2. An offshore geological map of Cornwall (after Evans, 1990)

However, in the Fal valley between Ruan Lanihorne and Grampound, and at Par and Pentewan, some of this infill resulted from mining activities in the upland areas.

Away from the estuaries, coastal morphology is geologically controlled. Resistant igneous rocks, such as granite and dolerite, tend to form the more prominent headlands, and the softer killas forms bays. The fine sands of the north-coast beaches usually contain a high proportion of fragmental shells and are used as a soil additive to correct acidity. Some of the south-coast beaches have a similar composition, but those at Par, Carlyon Bay, and Pentewan incorporate much mining waste derived from tin streaming and china clay working.

HISTORICAL REVIEW

Although there is little evidence of an interest in geology elsewhere in Britain until the early part of the 19th century, in Cornwall significant publications were appearing well before the end of the 18th century. This attention reflected the growing international importance of Cornwall as a source of Sn and Cu. The first work devoted entirely to the mineralogy of Cornwall was *Mineralogia Cornubiensis* (Pryce, 1778), and Rashleigh's manuscript catalogue of his mineral collection.

Pioneering Cornish scientists and engineers such as Humphry Davy, Richard Trevithick, William Henwood and Joseph Collins played a leading role in the early development of geology and mining engineering in Britain. In 1805, in a series of ten public lectures at the Royal Institution of Great Britain in London, Davy was able to assert to an influential audience the importance of geology to the miner, engineer and agriculturalist. Evidently a substantial body of unpublished knowledge on the mining geology of Cornwall was then in existence.

On 11 February 1814, Henry Boase, John Ayrton Paris and Ashendie Majendie founded the Royal Geological Society of Cornwall. At the first meeting held in the Union Hotel, Penzance, Dr Majendie delivered a lecture, on 'The veins of granite traversing the schisters at Mousehole'. Both Humphry Davy and Francis Lord de Dunstanville were associated with the inception of the society, and it was through the latter's friendship with the Prince Regent that the society gained royal patronage.

In 1839 Henry de la Beche, first Director of the organization now known as the British Geological Survey, published a monumental *Report on the Geology of Cornwall, Devon and West Somerset*. This early interest by the national survey was reactivated towards the end of the 19th century when detailed mapping of Cornwall at a scale of 1:10 560 was initiated. From this work, 12 one-inch (1:63 360) sheets were published with accompanying memoirs. They remain an important source of reference.

Notable papers were published by William Henwood and Joseph Collins within the *Transactions of the Royal Geological Society of Cornwall* between 1827 and 1874. These are an invaluable source of first-hand information on the mines and mining geology during that period. Collins was involved in the founding of both the Institution of Mining and Metallurgy and the Mineralogical Society, and much of his work was summarized in his book *Observations on the West of England Mining Region*, published in 1912.

After this early burst of activity few papers of significance appeared until the second half of the 20th century. However, Hendriks (1937, 1939) showed that most of the rocks of west Cornwall are of Devonian age, and demonstrated the importance of large-scale thrusting in this region.

Over the past 40 years new concepts in structural geology and stratigraphic palaeontology have been applied by many workers. The first phase of modern research was comprehensively reviewed in the 150th Anniversary Volume of the Royal Geological Society of Cornwall (Hosking & Shrimpton, 1966). Later, a great stimulus to research was given by the British Geological Survey when it returned to Cornwall. New 1:50 000 geological maps and memoirs for Falmouth (Sheet 352), Land's End (351 & 358), Bude (307 & 308), Boscastle (322) and Holsworthy (323) have been published. More recently the sheets for Tavistock (337) and Trevose Head & Camelford (335 & 336), have been revised by a team from the University of Exeter under contract to the British Geological Survey. At present the Survey has just completed the remapping of the Plymouth (348) and Mevagissey (353) sheets. Only revision of the Newquay (346), Bodmin (347) and Lizard (359) sheets is outstanding.

Apart from the Geological Survey memoirs, much of the more recent research is documented in the *Proceedings of the Ussher Society*. This society, formed in 1962, holds an annual conference at which many individuals engaged in geological research in South West England meet and read papers.

It is clear that our understanding of the geology of Cornwall is based upon the cumulative efforts of many dedicated workers. In the chapters that follow the references cited recognize the diverse contributions made in recent years. To all these geologists we owe a considerable debt of gratitude.

SUMMARY OF GEOLOGICAL DEVELOPMENT

Although present-day Cornubia occupies a mid-latitude position on the Atlantic margin of northwest Europe, such has not always been the case. The earliest evidence available indicates that some 500 Ma ago, the land area including the Cornubian basement lay at a latitude of about 70°S within the Gondwana continent. At that time, the Atlantic Ocean did not exist, and the global configuration of continental and oceanic areas was quite different from today. Cornubia then formed part of the northern margin of a continent that stretched south towards the South Pole. To the north of Gondwana lay the now-vanished Iapetus Ocean. It is against a pattern of movements of continents and oceans over the last 500 Ma that the geological history of Cornwall must be placed.

These movements are caused by processes that originate deep in the Earth's mantle. Here, convection currents, produced in part by heat liberated during the decay of radioactive elements, rise and cause segments of the more rigid outer layers of the Earth, known as plates, to move. Above rising convection currents plates separate and new ocean basins form, while convergence of plates produces the tectonic and volcanic activity characteristic of mountain building. Because new ocean floor is the initial building material of plates, plate margins where ocean basins form are known as constructive margins. Conversely, at sites of plate convergence, oceanic material descends into the mantle and is recycled, so these are called destructive margins. Where plates converge, but simply slide alongside one another, the margin is called conservative. Here plates are neither created nor destroyed. Over time, the character of a plate margin may change.

Not all plates carry continents, but where they do the trailing edges of continents separated by a new ocean basin are known as passive continental margins. Active continental margins are found where the edges of continents coincide with destructive plate margins. The early stages in the separation of a continent by the development of a new constructive plate margin will also produce intracontinental activity. At present, Cornwall is on the passive Atlantic margin of the European plate. The mid-Atlantic Ridge is the constructive plate margin separating the European and North American plates, and the western edge of the North American plate is, in parts, both a destructive margin (British Columbia) and a conservative margin (California).

The broad picture of the changes that have affected Cornwall is relatively straightforward. Firstly, there is the gradual northward migration of Cornubia against which all other changes must be viewed; and secondly, there is a history of episodic joining of continents and opening of ocean basins. These latter changes have resulted in Cornubia lying at different times at tectonically active and passive ocean margins, and in active and quiescent continental-interior positions. The palaeocontinental world maps for the Phanerozoic prepared by Smith *et al.* (1981) form an invaluable backcloth against which to set the complexities of relationships detailed in the following chapters.

It is now recognized that the Precambrian basement of Britain was formed by the accretion of a number of distinctive and possibly far-travelled terranes (areas of previously formed crust) during late

Precambrian to late Palaeozoic times. The Upper Palaeozoic rocks of Cornubia accumulated on the southern extension of the Avalonian terrane that forms the basement in southern Britain. This terrane, a fragment of the Avalonian Superterrane, consists of metamorphic and volcanic rocks forming part of a late Proterozoic magmatic arc system with associated sedimentary basins. The uplifted southern margin of this terrane, known as Pretannia, forms the Cornubian basement. It was deeply eroded through much of early Palaeozoic times, and provided sediment to the Welsh Basin lying to the north.

The Avalonian Superterrane was probably detached from the southern continent of Gondwana as a new ocean, the Rheic Ocean, opened farther south. Avalonia was thus located between the northern passive margin of the Rheic Ocean and the southern destructive margin of the Iapetus Ocean. The final closure of the Iapetus Ocean, in late Silurian times, led to the accretion of Avalonia into the basement of what is now northwest Europe. The Caledonide mountain chain of northern Britain was produced, and the Old Red Sandstone continent formed. This continent included the northern parts of continental Europe, Greenland, and much of eastern Canada and the Northwest Territories. Cornubia lay on the southern shelf of this great continent, but the influence of Caledonian deformation in this area is obscure.

After the closing of the Iapetus Ocean the Rheic Ocean also began to close, and the continental mass containing Cornubia was carried northwards. Britain lay in tropical latitudes south of the equator in the early Devonian, and across the equator in the late Carboniferous. In movements associated with the final closure of the Rheic Ocean, southern Britain became involved in deformations leading to the production of the E–W-trending Variscides. Thus the supercontinent of Pangaea, comprising most of the present-day continents, was assembled by Permian times. Within this landmass, Cornwall lay some 5°–10° north of the equator. A hot desert climate was initiated, and red-bed deposition was to last for 70 Ma. During this time northward drift persisted. Only when the Atlantic Ocean began to open to the west in the Jurassic Period did more humid conditions return. By then, Britain came to occupy its present mid-latitude, passive-margin position on the eastern shores of the Atlantic Ocean. Later phases of compression and tension that are associated with the reactivation of older structures in Cornwall have been linked to processes acting at a destructive margin at the southern edge of Europe. These eventually resulted in the formation of the Alps. Today, occasional earthquakes occur in Cornwall as movements along old structures accommodate stresses still associated with that destructive margin. However, these are infrequent and typically of very low magnitude. Active mountain building and volcanic activity are now completely absent. Only if Britain becomes part of a new active plate margin would this situation change. There is no evidence to suggest that this is likely to occur in the foreseeable geological future.

Although broad interpretations of the palaeogeographical evolution of Cornwall are facilitated by a knowledge of plate movements, more detailed appraisals depend upon geological data gathered not only in Cornubia, but throughout northwest Europe. Inevitably, many lines of evidence lie well beyond the scope of this book, and much has to be taken on trust. The reader is encouraged to refer to the commentary accompanying the palaeogeographical maps for the British area published by the Geological Society of London (Cope *et al.*, 1992). This monumental work provides the basis for the following paragraphs that review the evolution of Cornwall. *Table 1.1* summarizes the principal events in the geological history of the county.

Variscan deformation makes the palaeogeographical setting of the Upper Palaeozoic rocks of Cornubia unusually difficult to reconstruct. In particular, sequences have been disrupted within thrust sheets that show significant northward translation. *Figure 1.3* provides a generalized palinspastic reconstruction of Cornubia that relates the main sedimentary basins and their tectonic setting.

The base of the Devonian system is not observed in Cornubia. The oldest rocks, the sandy argillites of the Dartmouth Slate, are of fluvial/lacustrine aspect, and of late Lochovian–Pragian age. These sediments appear to have accumulated on the outer shelf bounding the oceanic Gramscatho Basin to the south. Possibly this shelf extended southwards from the uplands of the Caledonides lying hundreds of kilometres to the north. Later, in Emsian times, a marine transgression moved northwards across the shelf, first depositing sandy argillites of the Meadfoot Group. The Staddon Grits, represented at the top of this group, may have been produced by the reworking of previously deposited sediments, following activation of synsedimentary E–W-trending faults. Olistoliths within the mid to late Devonian clastic fill of the Gramscatho Basin show that at this time carbonates were being deposited on an area adjoining the southern margin of the basin, identified as the Normannian High.

By Eifelian times the Trevone and South Devon basins, lying on the shelf north of the Gramscatho Basin were subsiding, probably along fault-controlled boundaries. At first, these basins were barred from the Gramscatho Basin, and largely starved of southerly-derived coarse clastic sediments. However, flyschoid sediments of the Gramscatho Group soon onlapped the intervening shelf. This shelf carried local carbonate build-ups that became more extensive eastwards into Devon, where reefal complexes were established in Givetian/early Frasnian times. Periodically, carbonate turbidites were discharged into the basins lying to the north. The northern margin of both the Trevone and South Devon basins were

Table 1.1. SUMMARY OF THE GEOLOGICAL HISTORY OF CORNWALL

	Age (Ma)	Sedimentary history and principal geological events
QUATERNARY		
RECENT		Submerged forests, alluvium, beach deposits, peat; erosion
PLEISTOCENE	0.01	Glacial and periglacial: boulder gravel, erratics, head
hiatus		Interglacial: raised beach deposits, river terraces, periglacial and temperate erosion; weathering
TERTIARY		
PLIOCENE	2	*Onshore:* Erosion and temperate weathering. Locally St Erth Beds: near-shore sands and clays *Offshore:* Shallow seas with sandy carbonates in Western Approaches
MIOCENE		*Onshore:* Erosion and weathering. Locally St Agnes Formation: aeolian and colluvial sands and clays *Offshore:* Rise in sea-level. Shallow seas with sandy carbonates in Western Approaches
OLIGOCENE		*Onshore:* Erosion and subtropical weathering. Local lacustrine candle clays and sands *Offshore:* Restricted shallow sea in English Channel
EOCENE		*Onshore:* Deep weathering and erosion. Locally Bovey Formation: lacustrine pale grey to brown clays with silts, sands and lignite in Dutson Basin *Offshore:* Shallow sea with limestone in Western Channel
PALAEOCENE		*Onshore:* Erosion; deep weathering with possible silcrete production *Offshore:*Significant fall in sea-level. Limestones in Western Channel
MESOZOIC		
CRETACEOUS	66	*Onshore:* erosion and weathering, followed by Upper Cretaceous marine transgression and chalk deposition (subsequently removed by erosion) *Offshore*: Upper Cretaceous glauconitic limestone and chalk deposition Lower Cretaceous shallow-marine, brackish water and freshwater sandy clays and lignitic sands
	130–112	Minor igneous activity (Wolf Rock and Epson Shoal phonolites)
hiatus		Late Jurassic–early Cretaceous, widespread uplift and erosion

Table 1.1. SUMMARY OF THE GEOLOGICAL HISTORY OF CORNWALL (*continued*)

	Age (Ma)	Sedimentary history and principal geological events
MESOZOIC (cont.)		
JURASSIC	144	*Onshore:* erosion and weathering
	180	Start of supergene kaolinization (continued to Recent)
		Offshore: Middle–Upper Jurassic marginal marine/fluvial arenaceous sediments. Atlantic Ocean starts to open
		Lower Jurassic dark grey mudrocks and limestone
TRIASSIC	208	*Onshore*: hot desert erosion and deep weathering
	236	Cross-course mineralization and some wall-rock alteration
		Offshore: thick continental red-beds deposited in fault-bounded basins
		Late Triassic marine transgression
PALAEOZOIC		
PERMIAN	245	*Onshore:* hot desert erosion and weathering
	270	Main stage Sn–W–Cu mineralization
	280–270	Emplacement of quartz porphyry dykes (elvans)
		Offshore: intermontane basins containing continental red-beds and volcanic rocks.
		Later, crustal rifting produced fault-bounded basins
	286–270	Emplacement of later granite plutons
CARBONIFEROUS	286	
Stephanian	290–286	Emplacement of earlier granite plutons and extrusion of Kingsand rhyolites within continental red-beds
	296	Emplacement of lamprophyre dykes
Westphalian		Continued flysch deposition in Culm Basin, followed by deformation
Namurian		Flysch deposition progrades northwards into Culm Basin
		Main deformation of Trevone and South Devon basins. Flysch interleaved with developing nappes in east Cornwall
Dinantian		Widespread development of black mudrocks, cherts and submarine volcanism in north and east Cornwall. Flysch and shallow-water deltaic facies prograde northwards in the South Devon Basin
		Main deformation of Gramscatho Basin, and obduction of Lizard Complex

Table 1.1. SUMMARY OF THE GEOLOGICAL HISTORY OF CORNWALL (*continued*)

	Age (Ma)	Sedimentary history and principal geological events
PALAEOZOIC (cont.)		
DEVONIAN	360	
Upper–Middle		Intrashelf Trevone and South Devon basins established. Mudrock dominated sequences with submarine volcanism Gramscatho Basin: mainly flyschoid mudrocks, sandstones, and conglomerates with submarine (MORB) volcanism. Olistolithic blocks with Ordovician, Lower Devonian and early Eifelian faunas, indicate active tectonism
Lower		Rheic Oceanic established south of a mudrock dominated outer shelf
	408	Base of System not observed

Onshore and *Offshore* refer to the present shoreline

volcanically active, and provided a local source of magmatic and clastic basin infill during the Late Devonian.

By the early Upper Devonian, the Gramscatho Basin was closing rapidly, as nappes derived from the southern parts of the basin were generated by a northwards-advancing deformation front. Rapid uplift and instability at the tectonic front released submarine slides and mass flows. These produced olistostromes (Roseland Breccia Formation) at the margin of the basin, which carried olistoliths of Ordovician quartzite and Lower–Middle Devonian limestones. By Famennian times these sediments were caught up in the Carrick Nappe and transported northwards across parautochtonous outer-shelf deposits. Oceanic floor derived from the Gramscatho Basin was finally obducted as the Lizard Nappe in late Devonian to early Carboniferous times.

Meanwhile, to the north, pelagic sedimentation characterized the Trevone and South Devon basins. Purple and green argillites (Polzeath Slate) in the Upper Devonian were followed by deep-water, black argillites and volcanic rocks in the Lower Carboniferous. Within the South Devon Basin, intercalations of southerly derived flysch, and shallow-water deltaic facies bear evidence of uplift associated with the advancing deformation front. This front eventually invaded these intrashelf basins in late Viséan/early Namurian times, producing complex multiphase deformations. Flyschoid sediments generated in advance of the nappes were overridden and intercalated within the resulting nappe pile.

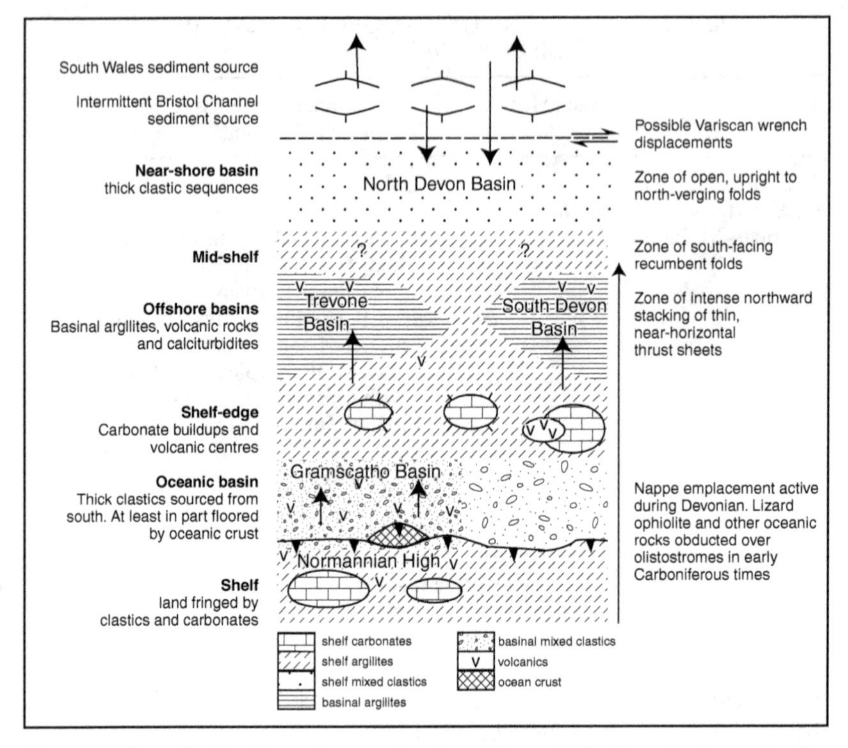

Figure 1.3. Generalized palinspastic reconstruction of Cornubia in Devonian times, showing the possible distribution of the main sedimentary basins and their tectonic setting (after Bluck *et al.*, 1992). Sediments from the North Devon Basin are not recognized in Cornwall (the Culm Basin developed north of the Trevone and South Devon basins in Carboniferous times)

To the north of this activity, the Culm Basin received great thicknesses of Upper Carboniferous (Namurian–Westphalian) flysch. The lower Namurian turbidites (Crackington Formation) are more proximal to the south, but by the upper Namurian the major supply was from the west and northwest along the axis of the basin. Although palaeocurrent directions in the Westphalian A (Bude Formation) are variable, surprisingly none come from the emergent area to the south. Relationships at the southern margin of the basin are unknown. Possibly, important strike-slip displacements may have taken place. The Culm Basin was deformed in the final major pulse of the Variscan orogeny in late Westphalian times. In Cornwall, the inverted limb of a major south-facing anticline was driven southwards on the Rusey Thrust across the earlier, northward-transported nappe pile.

By late Carboniferous times substantial uplift and N–S extension had taken place, probably accompanied by the rise of the granite batholith.

A profound climatic change took place as the Pangaean supercontinent drifted slowly northwards out of the equatorial region. In the Stephanian the area was already subject to severe hot desert erosion. At the same time, the development of acid magmatism was signalled by the eruption of acid volcanic flows (Kingsand Rhyolite). On the land, flash floods swept terrigenous material down to the surrounding lowlands, and alluvial fans built out across sand and gravel plains featuring playa lakes and dunefields. The main-stage Sn–W–Cu mineralization associated with the granite is dated within the early Permian.

Red-bed sedimentation continued into late Triassic times, with patterns of deposition being increasingly controlled by rifting. Cornubia became differentiated from other upland areas of northwest Europe by major graben identified below the Bristol Channel, Celtic Sea and English Channel. Now Cornwall lay some 15°–20° north of the equator, and the climate had ameliorated. There is evidence for high but very seasonal rainfall, possibly associated with a monsoonal regime. The cross-course mineralization of the Cornubian ore field is dated within the Triassic at 236 Ma.

A late Triassic marine transgression left Cornubia standing as a low-lying island within the Jurassic sea. The extent of the landmass is ill-defined, but included most of Devon and Cornwall and would have extended southwestwards as a ridge between the Celtic Sea basins and the Western Approaches Trough. Later, its area would have been increased considerably with a fall in sea-level in the Middle Jurassic. The general rise in sea-level recorded in Britain during late Jurassic times was not, however, seen in the Celtic Sea area. Rather, uplift led to the union of the Cornubian island with the Irish landmass. The area of emergent land reached its maximum extent in Early Cretaceous times when there was a marked fall in sea-level, and the British Isles became part of mainland Europe. The Wolf Rock phonolite gives evidence for minor volcanic activity at that time. Throughout this emergent period, Cornubia underwent deep weathering, and supergene kaolinization processes were initiated. Such conditions persisted into the late Cretaceous when Cornubia was eventually submerged by the Chalk sea. In early Palaeocene times a major regression followed, and subsequent erosion led to the complete removal of chalk deposits from Cornwall.

Although piecemeal in distribution, and occupying only a small area, more Tertiary systems are represented in Cornwall than elsewhere in Britain. These deposits are largely terrigenous, and accumulated locally on the deeply weathered Cornubian massif.

During most of the Palaeogene the palaeolatitude of Britain was about 40°N, and the climate was warmer than at present, with floras and faunas indicating subtropical/tropical conditions. Also, sea-level was much lower than at present, with shelf sea sedimentation being maintained only in the deeper parts of the Western Approaches

By late Palaeocene times, north Atlantic rifting was well established, and although rifting in the North Sea had finished, periodic connections with the early Atlantic Ocean existed along the line of the English Channel. During late Eocene–late Oligocene times, deep weathering products derived largely from Carboniferous shales, were transported by rivers into lakes. Some lakes filled topographic hollows (Beacon Cottage Farm Beds), but the more important and persistent basins were fault-controlled (Bovey Formation).

The Neogene is characterized by a significant rise in sea-level. Eustatic changes are known to have occurred, but, overall, the shoreline approximated to that seen now. It was during this time that the present latitudinal sequence of faunas of the western Atlantic was established, and their modern character progressively acquired. The terrestrial St Agnes Formation represents the only Miocene deposit known in Britain, while the shallow-water St Erth Beds accumulated during a Pliocene marine high-stand.

In Britain the Pleistocene is characterized by numerous cold and temperate stages, but in Cornwall the chronology is particularly difficult to unravel because the earlier record has been largely obliterated by the effects of the late Devensian ice sheet. The southern margin of this sheet probably reached the Isles of Scilly 18 ka Before Present (BP), and produced a periglacial environment in Cornwall and in the emergent offshore areas. At this time, sea-level would have fallen by as much as 120 m below Ordnance Datum (OD); the English Channel would have been emergent, and had a major southwest-flowing river draining much of the North Sea and its hinterland. Subsequently, the climate ameliorated, and by 13 ka BP the Devensian ice had largely disappeared. Sea-level rose rapidly, recovering to within a few metres of its present value by 6 ka BP. The final submergence of the offshore island of Scilly to create the present archipelago was not effected until the post-Roman period. Archaeological and historical evidence indicates that although the sea was rising on a unitary island about 2000 BC, 'submergence began in earnest during Norman times and was effectively completed by the early Tudor period' (Thomas, 1985). Late fault movements would appear to be responsible for this episode. The Holocene marine sediments around Cornwall are a mixture of terrigenous and bioclastic material, and include much reworked Pleistocene sediment.

The rocks of Cornwall have thus revealed a remarkable story covering the past 400 Ma of Earth history. The county, as we know it today, is evidently the product of diverse crustal and mantle processes that have generated, filled and deformed sedimentary basins. At the same time, dramatic changes in geographical position and climate were effected. As our understanding of these continuing processes increases, so it becomes possible to make short-term predictions about the future. The sum of geological experience will be much in demand as we seek to understand, and possibly control, the environmental changes going on around us.

Chapter Two

The Pre-Devonian Tectonic Framework

Britain is composed of a number of distinctive fragments of continental crust, known as terranes, that accreted during late Precambrian to Carboniferous times. Most of the Upper Palaeozoic rocks of Cornubia were deposited on the southern extension of the Avalonian terrane that forms the basement to much of southern Britain. This terrane was once part of the larger Avalonian Superterrane, which was dismembered during the opening of the North Atlantic Ocean. Fragments are now to be found on the eastern seaboard of North America, from New England through the Maritime Provinces to Newfoundland, and in parts of the British Isles, France and Belgium (Cocks, 1993; Cocks *et al.*, 1997).These rocks formed part of a late Proterozoic magmatic arc system and associated sedimentary basins.

In late Precambrian times, the southern margin of Avalonia represented in Britain was emergent, and subsequently supplied sediment to Lower Palaeozoic basins. This uplifted area has been identified as Pretannia, and probably forms the Cornubian basement, except perhaps for the region south of the Start–Perranporth Line (*Figure 6.1*). The latter area may belong to a separate terrane accreted by the closure of the Rheic Ocean in late Palaeozoic times. A small fragment of oceanic crust, known as the Lizard Ophiolite, was trapped between slices of continental crust in the process.

In Cornwall today there is only scattered or indirect evidence of a Precambrian crystalline basement. However, Precambrian schists and gneisses are known in Pembrokeshire. In addition, the adjoining sedimentary sequences of the Welsh Basin indicate a southern provenance, with detritus derived from Pretannia. This upland area consisted, at least in part, of mica schists or quartz-mica schists (Cope & Bassett, 1987). Geophysical investigations of the Bristol Channel and Celtic Sea areas (Brooks & James, 1975; Mechie & Brooks, 1984) have shown that the Precambrian basement of South Wales continues south and west, increasing in depth to about 7.5 km off the coast of north Devon. Unfortunately, good geophysical evidence for the presence of crystalline basement beneath South West England is lacking. Nevertheless, differences between the depositional framework of the Devonian rocks across the Start–Perranporth Line suggest that the Avalonian basement extends southwards beneath central Cornwall.

Farther south, the garnetiferous gneisses of Eddystone Rock and mica schists, which form the sea floor of Plymouth Bay, suggested to Brooks *et al.* (1993) that there is an extensive area of crystalline basement beneath the western part of Plymouth Bay and the Lizard peninsula. This is now supported by evidence from the Man o' War Gneiss (Sandeman *et al.*, 1997). Pre-Variscan cataclastic granitoid xenoliths in Devonian basic lavas near Land's End also led Goode & Merriman (1987) to propose that crystalline basement is present in the area, and possibly extends west as far as Haig Fras. Such rocks were almost certainly uplifted to the south of Cornwall during the Devonian, and supplied sediments to the Gramscatho Basin. Similarly derived sediment also appeared in the northward-prograding flysch within the South Devon Basin during the Carboniferous. This basement, which includes the Normannian High, appears to have been derived from the southern margin of the Rheic Ocean, and to have been uplifted during ocean closure.

Although much of the early history of Avalonia is speculative, it is probably correct that, at the end of the Precambrian, the superterrane lay at a latitude of about 70°S, on the northern edge of Gondwana. This megacontinent united the now separated continents of Africa, South America, India, Antarctica and Australia. To the north of Gondwana, the Iapetus Ocean extended to a latitude of about 20°S, where lay another major continental area, known as Laurentia. Laurentia was made up of most of North America, but also included the northwestern parts of both Scotland and Ireland. Between Laurentia and Gondwana, at a latitude of about 50°S, the small continent of Baltica was separated from Gondwana by the Tornquist Sea. Baltica was composed of present-day Fenno-Scandinavia and the Baltic region of Russia. This configuration of continental and oceanic areas persisted until the Ordovician (*Figure 2.1a*), although there was a general northerly movement of Gondwana and Laurentia, and Baltica moved somewhat to the west.

In Proterozoic and early Palaeozoic times, Gondwana showed early evidence of the continental separations that were to follow in the Ordovician. Crustal instability in northern Gondwana caused the development of Pretannia, an E–W-trending region of topographic highs and basins, with intervening shelf seas. This upland area of Precambrian basement is now recognized to have extended from Belgium in the east towards Newfoundland in the west. Details of the geology of Pretannia are vague, but in addition to mica schists and gneisses, extrusive rhyolitic lavas and intrusive acid-intermediate rocks are also known to be present (Cope, 1987). The structure of Pretannia was probably formed by late

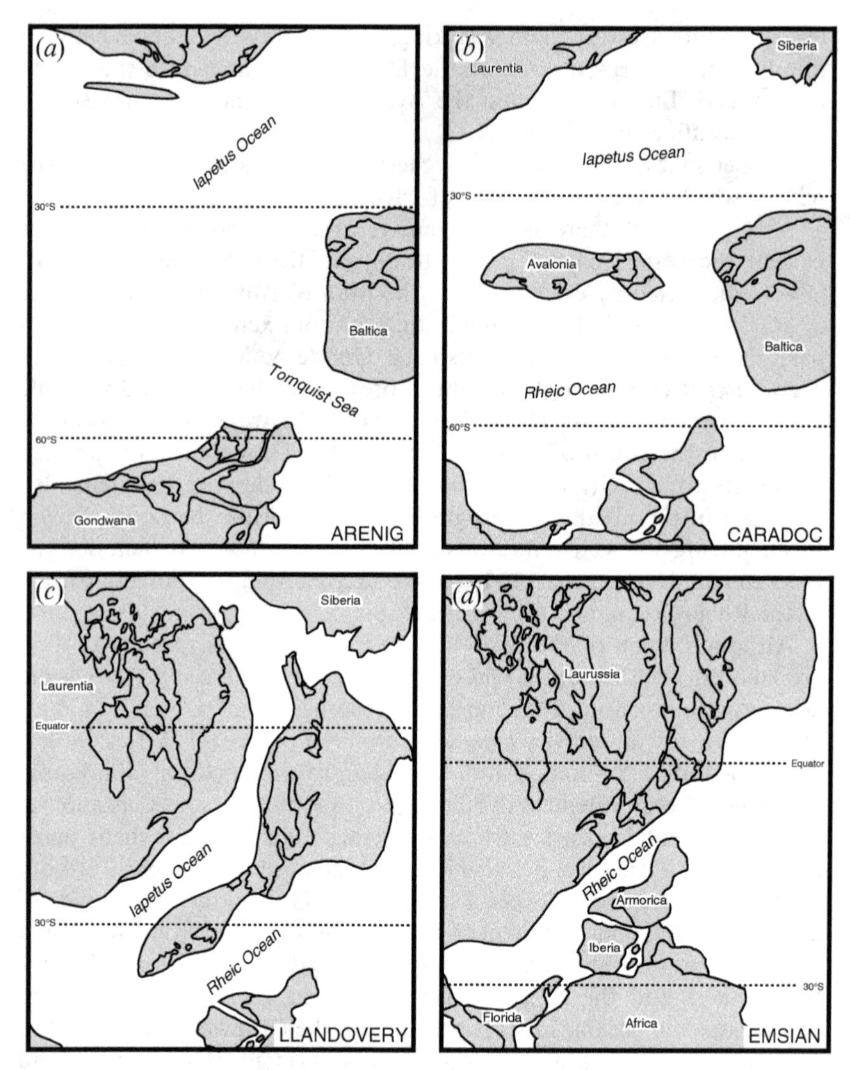

Figure 2.1. The palaeogeography of northwest Europe, (*a*) in mid-Arenig times with southern Britain attached to Gondwana, (*b*) in early Caradoc times, with Avalonia, including the London–Brabant massif, detached from Gondwana, (*c*) in late Llandovery times, showing the Avalonian fusion with Baltica and the narrowing Iapetus Ocean and (*d*) in mid-Devonian times, after the closure of Iapetus (after Cocks, 1993)

Pre-cambrian (Proterozoic) Cadomian deformations now seen to have affected northern France. During the Cambrian and Ordovician, the shelf areas around Pretannia received shallow-water marine sands and silts, while at greater distances basinal deposits were formed.

Pretannia separated from Gondwana within the Avalonian Superterrane during the Ordovician. (*Figure 2.1b*). The microcontinent moved northwards to a latitude of about 40°S as the Rheic Ocean opened behind it. At the same time, the Iapetus Ocean was closing and the Tornquist Sea narrowing. The timing of the opening of the Rheic Ocean is derived from palaeontological evidence. Direct evidence comes from the Lower Palaeozoic successions of Wales that document the history of the shelf seas and basins developed to the north of Pretannia. Indirect evidence comes from fragments of Ordovician rocks found in younger beds in east Devon and Cornwall. Of particular importance are the Ordovician quartzite pebbles found in the Triassic Budleigh Salterton Pebble Beds of east Devon. These contain fossils that can be grouped into two assemblages, one of middle Arenig age, the other of Llandeilo age (Cocks, 1993). Most of the fossils are brachiopods, but trilobites also occur in the Llandeilo assemblage. Brachiopods and trilobites also occur in quartzites of Ordovician age that are found as blocks within late Devonian tectonic and sedimentary mélanges in the Roseland and Meneage areas of Cornwall. Both sets of quartzites were originally deposited on a shelf well to the south of Pretannia. Although the quartzites from Cornwall are slightly older than the Llandeilo quartzites from Budleigh Salterton, they are of similar character. These shallow-water Arenig and Llandeilo assemblages have considerable palaeogeographical significance. Thus the Arenig fauna of Budleigh Salterton belongs to a faunal province including Brittany and eastern North America. This suggests that the Rheic Ocean had not then opened. The Llandeilo assemblage, on the other hand, is of central European character and distinct from that of eastern North America. Separation by the Rheic Ocean is indicated. It is difficult to be more precise about timing the origin of Avalonia and the Rheic Ocean, but Cocks (1993) considered a Llanvirn date to be most likely. The subsequent history of the Rheic Ocean involved the development of the Variscides and extended to its final closure in the early Permian, when the southern margin of this ocean accreted to the Avalonian terrane. Inevitably, profound differences occur between the rocks that formed at the southern edge of the Rheic Ocean and those that formed in basins on Pretannia. Variscan events are documented in the Upper Palaeozoic geology of Cornwall, but these events were superimposed on an ancient basement carrying structures that influenced later events.

The closing of the Iapetus Ocean towards the end of the Silurian led to the accretion of Avalonia to Laurentia. This event culminated in the formation of the Caledonides, with intense deformations in northern Britain and Ireland. Although some N–S-trending folds were formed, the Caledonides show mainly NE–SW trending structures stretching from northern Scotland to southwest Wales. Farther south, such Caledonian trends are thought to have been impressed upon parts of Pretannia. These are recognized through their influence on the later geological history of central southern England (Whittaker, 1985). However, they are not particularly evident in Cornwall. Probably only the development of NE–SW-trending mineral veins in west Cornwall, was controlled by Caledonide structures. More importantly in Cornwall, the E–W-trending Cadomian structures of Pretannia may have been reactivated during the formation of the Caledonides (Bois *et al.*, 1990). Reactivated Cadomian structures also appear to have been influential in controlling the sedimentary and structural history of basins and rises during the Devonian and Carboniferous.

Also of significance for Cornwall are the NW–SE-trending structures, developed during the Ordovician, that influenced the Devonian and Carboniferous history of South West England (Selwood, 1990). Indeed, these features have continued to influence the geology of the whole of southern England until the present. Today they appear as a series of NW–SE-trending transcurrent faults that cut across Cornwall and Devon (Dearman, 1963). The best known of these is the Sticklepath fault in Devon, but many are also present in Cornwall between Bodmin Moor and Dartmoor, and west of Bodmin Moor. They exercised a considerable control on the development of the drainage pattern of Cornwall and Devon, and the occasional minor earthquake caused by movement on these faults still occurs from time to time.

The origin of such important features must lie in a major geological event that produced deep-seated lines of weakness, which subsequently could be reactivated. The accretion of Avalonia to Baltica resulting from the closing of the Tornquist Sea (*Figure 2.1c*) provided such an event. Compressive closing was accompanied by considerable transcurrent movements along a NW–SE line. These transcurrent movements are likely to have been somewhat similar to those now occurring on the western seaboard of North America. Moreover, the coalescing of separate parts of continental crust that is still taking place along western North America, such as the addition of Vancouver Island to the main continent, may well parallel the coalescence of Avalonia and Baltica. If Avalonia was not one microcontinent, but consisted of several even smaller continental areas, of which Pretannia was one (Ziegler, 1990), the similarity becomes even more direct. Transcurrent faults produced in this type of environment are extremely deep-seated, extending through the crust, and can be expected to persist as planes of weakness long after suturing has ended. Later stress fields acting on these planes can give rise to vertical as well as transcurrent displacements.

Finally, the closure of the Rheic Ocean was a transpressive event, combining sinistral strike-slip movement with oblique compression as the Armorican and Iberian microcontinents slide past the Avalonian terrane of Laurussia (*Figure 2.1d*). The Start–Perranporth Line juxtaposing elements from the northern and southern sides of the Rheic Ocean, then came into existence.

Chapter Three

The Lizard Complex

The interpretation of the gabbro, amphibolite, serpentinized peridotite and associated rocks of the Lizard Complex has undergone radical reappraisal since the pioneering work of the Geological Survey. Originally the Complex was thought to have been formed by intrusive processes. Thus Flett (1946) recognized the cold diapiric ascent of low-density serpentinite, while Green (1964a) favoured the intrusion of hot diapiric serpentinized peridotite into gabbroic country rocks. In this latter view, the resulting high-temperature metamorphic aureole produced the amphibolites. However, the theory of plate tectonics has led to a fundamental change in interpretation. The assemblage found in the Complex is now recognized as a section of oceanic crust (Badham & Kirby, 1976).

It was during closure of the Gramscatho Basin that this part of the ocean floor became detached and displaced over low-grade metasediments lying to the north (Barnes & Andrews, 1986; Holder & Leveridge, 1986a). This process is known as *obduction*, and the rocks translated from the ocean crust are identified as an *ophiolite*. In undeformed ocean crust, an upward sequence of massive peridotite, layered gabbros, sheeted dolerite dykes, and near-surface basaltic pillow lavas and sedimentary rocks is found. The junction between the peridotite and gabbro is marked by a fundamental change in seismic velocities. This is distinguished as the Mohorovicic Discontinuity (Moho). Petrologically, however, the base of the crust is better defined as the base of the layered gabbros, and is known as the petrological Moho.

GEOLOGICAL SETTING

Overall, the Lizard Complex (*Figure 3.1*) consists of a number of fault-bounded slices which structurally overlie a series of low-grade, cleaved metasediments dominated by flysch but also including the sequence of mélanges, rudites and volcanic rocks variously known as the Meneage

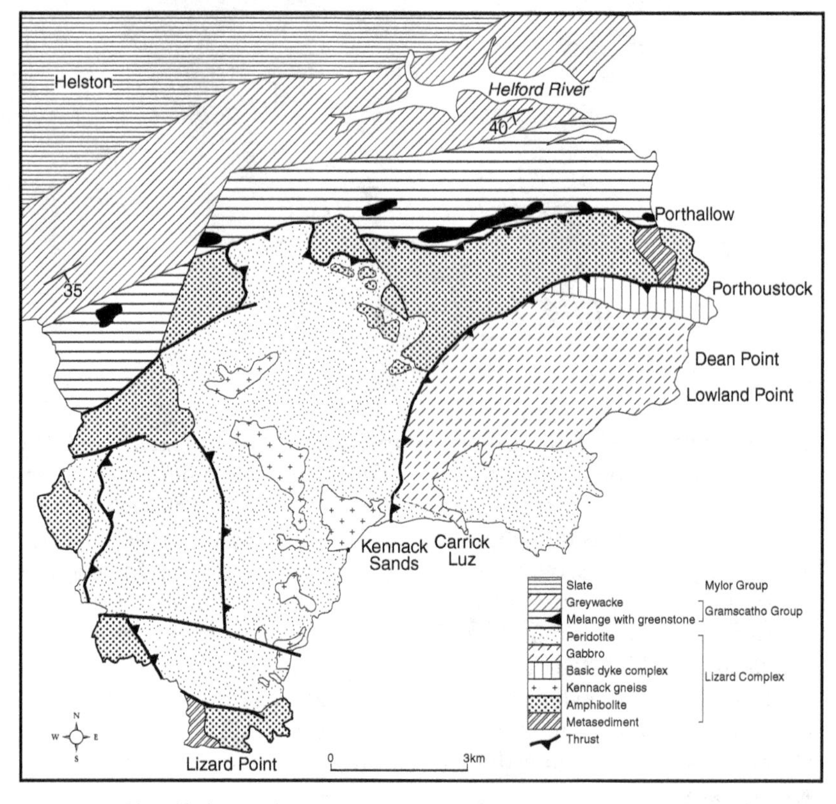

Figure 3.1. Simplified geological map of the Lizard Complex (after BGS 1:50 000 Sheet 359; Kirby, 1979)

Formation (Barnes, 1983) or the Carne and Roseland Breccia formations (Holder & Leveridge, 1986a). The presence of pillow lavas, breccias and tuffs with an oceanic geochemistry in this sequence has suggested a genetic link with the Lizard Complex (Floyd, 1984; Barnes, 1984). Similarly, the assemblage of pillow lavas, pillow breccias, gabbros and serpentinite of Nare Head in Roseland may be an extension of the Lizard Complex in the hanging wall of a large thrust fault (Hendricks, 1939).

The Lizard is dominated by peridotite, amphibolite and gabbro, with significant areas of granitic material and metasediment. The Complex is best described in terms of two tectonic units separated by the arcuate Kennack–Porthoustock Thrust. The eastern (upper) unit, sometimes referred to as the Crousa Downs structural unit, preserves a superb section through a piece of almost undeformed oceanic crust, in a northwards younging sequence of peridotite–gabbro–sheeted dykes. This is here named the Crousa Downs Ophiolite. The gabbro–peridotite contact at Carrick Luz [SW 755 165] is an early oceanic feature, interpreted either as an extensional shear zone generated in an oceanic spreading centre (Gibbons & Thompson, 1991) or as a fossil transform fault (Andrews *et al.*, 1995). The western (lower) unit shows a complicated juxtaposition of constituent lithologies produced by strong internal deformation and faulting. The Kennack–Porthoustock Thrust, which strikes N–S at Kennack Sands and swings around to strike E–W at Porthoustock, probably consists of a number of separate segments. Its steep inclination at Kennack Sands suggests a late, steep extensional structure downthrowing southeastwards. At Porthoustock it could either be a south-dipping extensional fault or a north-dipping reverse fault.

Geophysical evidence (Doody & Brooks, 1986; Rollin, 1986) suggests that the Complex is underlain at very shallow depths by material with low upper-crustal seismic velocities. This supports a thin-skinned interpretation of the structure.

THE CROUSA DOWNS OPHIOLITE

The ophiolite complex exposed in the coastal section from Coverack to Porthoustock, defines a northeast-dipping sequence from peridotite, upwards through gabbro into the root zone of a sheeted dyke complex. The expected transition into pillow lavas and oceanic sediments is truncated at Porthoustock by the Kennack–Porthoustock Thrust. This missing component may be represented by pillow lavas on Nare Head (*Plate 1*) and Mullion Island, and by the metasediments of the Old Lizard Head Series, which occur at Lizard Point and between Porthoustock and Porthallow.

Peridotite

This coarse-grained, porphyroclastic spinel-lherzolite is variably serpentinized leading to the rich variety of colours which has made the Lizard serpentine famous as an ornamental stone. Lherzolites are peridotites containing two pyroxenes in addition to olivine. The lherzolites typically exhibit a pervasive planar-linear fabric defined by flattened and elongated spinel, and blocky orthopyroxene grains. The fabric generally trends N–S (Flett, 1946) with steep dips. Occasional dunite pockets containing thin spinel layers can be seen, for example in Perprean Cove [SX 756 166]. Alteration to serpentine and talc is widespread and is due to subsequent low-temperature hydration which probably occurred via seawater ingress along fractures during cooling. Occasionally native copper is developed along some of these fractures, but not in sufficient quantities to be economic. Secondary magnesite is developed in occasional pockets of carbonated peridotite, so CO_2 was also an important component of circulating low-temperature fluids.

Gabbro

A series of randomly oriented, steeply-dipping gabbro sheets intrude the peridotite over a wide range of scales. They post-date the planar-linear peridotite fabric and are commonly, though not inevitably, undeformed. Good examples can be seen at Lankidden Cove. Steep dips suggest that the dykes acted as feeders to a magma chamber immediately above (Roberts *et al.*, 1993). These gabbro sheets are variably altered, with amphibole and sericite replacing clinopyroxene and plagioclase. Rounded, serpentinized olivine grains and pseudomorphs after olivine are present. Notably, olivine occurs as an interstitial phase crystallizing after clinopyroxene and plagioclase. The peridotite–gabbro contact is a complex transition zone more than 1 km wide in which the proportion of gabbro gradually increases. The gabbro ranges from olivine-rich troctolite to a feldspathic olivine-rich composition. It shows an extremely variable grain size, and is locally strongly deformed in a series of ductile shear zones defined by elongate plagioclase and pyroxene grains (flaser gabbro). The shear zones dip at moderate to low angles to the northeast with an east to northeast plunging mineral stretching lineation. Kinematic indicators consistently indicate a top-to-the-east sense of displacement (Gibbons & Thompson, 1991; Roberts *et al.*, 1993). All are cross-cut by a late set of basic dykes with Mid-Ocean Ridge Basalt (MORB) characteristics (Roberts *et al.*, 1993). This suggests that they were generated at, or near, a former oceanic ridge axis prior to obduction.

Farther north, the gabbro is generally compositionally homogeneous but texturally variable, displaying frequent abrupt and irregular internal contacts. Composition layering with a general steep E–W orientation is locally preserved (Kirby, 1978). The layering is due to variation in the content of olivine, plagioclase and rarely clinopyroxene. It is thought to represent differentiation in the lower part of ephemeral magma chambers beneath an ocean ridge crest. Distinctive zones of alteration are associated with the low-angle ductile shear zones. These are characterized by the development of oxide-rich gabbros which Hopkinson & Roberts (1995) have interpreted as the products of interaction with evolved Fe–Ti-rich, Si-poor melts which migrated along the shear zones during deformation. Similar gabbros have been identified in boreholes as a potential source of vanadium (Leake *et al.*, 1992).

South of Coverack, one large dyke-like body of strongly deformed gabbro mylonites runs along the coast from Carrick Luz to Spurnic Cove [SX 753 167], where it turns inland to join the main gabbro body. The gabbro approximates to a vertical body between 10–100 m wide, folded about a horizontal axis trending parallel to the strike. The mylonites display shear criteria indicating dextral displacement. The steep dips of the hanging-wall and footwall peridotite margins are more compatible with an interpretation of the mylonitic gabbro as the remnants of a leaky transform fault rather than a shear zone running along the petrological Moho as suggested by Gibbons & Thompson (1991, fig. 1).

Dykes

Occasional NNW–SSW-trending basic dykes up to 1 m thick are present throughout the section, but are most abundant between Dean Point [SX 806 205] and Porthoustock [SX 808 218]. The proportion of dyke to gabbro rises to a maximum of 50–70% on the coast, and in the West of England quarries [SX 809 215] near Porthoustock, constituting the base of a sheeted dyke complex. Using field relationships as a basis for initial classification, Roberts *et al.* (1993) showed that the dykes fall into three geochemically well-defined sets.

The earliest set (Set 1) comprises a small number of thin bodies, usually a few centimetres wide, which trend just east of north and dip shallowly westwards. This set occurs northwards of Coverack and is predominantly composed of plagioclase, green amphibole (after clinopyroxene) and abundant opaque minerals. The absence of chilled margins, and complex intrusive relationships with the gabbros, indicate that intrusion occurred soon after the gabbroic host had crystallized. Geochemistry suggests that they represent fractionated residual melts segregated from the gabbroic magma chamber. Low-angle, northeast-dipping ductile shear zones extensionally offset these dykes.

The second set (Set 2) comprises plagioclase-phyric metadolerite dykes which are predominantly observed towards Porthoustock and form the bulk of the sheeted dyke complex. These dykes are 0.1–2 m thick, trend NNW–SSE and typically dip moderately to steeply east-northeast. However, they become more irregular and comparable in form to Set 1 dykes southwards. Petrographically, a groundmass of plagioclase envelopes brown amphibole (after clinopyroxene) with occasional interstitial olivine. Although striking parallel to the ductile shear zones, these dykes dip more steeply and are deformed within the shear zones. Geochemically the dykes are tholeiites with MORB characteristics.

The third set (Set 3) comprises a series of sub-vertical, NNW–SSE-trending, dominantly aphyric dolerite dykes showing chilled margins. They cross-cut the peridotite, gabbros, plagioclase-phyric dykes and the NE–SW-trending shear zones. At Coverack they locally utilize the low-angle ductile shear zones which allowed transfer of extension (*Figure 3.2*). The dykes are 0.1–1.5 m in thickness and variably altered. Weakly metamorphosed samples preserve twinned plagioclase and clinopyroxene in sub-ophitic textures together with sparse olivine phenocrysts. Geochemically the dykes are tholeiites with more primitive MORB characteristics than the Set 2 dykes.

Plagiogranite

Intrusive felsic material is intimately associated with the basic dykes in a variety of ways. It frequently occurs at the base of the sheeted dyke complex, intermingled with dolerite in net-veined intrusions. Several examples can be seen on the north side of Godrevy Cove. Field relationships, such as colloform basic masses chilled against enveloping coarser-grained acid material which back-veins the basic masses along small fractures, strongly suggest the simultaneous existence of acid and basic magmas. Plagiogranites are a common feature of the top of the gabbro magma chamber in many ophiolites, and represent evolved differentiates of the gabbros. In this situation they could have become mixed with fresh pulses of undifferented magma from below, and ascended into the sheeted dyke complex to produce the net-veining. Support for this interpretation comes from the frequent observation that the plagiogranite forms thin selvages along the margins to the net-veined dykes. Plagiogranite veins also occur in the gabbros, and both cut and are cut by basic dykes.

THE WESTERN TECTONIC UNIT

The peridotites forming most of the Western Tectonic Unit are thrust over the amphibolites and metasediments forming the southernmost promontory around Lizard Point (Jones, 1997). These in turn overlie more coarse-grained gabbroic, amphibolitic and other more acid lithologies (Man o' War Gneiss) which occur at Lizard Point.

Amphibolites

About one third of the surface outcrop of the Western Tectonic Unit consists of amphibolites. The two varieties distinguished by Flett (1946) and Green (1964a) are now tectonically imbricated and it is not possible to map them as separate units. The Landewednack Hornblende Schists are well banded, occasionally garnetiferous, predominantly coarse-grained amphibolites occurring near Lizard Point and around Landewednack Church Cove [SX 715 127]. The Traboe Schists, composed of non-garnetiferous, predominantly coarse-grained amphibolites, are confined to, and make up most of, the northern and western amphibolite exposures. They are best seen in the coastal exposures between Porthoustock and Porthkerris [SX 807 218–SX 806 229].

Both types of amphibolite are cut by late NNW–SSE-trending dolerites. In some places strongly deformed early basic dykes are slightly discordant to the strong foliation in the Landewednack variety, whereas in the Traboe type, remnants of an original layering can occasionally be seen. Remarkably, serpentinized dunite layers are interdigitated with coarse amphibolite on the coast between Porthallow [SX 797 233] and Porthkerris. Schists of the Landewednack type are interbedded and/or infolded with psammitic and semi-pelitic schists of the Old Lizard Head Series near Lizard Point. This relationship is especially clear in exposures adjacent to the old lifeboat station [SX 701 115]. The Landewednack type corresponds to a mixture of the sheeted dyke complex and overlying volcanic extrusive components, whereas the Traboe type represents isotropic and cumulate sections of the magma chamber.

Peridotite

As in the Crousa Downs Ophiolite, peridotite is predominantly a more or less serpentinized spinel lherzolite with pockets of spinel-rich dunite. Where serpentinized it is either black or red according to the degree of oxidation of the magnetite released during the alteration process. In many places large porphyroclasts of orthopyroxene with a bright bronzy lustre (bastite serpentine) stand out. Good examples may be seen at Kennack Sands [SX 734 165]. A planar-linear fabric is inevitably present, usually with a steep orientation which has a predominant but more variable strike (Flett, 1946) than that in the Crousa Downs Ophiolite.

Kennack gneiss

A series of mixed acid/basic intrusive rocks intrude the peridotite at Kennack Sands and farther south towards Landewednack Church Cove. Small acid dykes and lenses also occur at Kynance Cove [SX 684 133] and larger areas of gneiss are mapped inland. The main body of the gneiss appears to follow a flat-lying contact between overlying peridotite and small areas of amphibolite (Landewednack type) which dips gently west and outcrops along the top of the cliffs. The tectonic nature of this contact (Styles & Kirby, 1980) has led to suggestions that it controlled the emplacement of the gneisses. Such field relationships can be seen at Kennack Sands (Flett, 1946). Late cross-cutting, sparsely porphyric (Set 3) dykes, which are possibly coeval with the migmatitic gneisses, may constitute the basic end-member of the series. Undeformed net-veined acid and basic dykes within peridotite become progressively more deformed as they join the main mass of the intrusion on the wave-cut platform to the east. The high-strain gradients suggest that synkinematic intrusion produced the gneissose fabric which generally characterizes these lithologies (Green, 1964b). The intimate, frequently net-veined interrelationships of the acid and basic fractions (and the presence of intermediate compositions with gradational contacts) clearly indicate the mixing of two end member magmas. The origin of the acid fraction is still conjectural.

Metasediments

Psammitic and semi-pelitic schists (Old Lizard Head Series) occur in the vicinity of Lizard Point and in a poorly exposed strip which trends south from Porthallow towards Porthoustock, where they terminate against the Kennack–Porthoustock fault. Although no longer exposed, there is also a lensoid body of mylonitized quartzite (Treleague Quartzite) sandwiched between the Crousa Downs Ophiolite and the Western Tectonic Unit. Flett (1946) reported that amphibolitized dolerite dykes intrude the quartzite, and Styles & Kirby (1980) noted microfabrics, suggesting high strains.

Many workers have noted the complex sequence of deformation events displayed in the psammitic and semi-pelitic metasediments. Their interdigitation with amphibolites is clearly an original feature, and both underwent amphibolite facies metamorphism during deformation. They now show signs of retrogressive metamorphism.

It is quite possible that these predominantly impure quartzo-feldspathic lithologies represent sediments originally deposited in an intracratonic basin floored by oceanic crust, now represented by the remainder of the Lizard Complex. The timing of the deformation and metamorphism is constrained by the absence of any major tectonic or thermal effects in the gabbros and dykes of the Crousa Downs Ophiolite and in cross-cutting (Set 3) dykes elsewhere. It must have taken place in an oceanic setting.

AGE

Although Lower Palaeozoic and Precambrian ages have been suggested, new isotopic and palaeontological evidence indicate that the Lizard Complex was generated late in the Upper Palaeozoic, probably in the Lower or Middle Devonian. Isotopic dating of the gabbro yielded an Sm–Nd age of 375±34 Ma. This is interpreted as the date of formation of the oceanic crust (Davies, 1984). A whole rock Rb–Sr isochron age of 369±12 Ma for the cooling of the Kennack gneisses (Styles & Rundle, 1984) is concordant with the gabbro age (the basic fraction of the gneisses at Kennack sands contains cognate xenoliths of gabbro) and together the ages span the Givetian to Famennian stages. The presence of Lizard Complex debris in the mélanges structurally underlying the complex (Barnes, 1983), from which fossils as young as Frasnian have been recovered (LeGall *et al.*, 1985), gives a lower age limit for ophiolite obduction.

U-Pb dating of zircon from the Man o' War Gneiss (Sandeman *et al.*, 1997), orthogneisses forming the lowermost structural unit at Lizard Point, demonstrates that the parental magmas crystallized in the earliest Ordovician (499^{+8}_{-3}) and that they must represent a sliver of basement incorporated into the thrust pile during obduction.

RIDGE-AXIS TECTONICS

The dip of the dolerite dykes in the Crousa Downs Ophiolite decreases with increasing age. Set 1 dykes are gently inclined, those of Set 2 are moderately inclined and those of Set 3 are subvertical. These orientations, the occurrence of top-to-the-east shear zones, and the gentle north-easterly dip of the ophiolite, are consistent with a simple model of listric faulting at an oceanic spreading centre.

Figure 3.2 indicates the possible sequence of development. Firstly (*Figure 3.2a*), Set 1 dykes were intruded as early flat-lying dolerite sheets in gabbro near the top of an early magma chamber, followed by Set 2 dykes and the initiation of extensional shear zones dipping towards the ridge axis. These were subsequently rotated in blocks upon a series of listric faults which flatten into a basal shear zone near the petrological Moho (*Figure 3.2b*). The amount of rotation (about 30°) suggests about 40% thinning of the oceanic lithosphere may have then taken place (Roberts *et al.*, 1993). An evolved Fe–Ti-rich, Si-poor melt migrated along the shear zones during pulses of extension to produce a hybrid oxide-rich gabbro (Hopkinson & Roberts, 1995). Finally, vertical Set 3 dykes cut the underlying tectonized peridotite, the shear zones and rotated gabbro blocks (*Figure 3.2c*). This suggests that the Set 3 dykes were either intruded along-axis, or propagated from a later gabbro magma chamber developed below the already thinned crust. The high structural level of Set 3 dykes is confirmed by the development of chilled margins even within the peridotite.

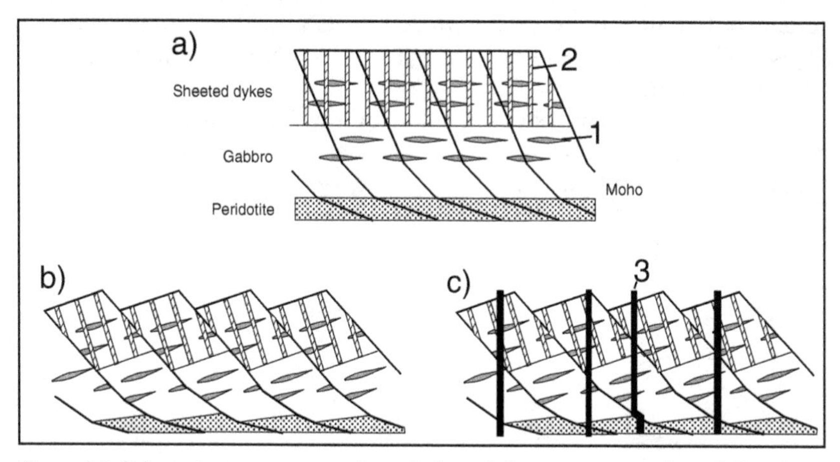

Figure 3.2. Schematic tectonomagmatic evolution of the east-coast section of the Lizard ophiolite: (a) generation of ophiolite stratigraphy and early flat-lying sheets (Set 1) cross-cut by Set 2 plagioclase phyric dykes; (b) activation of the extensional structures in response to continued spreading during a period of low magma production; (c) renewed magmatic activity with the injection of a series of aphyric dolerite dykes (Set 3)

The mix of lithologies in the Western Tectonic Unit is very strongly deformed in comparison to the Crousa Downs Ophiolite (Jones, 1997), yet there are occasional undeformed Set 3 dykes present in places. It is clear that most if not all of this deformation must have taken place close to, or at the palaeo-spreading axis, before the cessation of magmatic activity. Some stacking of the various units must have taken place, as the metasediments of the Old Lizard Head series lie structurally below the peridotites of the Western Tectonic Unit. On the other hand, Gibbons & Thompson (1991) suggested that some of the deformation is due to ductile extension at an oceanic spreading centre.

EMPLACEMENT/OBDUCTION

The present contacts of the Complex with the mélanges are controlled by late extensional faults. Deep seismic reflection data (BIRPS and ECORS, 1986; Le Gall, 1990) has imaged a structure inclined at 25–30° southwards, interpreted as a Variscan suture. This has carried the Channel Cadomian block over southern Britain. At the surface it corresponds to the Lizard Complex and Gramscatho Basin thrust system. Deformation accompanying the emplacement of the Lizard Complex over the Gramscatho Group mélanges is recorded by a development of a planar-linear fabric, and later spaced cleavages and accompanying north-northwest vergent folding (Rattey & Sanderson, 1984). None of these structural elements are developed in the overlying thrust sheet which carried the oceanic rocks northward. It seems to have acted as a cool, relatively competent sheet which had no thermal effect on the immediately underlying mudrock-dominated mélanges (Barnes & Andrews, 1984). There is no evidence of a hot-sole commonly seen beneath obducted ophiolites.

Emplacement occurred at the end of the Devonian or later as the Complex overlies sediments from which late Famennian palynomorphs have been recovered (Wilkinson & Knight, 1989).

Chapter Four

Devonian

'The Devonian was a System born from conflict ...'. With these words Martin Rudwick (1979) characterized a vehement scientific controversy which raged through much of the 1830s. The distinguished protagonists, Henry De La Beche, Roderick Murchison and Adam Sedgwick, drew into their dispute many of the leading scientists of the day, as attempts were made to resolve the interpretation of anomalous fossil evidence and complex field relations in South West England. At the outset, none of the participants could have anticipated the eventual solution: the creation by Sedgwick and Murchison in 1839 of a new system between the Silurian and Carboniferous, which included the continental rocks of the Old Red Sandstone and their marine equivalents in South West England. The complexities of the evolving arguments, the subtle and blatant shifts of opinion, together with the interactions between the principals were analysed with consummate skill by Rudwick (1985) in *The Great Devonian Controversy*, a volume recognized as an outstanding contribution to the history of science.

Despite the fact that the county boundary separating Cornwall from Devon has no geological significance, the problems at issue were recognized and resolved within Devon. It seems that the reported poor state of preservation of the fossils, and the structural complexity of Cornwall did not encourage travel to this relatively remote region.

Since those contentious times, the sedimentary rocks of the Devonian System have been clearly distinguished by their fossil contents. Besides giving an insight into the overall stage of evolution, these fossils also help to reveal the character of ancient environments. Continental conditions have only been recognized in the lowest parts of our Devonian sequence. Here we gain a glimpse of the spectacular evolutionary radiation of fish that was taking place at this time. Within fresh- and brackish-water sediments there is evidence of a community including jawless, armoured fish.

The marine fossil assemblages in Cornwall were derived from living communities formerly inhabiting a variety of depositional environments on the continental shelf and in the ocean beyond. Spire-bearing brachiopods dominated the benthos of shallow-water areas of sand and mud, together with strophomenoids, chonetoids, rhynchonelloids and a variety of bivalves. The diverse Middle Devonian communities associated with the stromatoporoid (sponge) reefs of the shallow-water carbonate complexes of Devon are not seen in Cornwall, although slipped blocks and turbiditic limestones, represented in the coeval basinal mudrock sequences of the county, indicate their presence on adjoining topographic highs. Frequent bottom anoxia meant that the sedimentary basins of Cornwall rarely sustained a rich benthos. Rather, the faunas were dominated by pelagic and planktonic species. It is these widely distributed forms, which evolved very rapidly in the Devonian, that provide a variety of schemes for the correlation of marine strata. Internationally recognized biozones based on ammonoid cephalopods (goniatites and clymenids), conodonts and thin-shelled 'fingerprint' ostracods are of particular importance in Cornwall. Such biostratigraphic dating methods allow geological events to be placed in relative order with considerable precision, but they do not tell us how long ago such events took place. Absolute dating depends upon radiometric methods, and these indicate that the Devonian spanned a period 408–360 Ma ago.

Despite the fact that South West England is the type locality for the Devonian, the stages that are used to characterize the sequence of marine faunas within the system take their names from continental Europe. The relatively simple tectonic settings at these localities aided faunal analysis and the characterization of biozones. It was not until the formulation of such schemes was well underway that any real progress could be made on the resolution of stratigraphic problems in Cornwall. Even now the poor preservation and sporadic occurrence of key index fossils continues to inhibit research in the county.

To facilitate international correlation in the Lower Devonian, the International Subcommission on Devonian Stratigraphy has determined that the Gedennian and Siegenian stages be replaced by the Lochovian and Pragian, respectively. These new stages, established in the Czech Republic, are broadly synonymous with the older ones but, in order to maintain clarity, the stages quoted below are those currently given in the literature.

It is generally agreed that, with the exception of parts of south Cornwall, Devonian sedimentation was initiated on Avalonian continental crust. At first, thick continental sedimentation extended across the region towards an ocean lying farther south. The progressive northern onlap of marine sediments across the continental shelf, which began in the late Lower Devonian, was accompanied by differential subsidence. It seems likely that this subsidence was basement-controlled, with movements on the deep-seated E–W and cross-cutting NW–SE structures. These allowed intrashelf (ensialic) extensional basins to develop and fill sequentially from south to north. In this setting, pulses of alkalic basaltic volcanism that punctuated basin development can conveniently be related to episodes of upper crustal extension along these structures.

Regionally (*Figure 4.1*), there is evidence for a basin with oceanic affinities in south Cornwall (the Gramscatho Basin), limited northwards by an outer shelf within which the Trevone Basin developed. Eastwards, along strike from the latter basin, vertical movements on deep-seated NW–SE basement faults created a topographic rise (Liskeard High) which separated and offset the Trevone Basin from the South Devon Basin. Later, and farther north on the shelf, the Culm Basin appeared, but there is no direct evidence in Cornwall for any Devonian sediments associated with this basin.

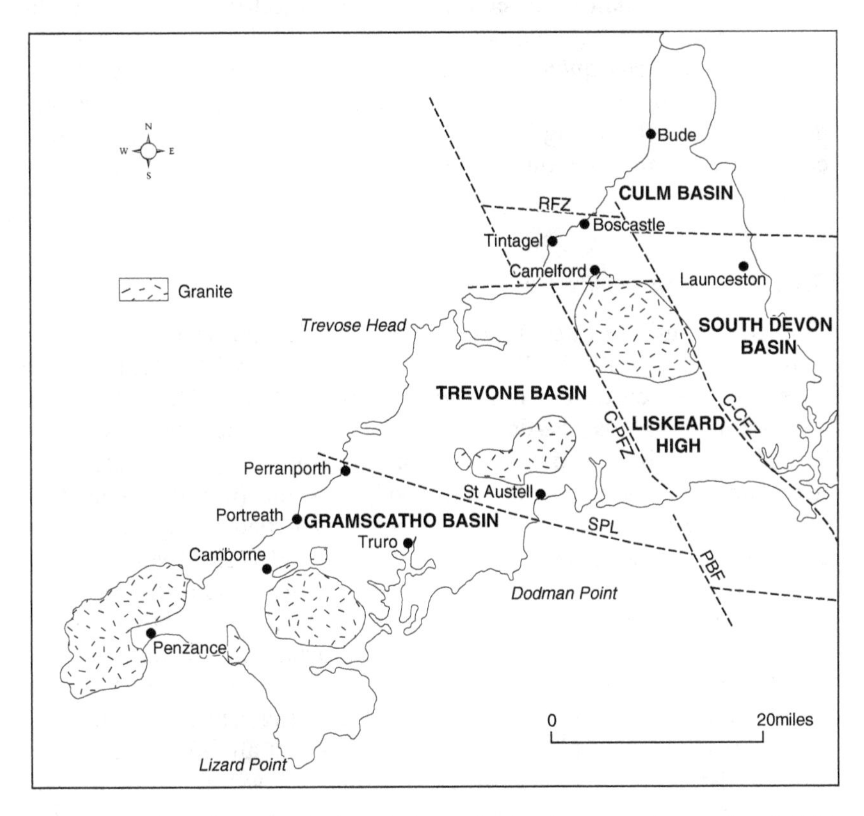

Figure 4.1. The sedimentary basins of Cornwall, showing controlling faults: C–PFZ, Cardinham–Portnadler Fault Zone; C–CFZ, Cambeak–Cawsand Fault Zone; SPL, Start–Perranporth Line; PBF, Plymouth Bay Fault; RFZ, Rusey Fault Zone.

LOWER DEVONIAN

The Lower Devonian succession is represented by the Dartmouth and Meadfoot groups. These are the oldest sedimentary rocks exposed in South West England and crop out in an approximately E–W-trending zone between Newquay and Torbay. Originally these rocks were included in the Grauwacke Group by De la Beche (1839), but subdivision into an older, non-marine, fish-bearing, mudstone-dominated succession (Dartmouth Slates), a sequence of marine mudstones (Meadfoot Group), and a series of marine sandstones (Staddon Grits), followed (Ussher, 1890). The combined succession was dated as Siegenian to Emsian on the basis of fish and brachiopod assemblages.

Lithostratigraphy, Sedimentology and Magmatism

The Lower Devonian succession in Cornwall is tectonically disrupted by a combination of approximately E–W and NW–SE-trending faults (*Figure 4.2*). Consequently, it is not possible to view the complete stratigraphy in any one continous section. The following account describes the lithostratigraphy in southeast Cornwall, where it is best constrained, and then outlines the characteristics of the succession sections farther west.

River Tamar–Portnadler Fault

The Dartmouth Group is exposed in the coastal sections between Bodigga Cliff [SX 276 541] and Portwrinkle [SX 359 538]. Although the base of the succession is not seen, it exceeds 4 km in thickness. The succession is mudstone-dominated, but sandstones become more common towards the top of the sequence. Barton *et al.* (1993) recognized five main lithofacies: (i) pale green muddy siltstone, (ii) purple mudstone or silty mudstone, (iii) pale grey, green or mottled green and mauve siltstone, (iv) pale greenish-grey sandstone, and (v) off-white to pale greenish-grey quartzite. The mudstones, muddy siltstones, and silty mudstones may exhibit colour mottling and form units between 1-2 m and, exceptionally, 30 m in thickness. Bioturbation is common. Thin conglomerates with erosive bases locally occur and contain flattened intraformational clasts of mudstone and fish fragments. The siltstones form structureless beds up to 2 m in thickness that are often interbedded with sandstones, typically less than 1 m thick. However, the sandstones may amalgamate into packets up to 4 m thick that locally exhibit erosional bases and low-angle cross-bedding. Fish remains found immediately below the Long Stone Thrust are of lower Middle Pragian age (Barton *et al.*, 1993). A mid–late Siegenian assemblage of corals,

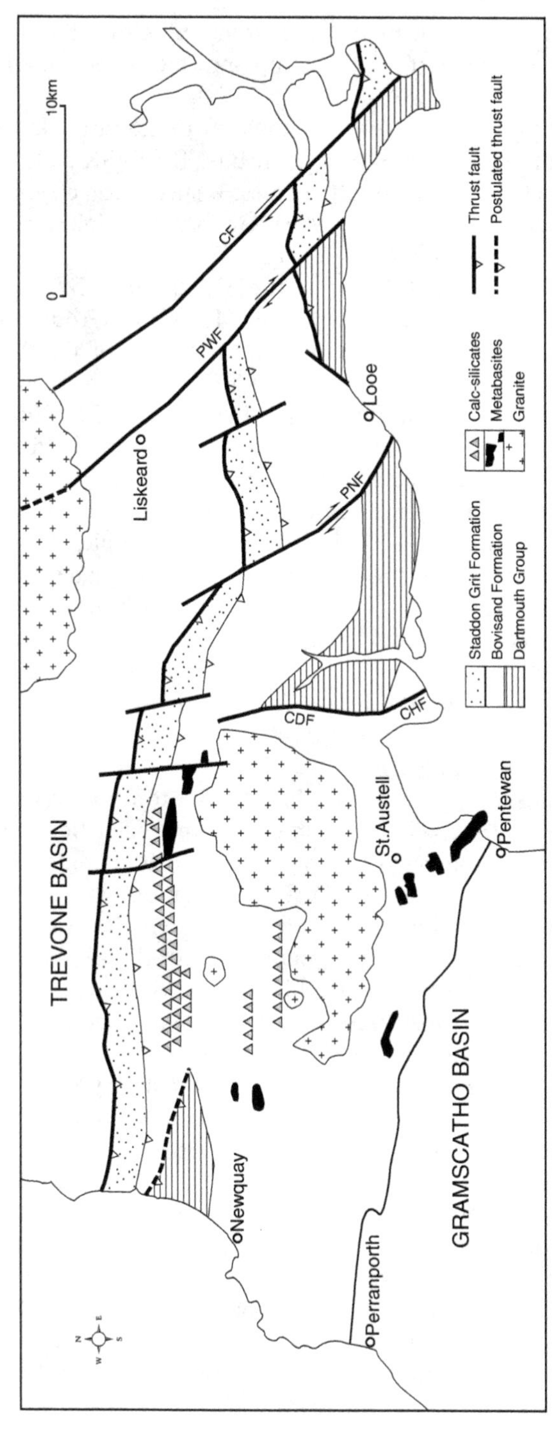

Figure 4.2. Distribution of Lower Devonian rocks in Cornwall: CDF, Castle Dore Fault; CF, Cawsand Fault; CHF, Coombe Hawne Fault; PNF, Portnadler Fault; PWF, Portwrinkle Fault (after Hobson, 1976; Barton *et al.*, 1993; British Geological Survey, 1994b)

bryozoans, and brachiopods recovered from Bull Cove [SX 4215 4845], close to the top of the Dartmouth Group, indicates a local marine incursion (Evans, 1981).

Crystal and crystal-lithic tuff horizons up to 2.5 m thick are locally developed within the sequence, such as at Bass Rock [SX 3230 5387] and The Pit [SX 3243 5393]. These tuffs possess a sheet geometry but exhibit intrusive contacts (Barton *et al.*, 1993). Gabbro and dolerite intrusions occur close to the Long Stone [SX 3375 5363].

A conformable contact between the Dartmouth Group and the overlying Bovisand Formation of the Meadfoot Group occurs at Tregantle Cliff [SX 386 528–SX 390 524], to the east of the Portwrinkle Fault. The most complete coastal section through the Bovisand Formation which occurs at Longsands Beach [SX 380 530] has been described by Barton *et al.* (1993). The lower 225 m of the formation is dominated by grey mudstone which is frequently bioturbated. Sandstone occurs as laminae and as lenses within mudstone–sandstone heterolithic sequences. Cross-lamination, ripples and climbing ripples, together with load casts are common lower in the sequence, whereas plane lamination is more prevalent in the higher beds. Thinly bedded bioclastic limestone horizons also occur higher within the sequence and include corals, crinoids and gastropods.

Above this succession, the Lower Longsands Sandstone Member is approximately 320 m thick and comprises a series of sandstone-dominated, thickening and coarsening-upwards cycles. Coarse bioclastic limestones up to 0.3 m thick may also occur close to the top of the coarsening-upwards cycles. The succeeding 150 m of the undivided Bovisand Formation is once again dominated by grey mudstone, and contains crinoids, shell fragments and solitary corals. The Upper Longsands Sandstone Member is approximately 100 m thick and similar to the Lower Longsands Sandstone, although cyclicity within the sandstones is not developed. These two members are likely to be equivalent to the Looe Grits, which are highly fossiliferous and have have yielded brachiopod assemblages of late Siegenian age (Evans, 1981).

The Staddon Grit Formation has an approximate thickness of 500 m. In southeast Cornwall the contact with the Bovisand Formation is faulted, but elsewhere it is conformable. The formation is exposed along the St Germans River and comprises thin to medium-bedded, fine to medium-grained sandstones, sometimes cross-laminated towards their tops, which are separated by mudstone partings (Barton *et al.*, 1993). The succession to the south of Liskeard is similar, although the sandstones are often somewhat coarser and more thickly bedded (Barton, 1994). Isolated occurrences of the Staddon Grit around Trevelmond [SX 2046 6351], to the west of Liskeard, yield a fauna indicative of the upper part of the Lower Emsian (Burton & Tanner, 1986).

Pentuan–Portnadler Fault

The Lower Devonian succession between the Portnadler Fault and the Castle Dore–Combe Hawne faults (*Figure 4.2*) is preserved within a large-scale southerly facing antiform. The Dartmouth Group is exposed in Lantivet Bay, where it occurs on the inverted limb of the antiform, and comprises grey, green and purple mudstones with thin to medium-bedded sandstones and rare basic igneous intrusions. There is a gradational boundary (Hobson, 1976) into the structurally underlying Bovisand Formation on the west side of Lantivet Bay [SX 1578 5106]. The Bovisand Formation largely consists of grey, locally calcareous, mudstones. At Great Lantic Bay [SX 1485 5077], thin to medium-bedded, fine to medium-grained quartz-rich sandstones, which can exhibit normal grading, plane lamination and/or ripple cross-lamination, together with load structures, are interbedded with mudstones and a heterolithic sandstone–mudstone sequence. Associated with the sandstones are thinly bedded, coarse bioclastic limestones which contain a crinoid, brachiopod and coral assemblage, plus rare intraformational conglomerates containing mudstone clasts and possible fish fragments. The combined succession is likely to correlate with one of the Longsands Sandstone members farther east. Brachiopod assemblages from sandstones around Fowey indicate a late Siegenian age (Evans, 1981).

West of the Castle Dore–Coombe Hawne faults, the Bovisand Formation crops out throughout St Austell Bay. Inland, calcareous sediments to the north of the St Austell granite have undergone contact metamorphism to form extensive areas of calc-silicate rocks described as calc-flintas by Ussher *et al.* (1909). A transitional boundary between the Bovisand Formation and the Gramscatho Group (Porthtowan Formation) occurs just to the north of Gamas Point, Pentuan [SX 0233 4720].

Magmatic activity within the Bovisand Formation is represented by numerous minor basic igneous intrusions and tuff horizons throughout St Austell Bay, with more substantial dolerites and local picrites inland. Between Spit Point [SX 0755 5244] and Fishing Point [SX 0679 5223], pale green tuff horizons are interbedded with mudstones and sandstones. They are largely bedding-parallel, but cross-cutting relationships locally occur suggesting an intrusive origin. A rhyolite sill of pre-folding age, just north of Hendra [SX 985 643], was described by Ussher *et al.* (1909).

North Coast

On the north coast (*Figure 4.2*), the Dartmouth Group crops out in Watergate Bay. Pale grey mudstones, with occasional thinly bedded, fine-grained sandstones and variegated purple and green mudstones predominate between Ontonna Rock [SW 8433 6575] and Horse Rock

[SW 8363 6418]. The succession to the south of Horse Rock is dominated by green and purple mudstones that contain fossil fish and possible phosphatic concretions. Fossil fish from horizons north of the Watergate Bay Hotel [SW 8413 6501] include *Althaspis leachi*, *Rhinopteraspis cornubica* and *Europrotaspis* sp., and an early to mid-Siegenian age is indicated (White, 1956). Palynological studies in Watergate Bay indicate an uppermost Gedinnian to lowermost Siegenian age for the Dartmouth Group (Davis, 1990).

A conformable transition into grey mudstones and thin to medium-bedded sandstones of the Bovisand Formation occurs at Whipsiderry Beach [SW 8320 6348]. Farther to the south, the Bovisand Formation largely comprises grey and green mudstones with subordinate thin to medium-bedded, plane and/or cross-laminated sandstones. Bioclastic limestones also occur. Approximately 40 m of marble, apparently of sedimentary origin, has been proved in a borehole 3 km north of Perranporth, and is associated with calc-silicate rocks similar to the calc-flintas exposed north of the St Austell granite (Goode & Merriman, 1977). Basic igneous intrusions occur in the southern part of the succession (Ussher *et al.*, 1909) where the boundary with the Gramscatho Group is not well constrained.

A thrust contact between the Bovisand Formation and the Staddon Grit Formation (British Geological Survey, 1994b) is exposed immediately east of Trenance Rock, Mawgan Porth [SW 848 647]. The Staddon Grit comprises fine to medium-grained sandstones interbedded with mudstones and is poorly fossiliferous. The boundary between the Dartmouth and Meadfoot groups and the Trevone Basin succession is formed by a thrust fault which transports the Staddon Grit northwards over the Meadfoot Group (Bedruthan Formation) to the southeast of Carnewas Island [SW 847 690].

Synthesis

Palynological evidence from Watergate Bay indicates that deposition of the Dartmouth Group had commenced by the late Gedinnian. The presence of a fresh- or brackish-water pteraspid fauna, suggested that the Dartmouth Group in Cornwall was deposited in a series of inland seas and lagoons (Reid & Scrivenor, 1906). More recent sedimentological studies essentially confirmed this view, and indicated that deposition took place within perennial lakes and distal alluvial fans or fluvial systems sourced from the east (Barton *et al.*, 1993).

Synsedimentary faulting appears to have controlled the occurrence of pebbly mudstones in south Devon and suggests an active tectonic regime (Smith & Humphreys, 1989, 1991). Faunal assemblages (Evans, 1981) and phosphatic concretions (Humphreys & Smith, 1989) indicate that marine incursions periodically occurred, particularly towards the top of the succession. Such incursions may also have been tectonically controlled. Tuff horizons and basic intrusive rocks in Cornwall, together with bimodal volcanicity in south Devon (Durrance, 1985), imply that the estimated 4 km of Dartmouth Group sediments were accommodated by subsidence associated with the early stages of continental rifting. Extension across E–W-trending faults in the pre-Devonian basement probably formed a series of half-grabens in which east to west axial transport of sediment was predominant. Although comparisons are frequently made between the Dartmouth Group and the Old Red Sandstone sequences in South Wales, there is presently no direct evidence to indicate that these successions form segments of a linked depositional system.

The mixed siliciclastic and carbonate shallow-marine conditions of the Bovisand Formation were established in mid-Cornwall by the late Siegenian (Evans, 1981), but are likely to be diachronous across the region (Evans, 1985). Fossil fish fragments, locally found within the Bovisand Formation (Ussher, 1907), could represent post-mortem transport of remains from a nearby non-marine environment (Evans, 1981), or fish tolerant of increased salinity. There are no recent detailed published accounts of the sedimentology of the Bovisand Formation in Cornwall. However, in south Devon the following depositional environments have been proposed: (i) a south-facing shelf dominated by high energy bipolar E–W tides (Richter, 1967), (ii) a storm-dominated shelf (Pound, 1983; Humphreys & Smith, 1989), and iii) a tidal flat and lagoonal model (Selwood & Durrance, 1982). Magmatic activity, both intrusive and extrusive, and in part bimodal, continued during the deposition of the Bovisand Formation and attests to the continued rifting of continental lithosphere. The change from the alluvial and lacustrine environments of the Dartmouth Group to the marine conditions of the Bovisand Formation may therefore be controlled by subsidence. However, Humphreys & Smith (1989) suggested that *in situ* phosphatic concretions within the Bovisand Formation may correlate with a major Devonian eustatic transgressive event.

The transition to the Staddon Grit had occurred by the lower Emsian in mid-Cornwall (Burton & Tanner, 1986), but is also likely to be diachronous across the region (Evans, 1985). In Devon, the following depositional environments have been suggested: (i) an offshore bar (Selwood & Durrance, 1982), (ii) a fluvial-dominated, low wave-energy delta (Pound, 1983), and (iii) coastal plain and shallow, sand-dominated ephemeral stream (Humphreys & Smith, 1989). Pound (1983) suggested that the Staddon Grit Formation might be sourced from ephemeral uplifts associated with E–W-trending fault zones, and so revived the Staddon ridge concept of Hendriks (1959). Such a feature would have formed a topographic high along the southern margin of the nascent Trevone Basin. A major ESE–WNW-trending basement fault zone, the Start-Perranporth Line (Holdsworth, 1989), runs approximately parallel to the southern limit of the Meadfoot Group. It probably marked the shelf break, to the south of which the Gramscatho Basin developed, but is unlikely to represent a Devonian terrane boundary. The Treworgans Sandstone Member of the Gramscatho Group is interbedded with the upper part of the Bovisand Formation and represents a sandstone-rich depositional system developed adjacent to this shelf break. It is possible that this formed part of a linked depositional system with the Staddon Grit (Hendriks, 1937).

GRAMSCATHO BASIN

The stratigraphy, age and interpretation of the low-grade metasediments in south Cornwall are topics of continuing controversy. The difficulties arise from a combination of poor biostratigraphical control, the widespread absence of distinctive lithologies, and considerable structural complexity.

The first detailed lithostratigraphy was proposed by Hill & MacAlister (1906). In this, the newly defined Grampound and Probus Series was assigned a Devonian age, while the Mylor, Falmouth, Portscatho and Veryan series were classified either as Ordovician or undivided Lower Palaeozoic. Later, Hendriks (1931, 1937) grouped the relatively sandstone-rich Falmouth, Portscatho, Veryan (part), Grampound and Probus series into the Gramscatho Beds, and suggested a Middle Devonian age on the basis of fossil plant material. Much later, further rationalization and formalization of the lithostratigraphy followed investigations by the Geological Survey (Leveridge *et al.*, 1984; Holder & Leveridge, 1986a; Leveridge *et al.*, 1990).

The dating of microfossils from parts of the succession previously regarded as barren has indicated Upper Devonian ages and implies that the Mylor Slate Formation is younger than most of the Gramscatho Group (Turner *et al.*, 1979; Le Gall *et al.*, 1985). In addition, interpolation from offshore seismic reflection profiles suggests that significant Variscan thrust faulting probably occurred at, or close to, the boundaries between certain lithostratigraphical units.

Lithostratigraphy, Sedimentology and Magmatism

The distribution of major lithostratigraphic units in south Cornwall (*Figure 4.3*) is controlled by thrusts that define a northern parautochthonous region, plus the Carrick, Veryan and Dodman nappes. The stratigraphy of each is described below and summarized schematically in

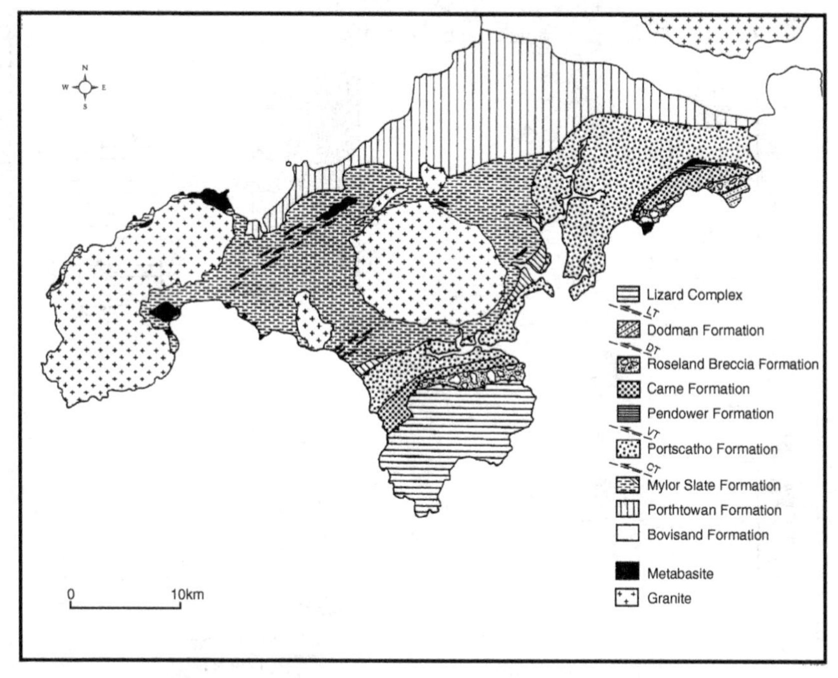

Figure 4.3. A simplified geological map of south Cornwall (after Leveridge *et al.*, 1990): LT, Lizard Thrust; DT, Dodman Thrust; VT, Veryan Thrust; CT, Carrick Thrust

Figure 4.4. Successions generally dip and young upwards to the southeast. There is no evidence of large-scale stratigraphical inversion. The lithostratigraphical nomenclature is largely after Leveridge *et al.* (1990), but the Gramscatho Group is here revised to include the Mylor Slate Formation, Roseland Breccia Formation and Dodman Formation.

Parautochthon

The lowest stratigraphical unit of the Gramscatho Group within the parautochthon is the Porthtowan Formation. Thickly bedded, medium to coarse-grained, disorganized sandstones, distinguished as the Treworgans Sandstone Member (Leveridge *et al.*, 1990), are interbedded with pale green mudstones and thinly bedded, quartz-rich sandstones of the Meadfoot Group just to the north of Gamas Point, Pentuan [SX 0233 4720]. This member, which represents deposition from high-concentration turbidity currents, can be traced west-northwestwards through a series of quarries, but thins markedly west of Ladock [SW 893 508]. Farther west, the lower parts of the Porthtowan Formation are largely mudstone-dominated, and were classified as the Perran Shales by Reid & Scrivenor

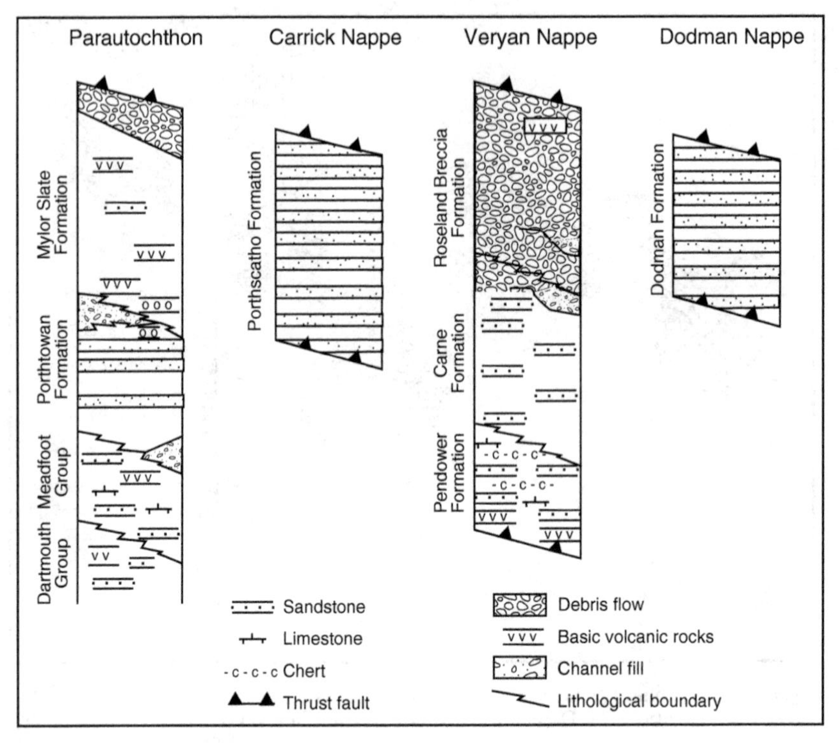

Figure 4.4. Summary lithostratigraphy and tectonic units within the Gramscatho Basin (after Holder & Leveridge, 1986a)

(1906). These beds include thinly bedded siltstone turbidites and occasional thin to medium-bedded sandstone turbidites. They are probably transitional with the Meadfoot Group in the vicinity of Carn Haut [SW 7499 5600], just north of Perranporth.

Exposures of the Porthtowan Formation along the coast from Perranporth to Lelant Towans [SW 5470 3825] provide a section through an upwards-younging succession. This comprises muddy siltstones, structureless or graded grey, green and purple mudstones, graded-stratified siltstones, and thin to very thickly bedded sandstone–mudstone couplets. Most are interpreted as turbidites, but some mudstones may represent hemipelagites. The sandstones are muddy and often occur in distinct packets that lack internal organization. They probably originated as unconfined sheet flows, and their occurrence may be controlled by proximity to point sources within the basin or variations in sediment supply to its margins. Proximity to a slope is indicated by up to 60 m of slide and slump deposits east of Fishing Cove [SW 5989 4281]. The traditional bipartite lithostratigraphical division into Gramscatho (sandstone-dominated) and Mylor (mudstone-dominated) successions is particularly difficult between Navax Point [SW 5924 4370] and Lelant Towans, as the two associations are more or less equally represented (Shail, 1989). At Black Cliff [SW 5542 3880] and Lelant Towans, medium to thickly bedded sandstones account for approximately 80% of the succession, and display sedimentary structures typical of turbidites. Limited slump fold data implies a depositional slope dipping towards the southeast. This is compatible with ripple cross-lamination, which indicates palaeocurrents from all directions except the southeast. This uppermost part of the Porthtowan Formation probably represents sandstone-rich channels cutting through slope-apron mudstone and slump deposits (Shail, 1989). Frasnian palynomorphs within the upper part of the succession have been reported by Leveridge *et al.* (1990) from Porthcadjack Cove [SW 641 447].

The Mylor Slate Formation is generally characterized by repetitive graded-stratified siltstones and graded mudstones, plus infrequent thin to medium-bedded, muddy fine sands and slump deposits. Assemblages of miospores and acritarchs recovered from lower parts of the succession at Mount Wellington Mine [SW 7758 4128] indicate a Famennian age (Turner *et al.*, 1979). Together with mine and borehole data, these suggest that the Mylor Slate overlies the Porthtowan Formation (Leveridge *et al.*, 1990). The Porthleven Breccia Member (Holder & Leveridge, 1986a) represents the uppermost parts of the formation and includes disorganized muddy gravels, gravelly muds and chaotic deposits. Many of the clasts are similar to the sandstone of the structurally overlying Portscatho Formation.

Metabasites (greenstones) account for approximately 10–20% of the Mylor Slate Formation. Most of these occur in the lower half of the formation to the west of St Ives and in a NE–SW zone between Penzance and Camborne. A further NE–SW zone occurs towards the top of the formation between Restronguet and Porthleven. High-level sills predominate, but there are also pillow lavas and rare volcaniclastic debris flows such as at Great Hogus [SW 5130 3040]. The contemporaneity of intrusive and extrusive events with sedimentation has been demonstrated by Taylor & Wilson (1975). Geochemical studies indicate intraplate tholeiitic affinities (Floyd, 1984).

Carrick Nappe

The Portscatho Formation displays a faulted contact with the underlying Mylor Slate and Porthtowan formations, and is the only lithostratigraphical unit to occur within the Carrick Nappe. The dominant facies types are similar to those of the Porthtowan Formation, although the proportion of sandstone is higher. These sandstones include poorly sorted, muddy sandstones with slurried tops that are often rich in mudclasts. Sandstone packets up to 30 m thick occur, particularly within the upper parts of the formation, and occasionally exhibit thinning and fining-upwards sequences. However, the more usual lack of channelling and internal organization suggest that sedimentation was largely in the form of unconfined sheet flows. Strained flute and groove marks can be seen at Pendower Beach [SW 8917 3784]. Minor slump horizons occur in Halzephron Cove [SW 6560 2178]. Pyritic mudstones are common, and suggest periodic anoxic conditions close to the basin floor. No volcanic rocks are observed. Fossil wood (*Dadoxylon*) recovered from the top of the Portscatho Formation and dated as Middle Devonian (Lang, 1929) has been reassessed as 'no older than lowest Upper Devonian' (Leveridge *et al.*, 1990). Miospore and acritarch assemblages recovered immediately south of the Loe Bar [SW 646 235], close to the base of the formation, indicate a late Famennian age (Wilkinson & Knight, 1989). Assemblages from structurally higher levels in the formation around Pennance Point [SW 804 306] and St Mawes Castle [SW 841 327] indicate a Frasnian age (Leveridge *et al.*, 1990), while those close to the top of the formation at Jangye-ryn [SW 659 206] indicate a mid–late Frasnian age (Le Gall *et al.*, 1985). The late Famennian age reported by Wilkinson & Knight (1989) brings into question the position of the Carrick Thrust, as it implies significant thrusting must have occurred to the south of Loe Bar. The abundance and diversity of miospore assemblages suggest a high terrestrial input to the basin. Lower to Middle Devonian (Wilkinson & Knight, 1989) and Ordovician (Le Gall *et al.*, 1985) miospores indicate the erosion of older successions.

Veryan Nappe

The Pendower Formation only occurs in Roseland, where it is exposed east of Pendower stream [SW 8982 3817]. Four main lithofacies associations occur: (i) structureless and graded-stratified siltstones plus varicoloured, graded and laminated mudstones; (ii) thinly bedded, coarse-grained siliciclastic sandstones; (iii) thinly to very thickly bedded calciclastic sandstone–mudstone couplets, and (iv) biogenic cherts. These are interbedded on various scales throughout the succession. Calciclastic turbidites are well developed to the east of Gidley Well Stream [SW 9080 3823]. Conodonts recovered from the limestones range from lower to middle Eifelian at the base of the formation, to uppermost Eifelian close to its top (Sadler, 1973). Partially boudinaged ochrous horizons interbedded with silicified black mudstones, which occur on Pendower Beach [SW 8995 3816], have Rare Earth Element chemistries of Mid-Ocean Ridge Basalt (MORB) type, and suggest contemporaneous basic volcanicity. Basic volcanic rocks are also exposed at the base of the Pendower Formation in quarry exposures at Tubb's Mill [SW 9620 4325]. These exhibit an enriched MORB-type geochemistry (Floyd, 1984).

The Carne Formation is dominated by fine to medium-grained, medium to thickly bedded structureless or planar-stratified sandstones, which often display considerable pre-lithification disruption by gravity-driven sliding and slumping. The succession was considered by Holder & Leveridge (1986a) to represent an association of slope and proximal submarine fan deposits. A conformable contact is observed with the underlying Pendower Formation in Gerrans Bay [SW 9089 3815], but elsewhere the base of the formation is faulted against the Portscatho Formation. There is no palaeontological data available for the Carne Formation, but since the upper parts of the underlying Pendower Formation have a latest Eifelian age (Sadler, 1973) a Givetian age is likely. In eastern Meneage, the Menaver Conglomerate (Flett, 1946) is exposed on the inverted limb of a large-scale fold and youngs upwards into a disrupted sandstone-dominated succession equivalent to the Carne Formation. The conformable contact between the Carne Formation and Roseland Breccia Formation is marked by the appearance of quartzite olistoliths at Pennarin Cove [SW 9125 3780] and quartz-mica schist/phyllite rudites at Little Perlea [SW 9513 4059].

The Roseland Breccia constitutes a major olistostrome (Barnes, 1983, 1984) deposited by mud-rich debris flows and large-scale slope failure processes. Individual quartzite olistoliths up to 100 m in length occur close to Carne [SW 9127 3805]. This association suggests a lower slope-apron environment in which periodic large-scale collapse was initiated by tectonic processes. Monomict and polymict conglomerates also occur locally in the form of channel fills. Monomict conglomerates include mafic amphibolites, quartz-mica schists and alkali-granite. They suggest very little mixing, and were probably derived from locally sourced fan-deltas close to the basin margin. Contemporaneous basaltic and acidic volcanicity has been documented within the upper parts of the Roseland Breccia. Virtually all fossil material is found within exotic blocks and must predate deposition. The quartzite olistoliths exposed in Roseland contain a trilobite and brachiopod fauna that indicates a Llandeilo age. These faunas have been correlated with the Grès de May inférieur (Petit May) of Normandy and the Grès de Kerarvail and schistes de Morgat of Finistère (Sadler, 1974a; Bassett, 1981). A synthesis of the conodont data recovered from limestone clasts has been presented by Austin *et al.* (1985). Most are Gedinnian to Emsian age, although a few Eifelian forms have also been recovered. Lower Eifelian ostracods occur within cherts in eastern Meneage (Cooper, 1987). The closest constraint on depositional age is provided by the Frasnian conodonts recovered from limestones associated with the Mullion Island pillow lavas (Hendriks *et al.*, 1971), but even these are likely to be somewhat older than the olistostromes. A Frasnian to Famennian age seems most probable, and suggests that the Roseland Breccia may be partially contemporaneous with the Mylor Slate.

Nare Head Greenstone

The Nare Head Greenstone occupies an onshore area approaching 1 km^2 and is about 100 m thick. A subhorizontal fault is observed to place pillow lavas of the Nare Head Greenstone over disrupted sandstones of the Roseland Breccia on the east side of Nare Head [SW 9240 3785]. Pillow lavas *(Plate 1)* and pillow breccias are superbly exposed northeast of Venton Vadan [SW 9180 3701] and are associated with massive dolerite sills and localized gabbros. The geochemistry of the Nare Head volcanic rocks indicates affinity with enriched MORB, suggesting development in a marginal oceanic basin (Floyd, 1984).

Dodman Nappe

The Dodman Formation is the sole lithostratigraphical unit contained within the Dodman Nappe (Leveridge *et al.*, 1984; Holder & Leveridge, 1986a). It comprises a series of phyllitic mudstones and sandstones which is similar to the Portscatho Formation, and was thrust over the Roseland Breccia.

Evolution of the Gramscatho Basin

The Gramscatho Basin has featured prominently in tectonic models for the Variscides of southern Britain as a consequence of its relatively internal position and close spatial association with the Lizard Ophiolite Complex. This section synthesizes stratigraphical and provenance evidence from the parautochthon and nappe units into a possible model for the Devonian evolution of the basin.

Gedinnian to Eifelian

Along the northern margin of the basin, the lacustrine and coastal mudflats of the Dartmouth Group were established by the late Gedinnian and succeeded by the mixed shallow-marine shelf of the Meadfoot Group by the late Siegenian. Contemporaneous basic volcanicity in both environments suggests that rifting of continental lithosphere was taking place. It is likely that the upper parts of the Meadfoot Group and lower parts of the Porthtowan Formation record contemporaneous deposition across a mudstone-dominated, south-facing shelf and slope, into the Gramscatho Basin. The transition to this basinal area was probably controlled by the Start–Perranporth Fault Zone. The sandstone-dominated upper parts of the Meadfoot Group exposed around Liskeard have a mid-Emsian age (Burton & Tanner, 1986) and might represent the deltaic or shallow-marine segment of a sediment transport system that eventually linked with the sandstone-rich slope apron of the Treworgans Sandstone Member (Hendriks, 1937). Lower Devonian limestone clasts contained within the Carne and Roseland Breccia formations imply that the southern margin of the basin was largely starved of siliciclastic sediment during this period. The mid–late Eifelian assemblage of radiolarian cherts, mudstones and lavas with MORB affinities preserved within the Pendower Formation is typical of small oceanic-rift basins such as the Red Sea or Gulf of California. The association suggests that continued rifting had, at least locally, brought about complete attenuation of continental lithosphere. Much of the Lizard Ophiolite Complex may have formed by this time. Calciclastic turbidites, together with their subordinate siliciclastic component, are likely to have been sourced from reefs developed on basement fault blocks within the basin.

Givetian to mid-Frasnian

Variscan convergence between South West England and the Armorican Massif was probably initiated around this time. Deep-marine siliciclastic sedimentation, represented by the undivided parts of the Porthtowan Formation, continued within the northern part of the basin, but input was minimal as a consequence of reduced or drowned topography. There is no evidence of contemporaneous volcanicity. Along the southern margin of the basin, thrust faulting brought about the break-up of the limestone platforms that had been the source of the calciclastic turbidites of the Pendower Formation. A major change in provenance followed, resulting in the siliciclastic sandstones of the Carne Formation (Holder & Leveridge, 1986a). Farther to the north the Portscatho Formation was deposited from unconfined sheet flows with no evidence of volcanicity. It is probable that the Portscatho and Carne formations, together with the mixed rudites of the Roseland Breccia may, at least in part, have formed segments of a linked depositional system developed in front of the northwards migrating thrust system.

Late Frasnian to Famennian

Renewed rifting, initiated close to the Frasnian–Famennian boundary, was heralded by a temporary increase in the proportion of sandstone within the uppermost parts of the Porthtowan Formation. As rifting progressed into the Famennian, there was a reduction in sandstone input and widespread basic intraplate volcanicity. The Mylor Slate Formation probably represents the progradation of a southeast-facing, mudstone-dominated slope apron. The same rift event was probably responsible for the generation of bimodal magmatism (Kennack Gneiss) within the partially assembled Lizard Ophiolite (Chapter 3), and bimodal volcanicity within the Roseland Breccia Formation.

Late Famennian to Tournaisian

Convergence appears to have resumed during the late Famennian. Major olistostromes in the Roseland Breccia were developed along the active thrust front at the southern margin of the basin. Basic volcanic and metavolcanic rocks became important source lithologies in the latter stages of closure, and attest to uplift and erosion of the basin floor. The Porthleven Breccia probably represents the erosion of the emergent sandstone-dominated Carrick Nappe (Leveridge & Holder, 1985). Following basin closure, the lithologies described above underwent thrust-related burial, deformation and regional metamorphism.

TREVONE BASIN

Intrashelf subsidence of the Trevone Basin commenced in the late Lower Devonian, and continued well into the Dinantian, giving a sedimentary fill in excess of 6000 m. At first, a common sedimentary sequence built up across the area, but in early Upper Devonian times the onset of volcanism led to the differentiation of the Pentire, Trevone, and Bounds Cliff successions. These successions (*Figure 4.5*) (Gauss & House, 1972; Selwood *et al.*, 1993) allow an E–W-trending half-graben basin to be modelled (*Figure 4.6*), in which the Pentire Succession is associated with the footwall scarp and the Trevone Succession with the hanging-wall dip-slope. The northern margin of the basin, which is characterized by the depositional instability and intense volcanic activity revealed within the Pentire Succession, is believed to have been controlled by E–W normal faulting linked to extensional reactivation of a basement thrust. Although the original geographical location of this margin is unknown, stratigraphical relationships between the Trevone and Pentire successions suggest that the Pentire Succession originated well north of its present position. The eastern margin of the basin approximates to the line of the Cardinham–Portnadler Fault Zone, which appears to be the surface expression of a NW–SE, deep-seated fracture that also controlled sedimentation. This fracture provides the western limit to the Liskeard High, separating the Trevone and South Devon basins (Selwood, 1990). The Bounds Cliff Succession may be represented as the transition between the basinal deposits of the Trevone Succession and the adjoining shelf.

The widespread distribution of the basinal Trevose Slates indicates that syn-rift sedimentation was well established by the early Middle Devonian. The conodont and palynomorph assemblages, goniatites

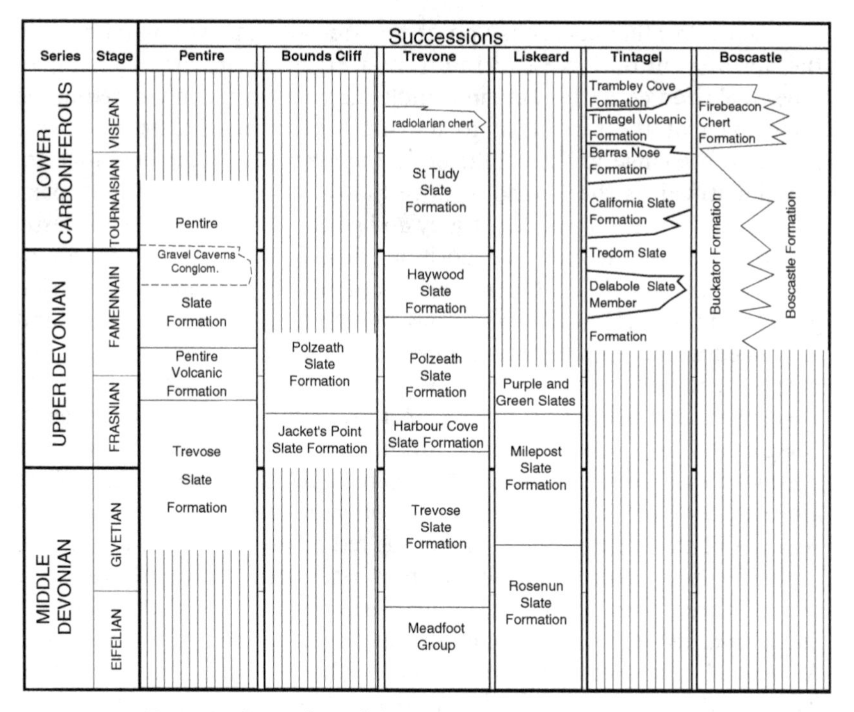

Figure 4.5. Successions of the Trevone Basin and adjoining areas

(House, 1963; Gauss & House, 1972) and infrequent horizons rich in styliolinids, all confirm a fully marine environment. About 4000 m of laminated, hemipelagic mudrocks were deposited over a period extending into the early Frasnian. Goode & Leveridge (1991) suggested a barred basin environment with rapid subsidence and fill, and that sporadic bottom-water anoxia was indicated by horizons of pyritous black slates. The limestone turbidites of the Marble Cliff Limestone, near the top of the formation in the Trevone Succession, suggest an intrabasinal carbonate rise lying to the southwest (Tucker, 1969). There is no evidence that the southern margin of the Trevone Basin was either tectonically active, or markedly differentiated. Within the proposed extensional basin setting, a hanging-wall dip-slope can be modelled, which was continuous with the submerged shelf bounding the Gramscatho Basin to the south. Holder and Leveridge (1986a) recorded the northward onlap of the basinal sediments (Porthtowan and Mylor Slate formations) from the Gramscatho Basin across this shelf, in a sequence generally coeval with the Trevone Succession. The depositional continuity between the Trevone and Gramscatho basins that is indicated is consistent with the barred basin model for the Trevone Basin.

Whereas continuous deposition of the Trevose Slate was maintained in the north (Pentire Succession) until mid-Frasnian or early Famennian times (Selwood *et al.*, 1993), minor tholeiitic volcanism commenced in the upper part of the formation to the south (Trevone Succession). This produced a change in the character of the associated sediments that are now identified as the Harbour Cove Slate Formation. The mudrocks changed from black to shades of grey and green, and falls of volcaniclastic material were associated with depositional instability and the generation of small-scale slump folds. Open marine conditions are indicated by rich goniatite faunas (House, 1963). Towards the top of the formation, and associated with evidence for a decline in volcanic activity, intercalations of purple slate occur. Such slates typify the succeeding Polzeath Slate Formation, and their appearance coincided with a widespread change to pelagic red-bed sedimentation throughout the Variscan basins of northwest Europe. The youngest faunas recorded from the Polzeath Slate are ostracods of the mid-Famennian *intercostata* Zone (Beese, 1984). Analysis of this facies has shown that the red pigment indicates a deficit in organic supply caused by oligotrophy associated with a decrease of sedimentary input from Caledonian source areas. In the Trevone Succession, red-bed deposition effectively ceased at the hypoxic or anoxic Annulata Event; a eustatic transgression with a worldwide representation (House, 1985). In Cornwall, this is recognized by the occurrence [SW 9988 7789] of *Posttornoceras inflexuoideum*, and clymenids within a black, ammonoid-bearing slate that marks the base of the Haywood Slate Formation. This newly recognized formation forms a disparate association of argillites, sandy argillites and volcanic rocks, some 100–200 m thick. The formation has been mapped east of Trevigo [SW9698 7937] (*Figure 6.3*) in normal succession above the Polzeath Slate, and as fault-bounded slices in the coastal section at Slipper Point [SW 9341 7951]. The formational base is marked by a change in the colour of the mudrocks from the uniform purple of the Polzeath Slate to a mixture of green, dark grey and black slates. Subordinate horizons of purple slate decrease rapidly in importance upwards. The type section, including the basal contact, is a continuous lane section south of the entrance to Trevathan Farm, St Endellion [SW 9988 7789]. These beds can be interpreted to signal a return to eutrophy associated not only with sustained transgression but also with greater clastic input and episodic volcanism. Entomozoacean ostracods from horizons lying above the black slates, which correlate with the Annulata Event, include *Richterina (Richterina) striatula*, *R. (R.) costata* and *Maternella dichtoma* from the *hemisphaerica-dichtoma* Zone of the upper Famennian. This age is in accord with a palynomorph assemblage recorded by Selwood *et al.* (1993) from near the top of the formation southwest of St Tudy [SX 5070 7542]. Volcanism is represented by lenticular lavas and tuffs that are particularly abundant east of the Allen Valley Fault (*Figure 6.2*). Volcanic rocks increase in importance upwards through the formation and into the overlying, largely Carboniferous, black mudrocks of the St Tudy Formation.

The Pentire Volcanic and Pentire Slate formations, which correlate with the Polzeath and Haywood slates (*Figure 4.5*) in the Pentire Succession, indicate rapid sediment accumulation under unstable conditions with much slumping. At first, the dark to mid-grey, often laminated, mudrocks typical of the Trevose Slate persisted into the Pentire Volcanic Formation. The extrusive volcanic rocks characterizing the latter formation are not older than mid-Frasnian, and appear about 20 m above the anhydrous alkalic dolerite intrusions marking the upper part of the Trevose Slate (Selwood *et al.*, 1993). Fine-grained, aphyric, alkali basalt lava flows and water-lain tuffs with pillow lavas predominate, with most of the basalts being vesicular or highly vesicular. The associated volcaniclastic rocks range from pillow breccias, through coarse-grained epiclastic tuffs, to fine-grained resedimented tuffs. The absence of red-beds in the Pentire Succession appears to reflect the high organic content within the sediments, which may be related to volcanic activity.

The main extrusive phase, characterizing the Pentire Volcanic Formation, opened with explosive magmatism. Volcaniclastic horizons in cliff sections at Sandinway Beach [SW 9360 8085] and Reedy Cliff [SW 9773 8098] give evidence of quench brecciation as the erupting lava interacted with seawater. Magma–water–gas–sediment interactions led to extensive fluidization of wet sediments as lava flows occurred. Sediment became entrained within the flows, and contacts frequently show the complex interpenetrations and clastic texture of peperites. Characteristically, the lower lavas in the sequence are either massive or not clearly pillowed. Pillows in the overlying horizons are highly variable in size, include frequent lobate and tube-like forms, and frequently show steep depositional dips. Cavity pillows ,which characterize the upper parts of flows, are much disrupted by fluidization that caused an intimate mixing of lava and sediment. Surface flow-cooling patterns can occasionally be recognized on flows. All of the pillows are highly vesicular, and show concentric patterns produced by gas boiling off and being entrained in distinct lines of bubbles during flow. A load-flattening fabric of vesicles characterizes pillows at the base of the pile. Minor amounts of chert and calcite, possibly precipitated by reaction between seawater and gas-rich magma, are sporadically represented between pillows. The overall physical appearance of the pillows more closely resembles pahoehoe lava toes, rather than typical deep-sea pillows (Selwood *et al.*, 1993). Such evidence is consistent with the low confining pressures associated with the development of peperites. The low-density pillows appear to have come to rest gently on or against each other, and do not show the characteristic 'wrap around' features of higher-density pillows.

Palynomorph assemblages from the green to grey mudrocks of the overlying Pentire Slate Formation (Selwood *et al.*, 1993) indicate an upper Famennian–Lower Carboniferous age, and establish it as the youngest formation in the Pentire Succession (*Figure 4.5*). Sheet-like and lenticular conglomerates, represented particularly towards the top of the formation, have been identified as the Gravel Caverns Conglomerate Member. These conglomerates are matrix-supported, and contain clasts of well-rounded siliceous mudstone, limestone, quartz-veined conglomerate, chert and a variety of locally derived igneous rocks. Although scoured and channelled bases occur commonly, more complex relationships are observed that involve considerable invasion of the enclosing mudrock by pebbles of the conglomerate. These occurred at both upper and lower contacts to give fluidization phenomena. It seems that significant dewatering processes operated soon after deposition. The scarcity of bioturbated horizons and the presence of planar slump beds suggest relatively rapid sedimentation. The sheeted conglomerates are interpreted as debris flows, and the channelled conglomerates appear to have been deposited from high-density turbidity currents.

Not only do the volcanic rocks of the Pentire Succession give evidence for shoaling at the northern margin of the Trevone Basin, but the diversity of the pebbles within the Gravel Caverns Conglomerate also suggests varied source areas, with rounding taking place in shallow, active environments before being released into the basin. In the volcanic setting proposed, the repeated steepening of slopes could be expected. This may have led to frequent detachment, sliding, and fragmentation of lithified and partly lithified sequences. Limestone turbidites from the Trevose Slate, derived faunas from conglomerates (such as *Pachypora*; Fox, 1905), and limestone-bearing megaclasts from the base of the Pentire Volcanic Formation (Austin *et al.*, 1992), all indicate derivation from a shallow-water carbonate platform source area.

The two formations identified within the Bounds Cliff Succession (*Figure 4.5*) include facies that indicate a depositional area marginal to that of the fully basinal Trevone Succession in Frasnian times. The thick, rapidly deposited, greenish-grey mudrocks that form the bulk of the Jacket's Point Slate Formation are similar to those of the Tredorn Slate Formation and their less metamorphosed equivalents (Kate Brook Slate Formation). Both characterize the outer-shelf environments westwards from the northern margin of the South Devon Basin to the Trevone Basin (Selwood & Thomas, 1993). The Jacket's Point Slates are, however, distinguished by the presence of abundant storm-generated sandstones, slumps, and intraformational conglomerates. The last two features indicate sedimentary instability, and were probably related to the steepening of slopes high on the shelf to basin margin. A transition to the shelf at the eastern margin of the Trevone Basin is probably indicated. The progressive onlap of pelagic red-bed sedimentation at this margin is recorded by the Polzeath Slate, which interfingers with, and overlies Jacket's Point Slate.

Considering the scale of volcanism in the Pentire Volcanic Formation, distal volcanic facies beyond the volcanic centres must have been widespread. However, the largely coeval and closely juxtaposed Polzeath Slate, represented in both the Trevone and Bounds Cliff successions, only carries evidence of insignificant volcanic activity. Considerable tectonic shortening appears to have taken place. Probably the Pentire Succession was displaced by the Polzeath Thrust (*Figure 6.2*) from a depositional area lying much farther north.

General models of extensional basins, such as that given by Leeder & Gawthorpe (1987), suggest that the sedimentary rocks of the Pentire Succession accumulated close to a growth-faulted northern margin of the basin, where lateral transport systems acted down the footwall scarp (*Figure 4.6*). The shallow-water megaclasts within the volcanic rocks appear to be peri-platform talus, while the conglomerate horizons can be interpreted as fan deposits at the foot of a gullied slope. The contrasting, fully basinal deposits of the Trevone Succession represent sedimentation on the hanging-wall dip-slope. The basin floor would have deepened northwards towards the modelled footwall scarp. The Bounds Cliff Succession appears to represent a marginal facies between the shelf and basinal Trevone Succession.

The stratigraphy reviewed above shows significant differences from that to be employed on the 1:50 000 Geological Survey Sheet 336 with 335, and described in the accompanying Memoir (Selwood *et al.*, in press). New palaeontological data collected since this survey was completed in 1991 has necessitated the major stratigraphic revision reviewed in this chapter.

It has been suggested that the sediments constituting the Tintagel Succession (*Figure 4.5*) accumulated on the outer shelf forming the northern margin of the Trevone Basin. The preserved record starts in late Devonian times with a thick, monotonous sequence of greyish-green slates referred to as the Tredorn Slate Formation. Local silty laminations, fine-grained sandstones and thin, lenticular limestones suggest rapid deposition. Macrofossils are restricted to thin horizons with polyzoa, and thick-shelled rhynchonellid and spiriferid brachiopods. This appears to be an opportunistic fauna only rarely able to colonize the muddy sea floor. Conodont faunas from the limestones belong to the polygnathid-icriodid biofacies and suggest an outer-shelf environment with some displaced shallow-water species.

Minor lithological variations recognized within the formation by early investigators have been abandoned by recent workers as metamorphic features, but the economically important roofing slate quarried at Delabole is distinguished as a member. These finely cleaved slates are uniform in colour and composition.

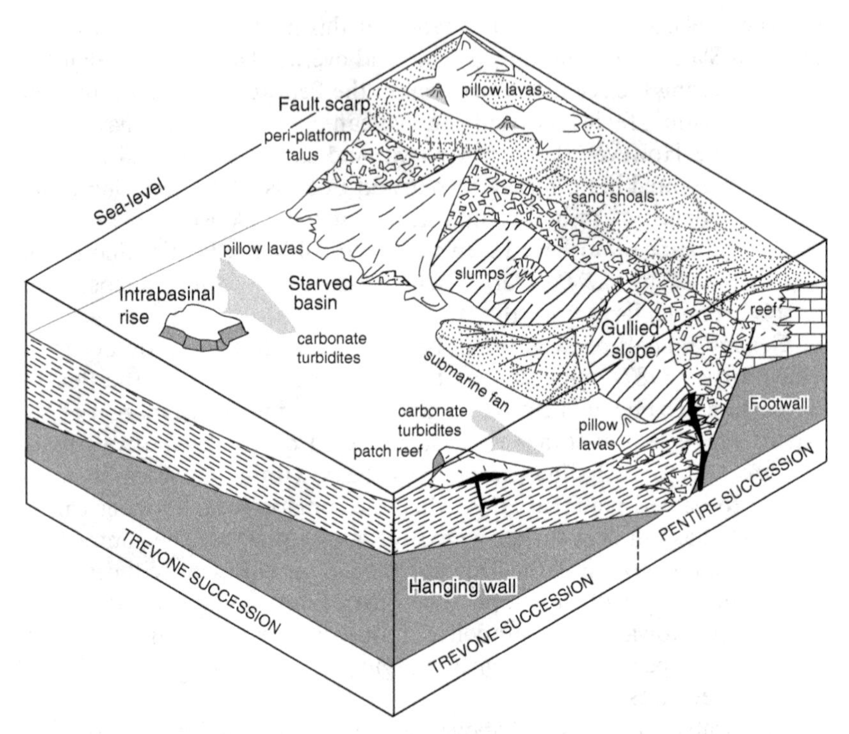

Figure 4.6. Half-graben model relating Upper Devonian facies developed in the Trevone Basin (after Leeder & Gawthorpe, 1987)

Selwood & Thomas (1993) have argued that the Tredorn Slate is coeval with, and lithologically identical to the Kate Brook Slate Formation deposited north of the South Devon Basin.

Igneous activity

The igneous rocks of the Trevone Basin fall within a distinctive alkaline magmatic province with enriched chemical features. This confirms the ensialic setting of the basin. Three principal dolerite suites have been identified by Floyd (Selwood *et al.*; 1998), which further distinguish the Trevone and Pentire successions.

The Pentire Succession is characterized by anhydrous, alkalic dolerite sills and dykes that are comagmatic with the later extrusive rocks of the Pentire Volcanic Formation. Major dolerite bodies can be seen at The Rumps [SW 9352 8125], in the lower cliff section of Com Head [SW 9395 8053], Lundy Bay [SW 9592 7996] and in headlands farther east. These intrusions are medium-grained, patchily vesicular, and profoundly altered. Flow-cooling structures indicate near-surface intrusion into the wet basinal argillites of the Trevose Slate Formation, which were adolinized and fluidized in the process. All of the intrusions have been carbonated to a greater or lesser extent. The Com Head intrusions are remarkable for the development of pipe vesicles. These cooling structures extend inwards from both the upper and lower intrusive surfaces, and may be developed through zones up to 1m thick. Thin (about 1.5 m) porphyritic basic sills form a useful marker horizon near the top of the Trevose Slate. These sills are characterized by a central zone consisting of highly altered, ferromagnesian and feldspar megacrysts, and rock fragments. The ascending magma appears to have passed through a crystal cumulus pile at depth, and has entrained part of the crystal mush.

The Trevone Succession shows two distinctive suites. These are an older group of anhydrous, tholeiitic dolerite sills intruded near-surface within the Trevose Slate (such as at Park Head [SW 840 708], Dinas Head [SW 847 862], and Merope Rocks [SW 860 766]), and a later (Famennian) group of near-surface, hydrous, alkalic dolerites characterizing the Polzeath and Haywood Slate formations. The latter suite includes the well known minverites of Trevose Head [SW 850 766], Cataclews Point [SW 923 7601], St Saviour's Point [SW 873 762], and Rock Quarry [SW 931 758], described by Reid *et al.* (1910) and Dewey (1914).

LISKEARD HIGH

South of Bodmin Moor, Burton & Tanner (1986) demonstrated the existence of a persistent shelf area between the South Devon and Trevone basins. The shelf that was limited westwards by the St Teath–Portnadler Fault Zone and eastwards by the Cambeak–Cawsand Fault Zone.

To the north, in the area presently occupied by the granite of Bodmin Moor, it appears that the outer-shelf facies characteristic of the northern margin of the South Devon Basin extended onto the high. These beds, identified as the Tredorn Slates, suffered contact metamorphism from, and were eventually displaced northwards in the Tredorn Nappe by, the rising pluton (Selwood & Thomas, 1993).

The character of shelf sedimentation changed southwards across the Liskeard High, where a persistent, shallow-water shelf area was maintained until the end of Givetian times (Burton & Tanner, 1986). Around Liskeard, open, shallow-marine shelf conditions, characterized by the Staddon Grit Formation, persisted well into the Emsian, when the eruption of tuffs and lavas in the succeeding Tempellow Slate Formation signalled the activation of the boundary faults defining the high, and the start of subsidence of the adjoining basins. Rich Middle Devonian benthonic faunas from limestones in the succeeding Rosenun Slate Formation indicate that open-shelf conditions were maintained on the high, but there is no evidence for the establishment of the reefal facies characterizing the shelf south of the South Devon Basin lying east of the River Tamar. The environment changed markedly during the deposition of the Milepost Slate Formation. Shallow shelf argillites persisted into the lower part of the formation, where they are accompanied by volcanic and volcaniclastic rocks, including scoriaceous lapilli, which suggest subaerial activity. The upper part of the formation consists of deeper-water, grey argillites with a pelagic fauna. By early Upper Devonian times, pelagic red-beds characteristic of the adjoining basins had started to lap onto the high. Although the succeeding strata are not observed, Selwood & Thomas (1993) speculated that the high could have carried shallow-water clastic sediments (Boscastle Formation) northwards from the southern margin of the South Devon Basin.

SOUTH DEVON BASIN

In Cornwall, the western part of the South Devon Basin lies between the Tamar Fault Zone (Turner, 1984) and the Cambeak–Cawsand Fault Zone. The character of the area is greatly influenced by movements on these fault zones which were active during the Variscan orogeny and which exerted a control on the thrusting and evolution of the nappe stratigraphy (Selwood, 1990). Furthermore, this structure may also have influenced deposition, for significant stratigraphic differences appeared across it during the Carboniferous. Most importantly, the thick Middle Devonian limestones characteristic of the Plymouth district are missing in Cornwall.

Basinal sedimentation was initiated in late Lower Devonian times, following the paralic sedimentation recognized in the southern part of east Cornwall. The Middle and Upper Devonian rocks (*Figures 4.7 & 4.8*) consist either of uniform basinal slate or lithologically complex limestone–slate rise sequences. The latter are only locally developed and often condensed. All successions pass conformably upwards into the Tournaisian.

The biostratigraphy of the area is based upon ammonoid, ostracod, and conodont faunas. The last have not only revealed the complexity of the rise facies, but also have provided a useful palaeoenvironmental control (Stewart, 1981a, b). The formations are discussed in relation to a nappe stratigraphy elaborated in Chapter 6.

Rise Facies Association

Rise facies association sequences of Famennian to latest Tournaisian age are revealed at the lowest structural level (autochthon) through fenster (tectonic windows). They are also intercalated with Lezant Slate in the Petherwin Nappe, and outcrop in the floor of the River Inney and north

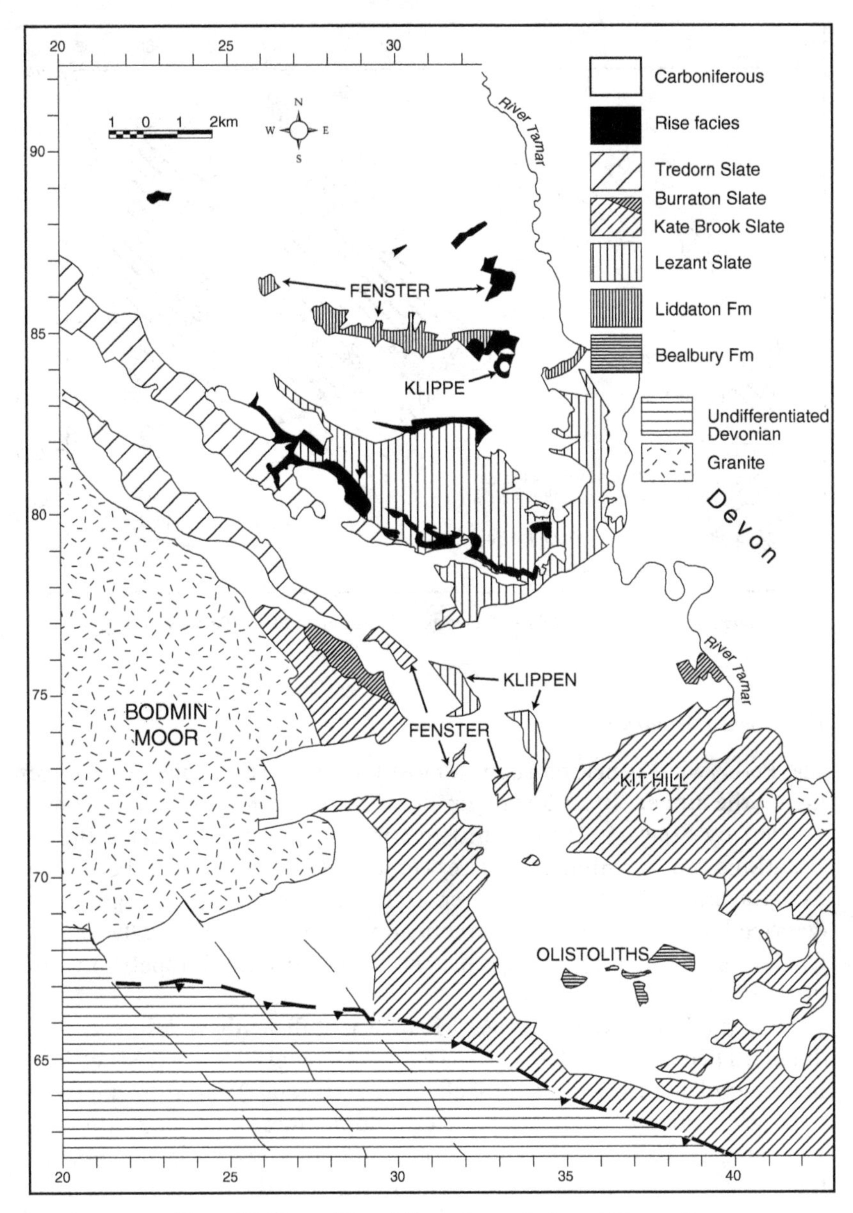

Figure 4.7. Disposition of Devonian rocks in east Cornwall

of South Petherwin (*Figure 4.8*). The two main units, the Petherwin and Stourscombe formations, are overlain by, or interfinger with, the Yeolmbridge Formation.

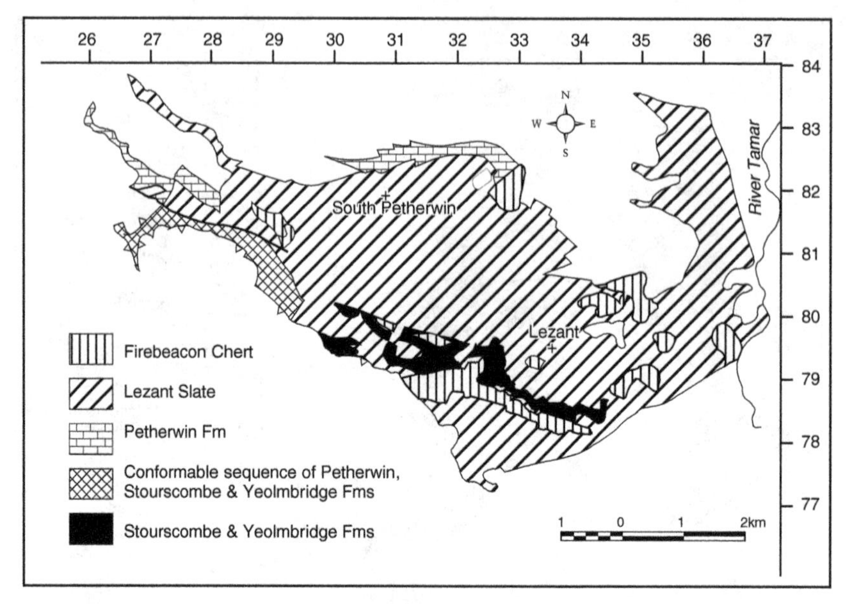

Figure 4.8. Disposition of rise facies association units of the Petherwin Nappe

Petherwin Formation

The Petherwin Formation spans almost the entire Famennian and ranges up to the upper Tournaisian *anchoralis-latus* conodont zone. It was divided by Stewart (1981a) into five members which reflect different situations on a submarine rise (*Figure 4.9*). The West Petherwin Conglomerate Member is, however, exclusively Tournaisian in age, and is described in Chapter 5. The component members are tectonically isolated from one another in a sequence showing structural repetition and inversion.

Only the Landlake Limestone Member is sufficiently widespread to be differentiated on the 1:50 000 map (British Geological Survey: Sheet 337). It is a condensed sequence, probably not exceeding 25 m in thickness, of dark grey calcareous slates with crinoidal limestones. The latter are lenticular, up to 0.5 m thick, and richly fossiliferous, yielding abundant thick-shelled brachiopods and conodonts. Mostly the sequence is of Famennian age, but at Trenault [SX 2625 8292], a highly condensed sequence only 2 m thick includes much of the Famennian (*velifer* Zone upwards) and ranges into the basal Carboniferous *sulcata* Zone.

The Cephalopod Limestone Member is known only from a quarry at Landlake [SX 3286 8229] infilled last century. It is remarkable for a rich Famennian clymenid fauna carried in nodular and thin-bedded limestones that range from the upper *Platyclymenia* Stufe to *Clymenia* Stufe

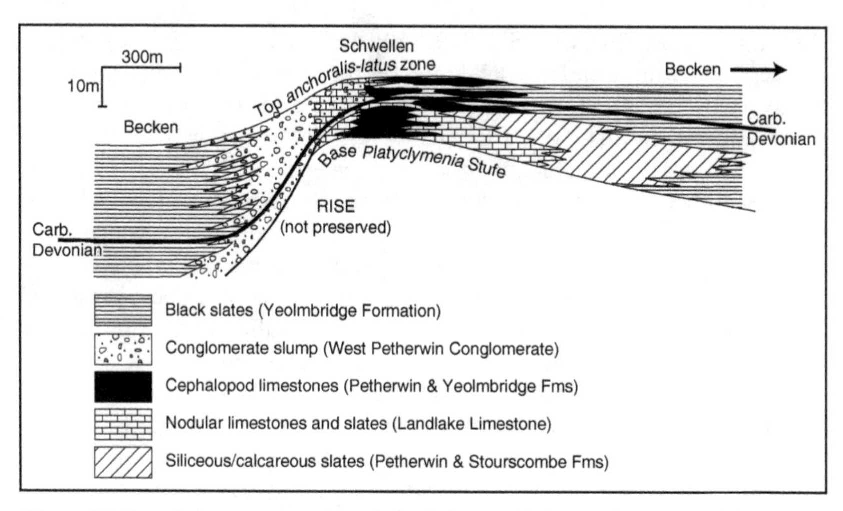

Figure 4.9. Speculative reconstruction of rise facies association environments of deposition

(Selwood, 1960). Conodont faunas from loose blocks indicate a range up into the late Famennian middle *costatus* zone (lower *Wocklumeria* Stufe), and considerable reworking, including elements indicative of a biofacies (upper *styriacus* Zone), suggesting water depths of only a few decimetres.

The Gatepost Sandstone Member is a fine-grained calcareous sandstone associated with a decalcified cephalopod limestone (Lower *Clymenia* Stufe). Although known only from another infilled quarry at Landlake [SX 3258 8216], it is significant because identical lithologies are associated with rise sedimentation elsewhere in the Variscides.

The pale grey, silty slates of the Landerslake Slate Member have yielded a rich shelly fauna of spiriferid and chonetid brachiopods. Conodonts are rare, but indicate a Famennian age range within the *velifer* to *styriacus* zones. A sedimentary environment marginal to a rise is indicated. The Landerslake Slate passes conformably upwards into the Strayerpark Slate of the Stourscombe Formation near Landerslake Quarry [SX 2665 8110].

Stourscombe Formation

The Stourscombe Formation outcrops as a series of tectonically intercalated sheets within the Petherwin Nappe (*Figure 4.8*) between Lezant and the River Inney. The lithologies consist of dark grey to green slate, with nodular and lenticular limestones that are frequently decalcified or silicified. Although the relationship with the Petherwin Formation is not clear, the two formations are partly coeval. Both pass conformably upwards into the Yeolmbridge Formation. Only west of

Lewannick [SX 263 811–SX 270 810] does a complete, conformable succession from Landerslake Slate (Petherwin Formation) pass into Strayerpark Slate (Stourscombe Formation) and up into Yeolmbridge Formation (Stewart, 1981b). The whole sequence is folded into a recumbent northwards facing anticline (Stewart, 1981b). The Strayerpark Slate Member is a thin, probably condensed, sequence of green, siliceous slates with abundant cephalopod-bearing limestone nodules which spans the *costatus* conodont zone. The youngest conodont faunas indicate the middle *costatus* Zone. A setting marginal to the rise could be represented.

Elsewhere, the Stourscombe Beds have been differentiated by Stewart (1981a). These are black slates, locally carrying silicified nodules with a remarkable late Famennian (upper *Wocklumeria* Stufe) clymenid and trilobite fauna. They are younger than the Cephalopod Limestone Member of the Petherwin Formation, and appear to have formed on the rise slope. West of Lewannick the Stourscombe Beds are replaced by the Yeolmbridge Formation (*Figure 4.9*).

Yeolmbridge Formation

The Devonian–Carboniferous boundary lies within the Yeolmbridge Formation (Selwood, 1960). Although largely of basinal facies this formation includes limestones of rise facies. The beds consist of dark grey to black, often silty laminated slates containing nodular limestones and rare calcareous sandstones. The Yeolmbridge Formation represents the final phase of drowning of the pelagic rise facies by deepening waters, and also increased basin anoxia. Within the Greystone Nappe, the base of the formation is diachronous (*Figure 4.9*). The oldest conodont faunas recorded from the Lewannick area above the Strayerpark Slate are middle *costatus* Zone in age. Elsewhere, the base of the formation lies in the upper *costatus* Zone.

In the north of the area the Yeolmbridge Formation is developed in the autochthon beneath overthrust Crackington Formation (*Figure 4.7*). Here, it is lithologically indistinguishable from that in the Greystone Nappe, but conformably overlies the Overwood Slate Member of the Liddaton Formation. These slates are of rise-slope facies, and the locally developed limestones at the base of the Yeolmbridge Formation reflect this setting.

Regional Significance

All the rise facies association sequences have restricted outcrops and, in east Cornwall, are generally part of tectonically allochthonous units of the Petherwin Nappe (Isaac *et al.*, 1982). No autochthonous rise facies sequences equivalent to the Whitelady Formation in Lydford Gorge, west

Devon (Isaac *et al.*, 1983), are known in east Cornwall, although the rise-slope deposition inferred in the Yeolmbridge Formation and Overwood Slate is likely to relate to it. *Figure 4.8* reveals a complex outcrop pattern with rise facies units tectonically intercalated at two main levels within the Petherwin Nappe (see Chapter 6). The disposition of the rise units suggests that a laterally extensive rise has been tectonically telescoped along with the Lezant Slates. The stratigraphic relationship of the Petherwin Formation to the Stourscombe Formation is unknown; the two sequences are never in stratigraphic contact, but the latter appears to have developed in deeper water marginal to the rise.

Rise limestone deposition continued from early Famennian to latest Tournaisnian times, but not all the sequence is either preserved or exposed. Unfortunately, no evidence of the origin of the rise survives. However, the occurrence of the trilobite *Asteropyge* from the Cephalopod Limestone Member (Reid *et al.*, 1911), and old quarry records which allude to massive bedded limestones occurring beneath the cephalopod limestones (Phillips, 1841), suggest a Frasnian limestone sequence of possible reefal origin beneath the Cephalopod Limestone.

A schematic representation of the rise facies association sequences of the Petherwin Nappe prior to deformation is shown in *Figure 4.9*. The tectonic significance of these sequences is discussed in Chapter 6.

Basinal Facies Association

The greater part of the Middle and Upper Devonian of east Cornwall consists of argillaceous metasedimentary sequences, mainly slates, with subordinate siltstones, sandstones and volcanogenic lithologies. Due to the deformed state of the sequences, poor exposure and regional metamorphism (see Chapter 7), the biostratigraphy and sedimentology of these units is poorly known.

The basinal slates are mainly of Upper Devonian age, lithologically monotonous, and mostly more than 200 m thick. However, it is recognized that there is a diversity in microfacies within formations. Since the formations appear at all tectonic levels in the nappe stratigraphy, they are largely isolated from each other. Only the Lezant Slate of the Petherwin Nappe is unequivocally associated with age-equivalent rise facies formations, although the Overwood Slate Member of the Liddaton Formation, in the autochthon, is known to pass laterally into rise lithologies of the Whitelady Formation (Isaac *et al.*, 1983) in west Devon. The Yeolmbridge Formation provides some continuity between age-equivalent rise and basinal facies associations between tectonic units. These beds record the spread of the black shale facies at the Devonian–Carboniferous boundary, that is widely recognized throughout the Variscides.

Kate Brook Slate Formation

The Kate Brook Slate Formation crops out in the autochthon and occupies two major inliers, one around the granite outcrops of Kit Hill and Gunnislake, the other running from Bathpool [SX 285 747] south to St Ive [SX 310 672] and beyond (*Figure 4.7*). The formation constitutes a thick, monotonous sequence of greenish-grey slates, including calcareous sandy lenses, beds of siltstones, and very fine-grained sandstones. A fauna of thick-shelled, coarse-ribbed rhynchonellid and spiriferid brachiopods is widespread, but of sporadic occurrence. It seems that this fauna was only rarely able to colonize the sea floor. Conodonts of the polygnathid-icriodid biofacies suggest an outer-shelf environment with some displaced shallow-water species. To the south, impersistent intercalations of purple and buff micaceous sandstones may include a sparse nektonic fauna of ostracods and conodonts of late Famennian age. No benthic elements are represented, and a deeper depositional environment is inferred.

Tredorn Slate Formation

The monotonous greenish-grey slates of the Tredorn Slate Formation are lithologically similar to the Kate Brook Slate Formation, but are separated from them by the Greystone and Petherwin nappes. The Tredorn Slate is unambiguously allochthonous.

The Tredorn Slate can be traced westwards from Lewannick to the extensive outcrops of the north Cornwall coast (Freshney *et al.*, 1972; Selwood & Thomas, 1993). In east Cornwall the slates are intruded by numerous dolerite sills of various sizes. The largest are in excess of 2 km long and up to 60 m thick, and include an ultrabasic complex of peridotite with subordinate gabbro and dolerite (Chandler *et al.*, 1984; Floyd *et al.*, 1993) at Polyphant [SX 262 820].

Faunas from occasional localities indicate upper Famennian ages. A diverse conodont fauna from Strayerpark [SX 2654 8163] (Stewart, 1981a) indicates the *marginifera* Zone, while ostracods from Trevell Quarry [SX 2589 8110] indicate the upper *Clymenia* to *Wocklumeria* Stufe.

Selwood (1961) united the Woolgarden Slate Formation into the Tredorn Slate. This usage has been followed by most workers, but the Woolgarden Slate is differentiated on Sheet 337 (British Geological Survey, 1994a).

Burraton Formation

The Burraton Formation consists of grey to black micaceous slates, siltstones and sporadic fine-grained sandstones, which are locally interbedded with purple, buff and green slates. At the type locality of East Burraton [SX 4123 6726], the formation lies within the Kate Brook Slate and yields an upper Famennian ostracod and clymenid fauna. However, southeast of North Hill [SX 273 767] it succeeds this formation and forms part of a black shale sequence that marks the passage into Dinantian strata.

Lezant Slate Formation

The Lezant Slate Formation is tectonically intercalated with rise facies association units within the Petherwin Nappe. It occupies a lozenge-shaped tract of country centred on South Petherwin [SX 310 820] and Lezant [SX 338 792]. In the east it abuts the Tamar Fault Zone, where it is seen to overthrust the Greystone Formation in Greystone Quarry [SX 365 805]. The formation comprises greenish-grey slates which locally carry highly fossiliferous silty horizons yielding spiriferid and rhynchonellid brachiopods, ostracods and ammonoids. The last indicate the presence of the upper Famennian *Platyclymenia* Stufe in Greystone Quarry. Mostly, however, the slates are barren, and grade into grey slates with thin siliceous lenses.

Liddaton Formation

The Liddaton Formation is principally developed in a series of major tectonic inliers extending from Lydford in Devon to the Tamar Fault Zone. West of this zone, into east Cornwall, the unit is only present north of Launceston in narrow fenster beneath the Crackington Formation in the valleys of the River Ottery and the River Kensey *(Figure 4.7)*. Overall, the sequence consists of dark grey to greenish-grey slates with subordinate thin siltstones, and rare fine-grained sandstones, particularly towards the top of the formation. An upward conformable passage is effected into the Yeolmbridge Formation. Considerable lithological diversity is, however, recorded. Thus in the Kensey valley, the formation is represented by black slates carrying horizons of silicified and dolomitized limestones nodules, with many thin seams of brown-weathering chert. A sparse ammonoid fauna indicates a late upper Famennian age. These beds are interpreted as a basinal facies, carrying sandstones of distal turbiditic origin. However, a rise-slope sedimentary environment is also indicated by the Overwood Slate Member. The type locality of this member lies in the Ottery valley west of Yeolmbridge [SX 3075 8722]. These green slates may show a fine lamination, and carry silicified and decalcified limestone nodules. Abundant fossils, including trilobites, pelagic ostracods and conodonts, indicate that this member ranges up into the upper *costatus* zone, before passing conformably into the Yeolmbridge Formation.

Purple, Green & Grey Slate

A tectonically defined slice of purple, green and grey slate, of Middle and Upper Devonian age, extends into east Cornwall from the Plymouth area. The succession is best developed in sections on the west bank of the River Tamar below Cargreen [SX 435 626], and a complex tectonic setting is revealed.

Relationships in Devon indicate that these beds are associated in a reefal setting. Here grey Middle Devonian slates are recognized as the basinal equivalents of a shelf carbonate complex lying to the south. The purple and green slates are younger and first appeared in the deeper waters adjoining reefal carbonates in early Frasnian times. Following the termination of active reef growth in middle Frasnian times they spread widely across the area. This sedimentary regime, which persisted into the early Tournaisian, most probably applied in east Cornwall. Although reefal carbonates are not recognized, the Middle Devonian carbonate turbidites reported at Neal Point [SX 437 614], were more likely derived from an adjoining carbonate platform.

The Bealbury Formation, known only from olistoliths within the Carboniferous Crocadon Formation, forms a related rise-slope association. These greenish-grey, purple, and bluish-grey slates and sandstones were largely documented from temporary exposures (Whiteley, 1981). Siliceous nodules, and the enclosing slates, yielded a rich fauna of ostracods, trilobites and ammonoids of late Famennian age.

Regional Significance

Although now tectonically dismembered and variably displaced by thrusting, the basinal formations of east Cornwall formed a linear facies depocentre through the Upper Devonian. The facies change from north to south. The Kate Brook Slate of the autochthon probably represents the outer, mud-dominated shelf environment developed at the northern margin of the South Devon Basin. The associated thick-shelled, strongly ribbed brachiopod fauna, recalls the near-shore fauna of the Rhenish magna facies described in beds of similar age lying south of the Old Red Sandstone Continent in Europe. The Kate Brook Slate interdigitates southwards with purple and green slates. The latter beds, together with the rise deposits described earlier, characterize the Hercynian magnafacies recognized in more offshore situations in the European Variscides.

Chapter Five

Carboniferous

The Carboniferous System was first recognized by Conybeare & Phillips (1822) in their description of the geological map of England and Wales, produced by William Phillips in 1821. This was the first formal usage of the term *System* to define a group of strata with similar lithological and palaeontological characters. The name selected referred to the occurrence of coal seams, although it was soon recognized that such deposits were largely limited to the upper part of the system. At this time, the highly deformed rocks of Cornubia were undivided, and correlated with the Greywacke of Wales. This view, which made these rocks of pre-Carboniferous age, was accepted not only by Henry De la Beche, the principal worker in South West England, but also by the geological establishment. Such accord, was shattered in 1834, when John Lindley, in correspondence with William Lonsdale, identified Coal Measure plants in a collection made by De la Beche from the Greywacke near Bideford in north Devon. While De la Beche accepted the downward extension of the range of plants into the Greywacke, and later (1839) distinguished a Carbonaceous Series from it, Roderick Murchison, the author of the newly created Silurian System, profoundly disagreed. Some of the most convoluted and bitter arguments recorded in the history of geology followed. These were eventually resolved by the creation of the Devonian System (Rudwick, 1985). The lower part of the Greywacke in South West England was then referred to this new system and the upper part to the Carboniferous.

Across Britain, rocks of Carboniferous age show a remarkable range of facies, having diverse floras and faunas. These have allowed a number of biostratigraphic schemes to be established, the most successful of which reflect the occurrence of rapidly evolving, widely distributed organisms. In marine rocks, sequences of goniatites, conodonts, and to a lesser extent foraminifera, are commonly used. However, coral–brachiopod biozones have largely fallen into disuse as facies control on the distribution of these faunas came to be recognized. The rapid diversification of land plants, including lycopods, seed ferns, calamitids and gymnosperms, recorded in the Upper Carboniferous, is important biostratigraphically, with palynomorphs proving increasingly useful. Biozones represent time intervals of only a few million years and afford great precision in the correlation of strata. Stages are used for the international correlation. These chronostratigraphic units refer to the rocks that accumulated during the time interval represented by a group of biozones.

Within South West England, Carboniferous fossil assemblages consistently indicate a distinctive, largely offshore sedimentary province to which the epithet *Culm* has often been applied. This term, introduced by early workers, refers to the poor quality coals occurring around Bideford. Since these beds have a limited distribution and belong to a deltaic facies characterizing the Coal Measures, this usage has been abandoned in lithostratigraphy. Nevertheless, the term is so deeply entrenched, that it is still used informally. In particular, *Culm Basin* is widely used to identify the major Carboniferous depocentre lying north of the Rusey Fault Zone (*Figure 6.1*).

The *Bathyal Lull* (Goldring, 1962) is used to identify an episode of reduced sedimentation, recognized throughout the Variscides. This interval was initiated in late Devonian times and attained its maximum development in the Lower Carboniferous. Across the region it is marked by the occurrence of deep-water, black mudrocks with interbedded volcanic rocks. In Cornwall, Lower Carboniferous rocks are only found in the complex tectonic setting existing south of the Rusey Fault Zone. However, to the north of this structure, such rocks probably underlie thick clastic sequences of Upper Carboniferous strata. In Upper Carboniferous times volcanic activity ceased and flyschoid sediments filled the sedimentary basins.

Although the relative dating established by biostratigraphic methods is of great importance in the detailed sequencing of events and in the reconstruction of palaeogeography, absolute dates have to be approximated by radiometric methods. These show that the Carboniferous spanned a period 360–286 Ma ago, and indicate that in South West England the lowest part of the post-orogenic New Red Sandstone may be of Carboniferous age. Red-beds of presumed Stephanian age are associated with the Kingsand rhyolite in Cornwall.

SOUTH OF THE CULM BASIN

South of the Rusey Fault, Carboniferous strata form part of the nappe terrane of South West England. In this complex setting, biostratigraphic data have proved invaluable in unravelling the geological history of the area. Detailed correlation is largely based on conodonts, but goniatites are also helpful in the Tournaisian (*Figure 5.1*).

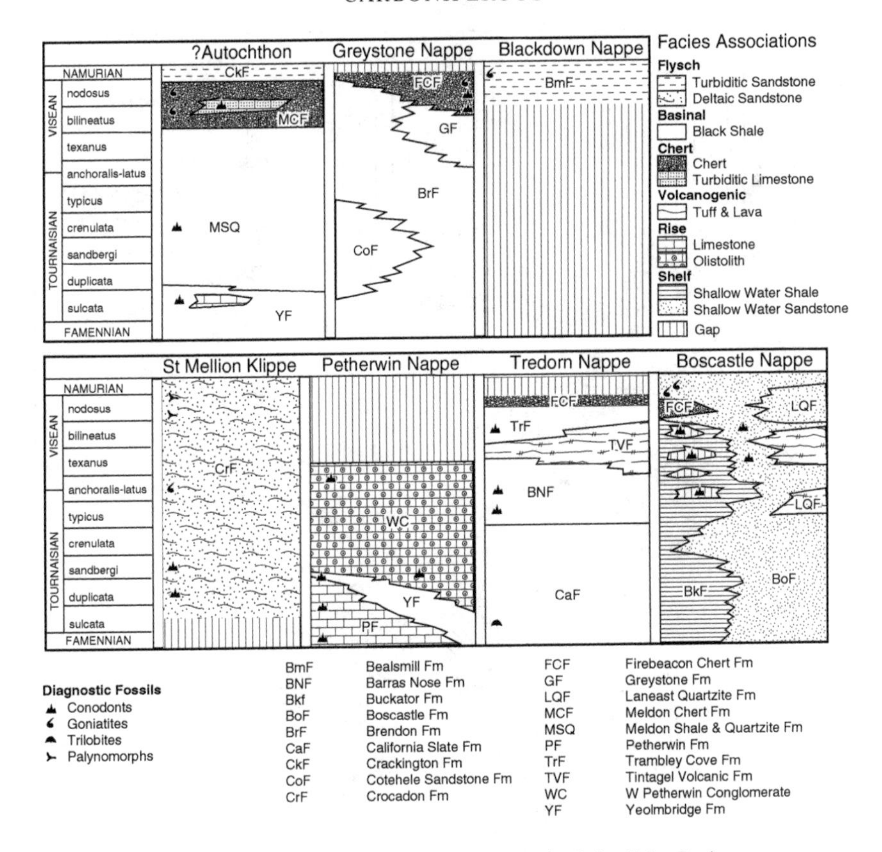

Figure 5.1. Carboniferous successions south of the Culm Basin

Strata at the Devonian–Carboniferous boundary mark a dramatic change in basin configuration and concomitant depositional environments. A deepening of the basin with attendant increased anoxia, is inferred by onlapping of black slate facies onto the rise margins. Characteristic lithologies and faunas allow six facies associations to be recognized within the nappe-bounded successions (*Figure 5.2*) of east Cornwall. Within these associations the system boundary is only preserved in the Yeolmbridge Formation and in the rise facies of the Petherwin Nappe.

Rise Facies Association

Although pelagic rise carbonate deposition persisted from the Devonian within the Petherwin Nappe, such limestones are rare. Deposition marginal to the rise was, however, manifested as spectacular polymict conglomerates and slumps of the West Petherwin Conglomerate.

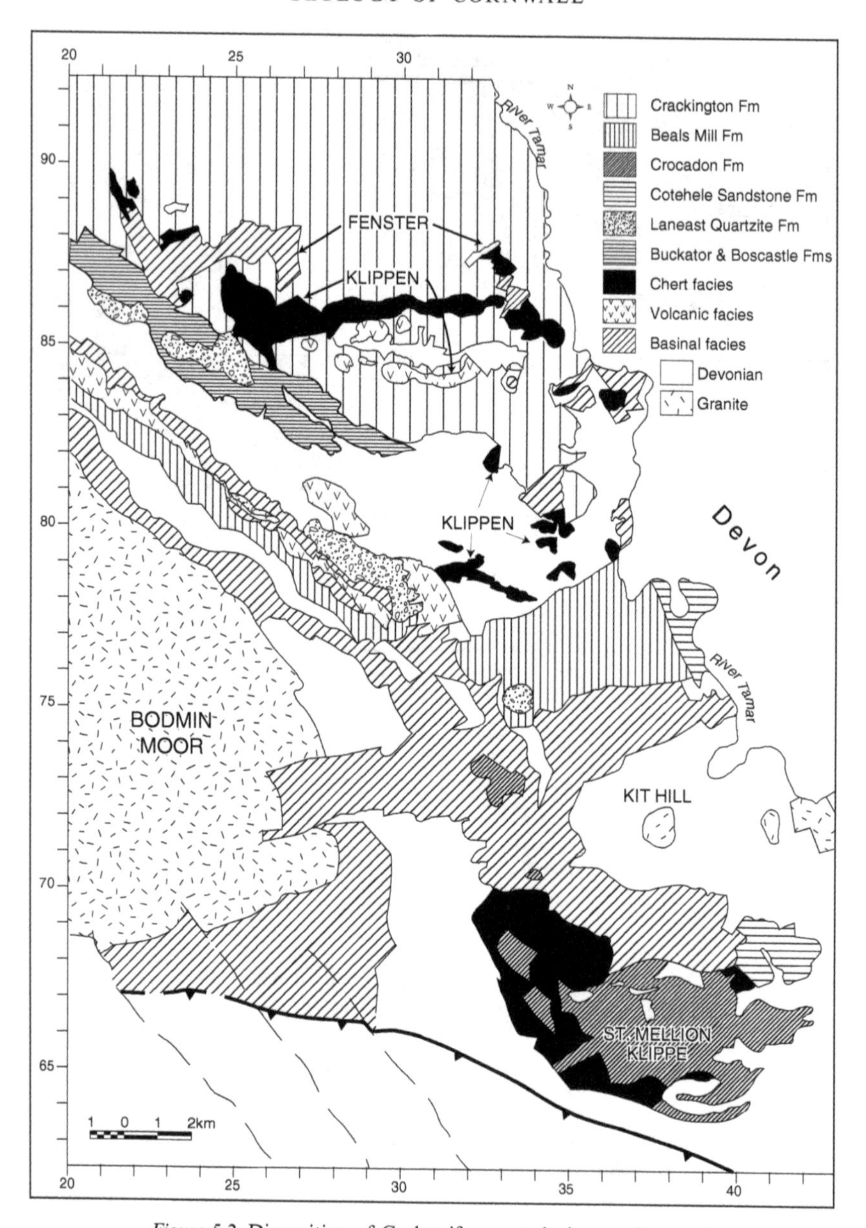

Figure 5.2. Disposition of Carboniferous rocks in east Cornwall

The West Petherwin Conglomerate ranges from the lower Tournaisian (upper Gattendorfia Stufe) up into the *anchoralis-latus* Zone. Clasts include abundant Tournaisian limestones and shales, and upper Famennian trilobite-bearing limestones not known from the present outcrop of the Petherwin Formation. This conglomerate was deposited as debris flows and slumps on the margins of the Petherwin Rise which were probably fault-controlled (*Figure 4.9*).

Further evidence that rise limestone deposition was continued into the lowest Carboniferous is provided by the local development of pelagic limestones within the black slates of the Yeolmbridge Formation (see Chapter 4). These limestones contain goniatites and trilobites characterizing the Gattendorfia Stufe and conodonts from the *duplicata* Zone.

Shelf Facies Association

The Buckator, Boscastle and Laneast Quartzite formations (Freshney *et al.*, 1972; McKeown *et al.*, 1973; Selwood *et al.*, 1985) form a stratigraphically contiguous unit of Famennian to Viséan age. These beds constitute the Boscastle Nappe that outcrops from the coast near Boscastle to the east Cornwall area.

Buckator Formation

The Buckator Formation, which occurs as a tectonically bounded unit within the Boscastle Nappe, consists of greenish-grey slates with thin calcareous siltstones, rare thin limestones and thin sandstones. Locally the slates are bluish-green in colour and show a thin siltstone banding. Siliceous nodules are common.

Conodont faunas from the coastal outcrops (Freshney *et al.*, 1972; Selwood *et al.*, 1985) yield *typicus–bilineatus* Zone ages. Inland at Trenault [SX 263 830], a fauna typical of the *styriacus–velifer* zones shows that the formation ranges well down into the Famennian. A poorly preserved Viséan coral–brachiopod fauna with nuculid bivalves, gastropods, bryzoa and crinoid ossicles (Freshney *et al.*, 1972) indicates a shallow outer-shelf environment of deposition. This is supported by the Famennian conodont fauna, which includes species of the shallow-water pandorinellid biofacies, and *Icriodus pectinatus* characteristic of intertidal and subtidal environments.

Boscastle Formation

The Boscastle Formation consists of interbedded sandstones and mudrocks. The sandstones are quartzose and bedded on the decimetre scale. They may have sharp or gradational bases and tops, and show no obvious internal grain-size trends. Generally these beds are structureless, but infrequent cross-laminated tops include bundling diagnostic of wave ripples. A storm-dominated, muddy inner-shelf environment of deposition is envisaged.

Although bioturbated, bioclastic limestones are noted from coastal sections, the only diagnostic fauna is from east Cornwall, at Trenault, where an inverted sequence yielded *texanus–bilineatus* Zone conodonts. A single goniatite of P_{2a} age from near Gimble Mill [SX 2456 8395] suggests that the formation ranges into the Lower Namurian. Overall, equivalence with the Buckator Formation indicates that most of the sequence is of Dinantian age.

Laneast Quartzite Formation

This formation consists of thick-bedded, coarse-grained, quartz-cemented, quartzose sandstones with subordinate mudstones. Sandstone beds may reach 2 m in thickness, but are generally structureless. Some, however, show parallel lamination and, rarely, primary current lineations. The sedimentology of the sandstones indicates an inner-shelf, possibly shoreface, environment of deposition. The grey to black mudrocks contain abundant plant debris, but no diagnostic fauna or flora has been recorded. However, as the formation is closely associated with the Boscastle Formation, a Dinantian age is favoured.

Regional Significance

Although the shelf facies association is known to thicken westwards, true stratigraphic thicknesses are unknown. It seems that a tectonic thickness in excess of 1000 m is likely for the Boscastle Formation. The sedimentology and supporting faunal evidence indicate that these units represent deposition on a stable subsiding shelf. Water depths were probably less than storm wave base (about 100 m), and may have been less than 10 m for the deposition of the Laneast Quartzite. Formations of the shelf facies association appear to represent a tectonically dismembered conformable sequence, without the major periods of synsedimentary deformation evident in other Dinantian sequences in east Cornwall and west Devon (Isaac *et al.*, 1982).

Basinal Facies Association

The Carboniferous basinal facies association evolved from the Famennian basinal slate sequences, but the lithologies represented are more diverse. They incorporate significant proportions of non-argillaceous lithologies, mainly cherts, sandstones, volcanogenic units, and major horizons of synsedimentary deformation. The association is entirely of Dinantian age and is most fully developed within the Greystone Nappe (Brendon and Greystone formations) and Tredorn Nappe (California Slate Formation, Tintagel Volcanic Group). Black slates are also present at the top of the Yeolmbridge Formation (see Chapter 4), both in the Petherwin Nappe and in beds sporadically exposed at the lowest tectonic level (probably autochthon). In the latter setting, basinal sedimentation was continued up into the Meldon Shale and Quartzite Formation. This is a unit of dark grey to black mudrocks with thin quartzitic sandstone beds, thought to have accumulated under anoxic basinal conditions. Although the formation is unfossiliferous, a more extensive outcrop in Devon (Edmonds *et al.*, 1968) suggests a Tournaisian to early Viséan age.

Brendon Formation

The Brendon Formation is a thick (750 m) sequence of black slates with subordinate thin siltstones, sandstones (both orthoquartzites and greywackes), tuffs, lavas, and dolerite sills. Around Newbridge [SX 347 680], extensive tracts of chert are differentiated as the Newton Chert Member. Although Famennian faunas occur, the formation is considered to be mainly Dinantian. The youngest faunas (Whiteley, 1981) recorded are Viséan goniatites (P_1–P_2 boundary) from Axford Hill [SX 3608 6775].

Greystone Formation

The Greystone Formation mainly subcrops beneath the Petherwin Nappe in east Cornwall, but is well exposed in Greystone Quarry [SX 365 805]. It consists mainly of dark grey to black siliceous slates and siltstones with subordinate radiolarian cherts, sandstones and volcanogenic lithologies, especially pillow breccias and lavas (Turner, 1982). Rare conodonts indicate a probable Tournaisian age. The whole unit is pervasively intruded by dolerite sills which may (Chandler & Isaac, 1982) or may not (Floyd *et al.*, 1993) have some MORB affinity. Many dolerites are non-vesicular and intruded into partially lithified sediments (Turner, 1982; Isaac *et al.*, 1983). This suggests intrusion at a shallow level in the sediment column, but under pressure conditions involving a substantial water column in addition to the sediment load.

In the absence of limestones (except in olistostromes), the association of abundant radiolarian cherts and fissile, finely laminated black mudrocks, with an abyssal trace fossil assemblage, suggests an extremely deep-water basinal environment of deposition.

California Slate Formation

This distinctive unit of black slates that immediately overlies the Tredorn Slate was delineated by Selwood & Thomas (1993) as a mappable unit within the Tredorn Nappe. The slates are pyritic and locally silicified, and may be interbedded with thin cherts and tuffs. The type locality at

California Quarry [SX 0902 9084] yielded *Gattendorfia* Stufe trilobites (Selwood, 1961). Elsewhere, at Helsbury Quarry [SX 088 791] thin micaceous sandstones preserve a shelly fauna indicative of a late Devonian age. The California Slate Formation represents the culmination of outer-shelf submergence by basinal black argillites initiated in late Devonian times.

Tintagel Volcanic Group

The Trambley Cove and Barras Nose formations (Freshney *et al.*, 1972; McKeown *et al.*, 1973) include units of dark grey to black, pyritous slates with laminated siltstone horizons and interbedded tuffs. Lenticular limestones are sporadically developed in the former, and silicification in the latter appears to be related to deformation. Conodont faunas (Austin & Matthews, 1967) indicate an *anchoralis-latus–texanus* Zone age for the Barras Nose Formation and a possible *bilineatus–nodosus* Zone age for the Trambley Cove Formation.

Regional Significance

The widespread Carboniferous basinal facies association reflects a significant increase in the depth of depositional environments. This is associated with a major transgression identified in early Carboniferous times throughout the Variscides. Detailed environmental interpretations are speculative, but deposition in small, possibly isolated, deep-water and tectonically unstable depocentres in the Hercynian foreland basin lying in front of the advancing deformation front is envisaged.

Chert Facies Association

Chert facies of uppermost Tournaisian–Viséan age are widely distributed throughout South West England. Although recognized by a variety of local names, these sequences appear to have been deposited contiguously. The Firebeacon Chert and Meldon Chert formations are the principal developments, although the facies is also present in the Brendon Formation (Newton Chert). The chert facies association is a manifestation of basin-wide, deep-water environments of deposition. Local trace fossil assemblages indicate abyssal water depths (1000s of metres). The abundance of bedded radiolarian cherts and the absence of pelagic limestones (except in olistostromes) attest to deposition below the carbonate compensation depth.

Firebeacon Chert Formation

This distinctive formation consists of well-bedded, dark grey to black cherts with subordinate black siliceous mudrocks. Bedding may range up to decimetre-scale but lamination is also common. The cherts and shales are rich in radiolaria, although these are generally indeterminate. The matrix of the cherts is entirely recrystallized to microcrystalline quartz (chalcedony). Conodonts from a few localities indicate the *texanus* and *bilineatus* zones. A mid-upper Viséan age is generally inferred by association with other interfingering or conformable dated sequences.

The main outcrops of Firebeacon Chert appear in klippen capping hills to the north of Launceston, and extending westwards to the coast. Farther south, numerous small klippen or fault-bounded outliers occur on the Petherwin Nappe. The original sequence from which these cherts were derived is yet to be determined.

Meldon Chert Formation

The Meldon Chert Formation contains prominent radiolarian chert horizons, but is distinguished by the presence of limestone turbidites that are frequently replaced by chert. The formation is limited to small tectonic inliers (fenster) near Launceston that expose the autochthon. Conodonts indicate a *bilineatus* Zone age.

Regional Significance

Deposition of bedded radiolarian cherts reached maximum development in the upper Viséan in both shelf (Boscastle Nappe) and basinal (Greystone Nappe) settings. This range of depositional environments is consistent with modern analogues which occur in water depths from less than 500 m to thousands of metres. However, the prolific plankton does indicate high nutrient levels in surface waters and favourable palaeocirculation patterns. An oceanic setting is not required.

Flysch Facies Association

The term flysch is here used to denote thick, turbidite sandstone-dominated sequences. Flysch units with ages ranging from Tournaisian–Lower Namurian occur intercalated at several levels in the tectonic stratigraphy.

Four lithostratigraphic units are dominated by turbidite facies in east Cornwall: the Crackington, Bealsmill, Crocadon and Cotehele Sandstone formations. Although accurate determination of stratigraphic thickness is impossible, all these formations are likely to be several hundreds of metres thick. Together they indicate flysch deposition over a period of 30 Ma. These formations were deposited in spatially discontinuous and temporally disconnected basins before an advancing deformation front. Their present-day disposition reflects subsequent displacement and deformation.

Crackington Formation

The Crackington Formation crops out principally in the Culm Basin, but it extends southwards into the thrust and nappe terrane where it is interleaved in the tectonic stratigraphy. In this area it consists of dark grey to black mudrocks interbedded with thin siltstones, and subordinate sandstones of early Namurian age. The formation is here dominated by distal turbidite facies, hemipelagic shales, and occasional slumps and proximal turbidites.

Bealsmill Formation

The Bealsmill Formation occupies extensive tracts of generally high country between Dartmoor and Bodmin Moor and constitutes the Blackdown Nappe. The formation is typically inverted and underlain by the Blackdown Thrust, with a thrust zone more than 10 m thick of tectonic gouge and mélange derived from the formation itself. Quartzose sandstones with a low feldspar content prevail in most sequences, although locally the sandstones are coarse and feldspathic. These are proximal turbidites but may be interbedded with subordinate distal turbidites and occasional thick-bedded amalgamated turbidite sandstones up to 30 m thick. Near Woodabridge [SX 347 768], metre-thick, matrix-supported conglomerates have been recorded by Witte (1983). In general, bedding is on a decimetre to metre scale. Compared to the Crackington Formation, the Bealsmill sequences are more proximal, thicker bedded, and contain a higher proportion of sandstone. Faunal evidence for the age of the Bealsmill Formation is scant. However, limestone pebbles from the conglomerates at Woodabridge have yielded Upper Famennian, early Tournaisian and Lower Namurian ammonoids. These indicate that the formation is, in part, post-Lower Namurian (Selwood, 1966).

Cotehele Sandstone Formation

The Cotehole Sandstone Formation is represented in a series of klippen around Cotehele [SX 423 686], and is also infaulted within the Tamar Fault Zone south of Dunterton [SX 377 794]. The sandstones are fine to medium-grained greywackes, and often extremely micaceous. Beds are commonly structureless, show sharp erosive bases, but rarely fine upwards. The interbedded shales are subordinate dark grey to black in colour and extremely micaceous. Tight isoclinal folding and apparent repetition of sequences may be tectonic, but synsedimentary folding during slumping is also possible. No faunas have been recorded, but a Tournaisian age is inferred from the lateral transition into the Brendon Formation.

Crocadon Formation

The Crocadon Formation occurs as klippen about St Mellion [SX 389 655]. It is a Tournaisian to Namurian sandstone-dominated unit which includes mappable rafts of Upper Famennian green and purple slates of the Bealbury Formation (Whiteley, 1981), and of Viséan chert (Matthews, 1969; Selwood & Stewart; 1985). These exotic enclaves are up to 1 km long and are interpreted as sedimentary olistoliths.

The sandstones are medium to very coarse-grained, locally conglomeratic, feldspathic, and exhibit features typical of turbidites. Sedimentary structures such as cross-bedding indicate the presence of proximal turbidites, although more evolved, classical turbidites are also common. Interbedded pale grey shales and siltstones are often rich in plant debris. The occurrence of burrowing, and a record of an *in situ* rootlet bed, suggest extremely shallow-water conditions. Whiteley (1984) inferred a deltaic environment of deposition, and this is entirely consistent with the overall predominance of proximal turbidite facies.

The olistoliths require a source of both approximately contemporary and older lithified sediments to be incorporated by gravity sliding into the Crocadon Formation during deposition. Derivation from nearby submarine fault-scarps active during deposition is possible, but slumps or slides down a steep paleoslope cannot be ruled out.

Conodont faunas from the Viverdon Down area (Whiteley, 1981) are diagnostic of the *sandbergi* Zone, whilst a cephalopod fauna from the same area (Matthews, 1970) indicates a late Tournaisian date equivalent to the *crenulata* conodont zone. An impoverished miospore assemblage also suggests a probable range up into the Namurian (Whiteley, 1981).

Regional Significance

Whether the isolated outcrops of the Bealsmill and Crackington formations represent a formerly continuous sheet overlying earlier nappes is unknown. The Bealsmill Formation is a more proximal turbidite facies association than the Crackington Formation lying farther north, and a southern provenance seems likely. It is possible that these formations were deposited on top of an evolving nappe sequence, forming a passive roof to the duplex beneath (*Figure 6.6*), which subsequently became involved in later thrust displacements. The variable thicknesses of flysch could reflect irregular minor basins on an evolving topography.

The Cotehele Sandstone represents turbidite sandstone deposition in local depocentres, and is associated with the basinal Brendon Formation. Shedding of terrigenous siliciclastic debris from uplifted and higher-grade metamorphic terrains is indicated by abundant, coarse white mica and polycrystalline quartz showing ductile deformation textures. The

Crocadon Formation represents a more proximal environment with turbidites and spectacular gravity sliding.

Together the flysch units comprise a continuum of turbidite facies deposition onward from earliest Carboniferous times in a range of very shallow- to deep-water environments. All could have been deposited in minor, spatially and temporally disconnected depocentres, in advance, or on top of the nappe front.

Volcanogenic Facies Association

Volcanogenic rocks are sporadically developed throughout Dinantian successions, but are absent from the Upper Carboniferous. The principal volcanogenic unit is the Tintagel Volcanic Formation of the Tredorn Nappe, which occupies a sinuous outcrop pattern from the northeast margins of Bodmin Moor through to the north Cornwall coast (Freshney *et al.*, 1972). Detailed mineralogical and geochemical studies were given by Floyd *et al.* (1993). Lithologically the unit consists of variably but intensely deformed metabasic volcanogenic rocks, dominantly agglomerate, tuffs and subordinate carbonate-rich lavas, with possible minor intrusives. Some basalt clasts exhibit quench textures typical of rapidly chilled submarine lavas, but no pillow lava is present. Overall, clasts are flattened and show a pronounced NNW-trending orientation. This is the tectonic transport direction. The dating of the Tintagel Volcanic Formation is constrained by conodont dates from the Trambley Cove Formation above and the Barras Nose Formation below, indicating an *anchoralis-latus–bilineatus* Zone age for the volcanic rocks.

The isolated klippen of volcanic rocks dominated by lavas are widely distributed at Launceston, where interbedded sediments have yielded *texanus* Zone conodonts. The extensive volcanogenic units occurring in the Greystone Nappe of west Devon (Chandler & Isaac, 1982) are similar in age to the Tintagel Volcanic Formation, but are less extensively developed in east Cornwall.

CULM BASIN

Following the thin successions accumulated during the Bathyal Lull, continued subsidence maintained an E–W trough through north Cornwall and north Devon whose fill is known as the Culm Basin. Its southern margin may have extended westwards from north Dartmoor, possibly not far south of the Rusey Fault Zone (*Figure 4.1*). There is little evidence for the position of the northern margin, which may have lain in north Devon. Within this trough the Upper Carboniferous was marked by a return to rapid sediment supply and a thick sequence of flyschoid sediments accumulated well into Westphalian times. Only these Silesian sediments crop out in the Culm Basin of Cornwall.

Flysch Facies

In Cornwall the Upper Carboniferous consists almost entirely of interbedded sandstones and shales. A division is made into an older Crackington Formation and younger Bude Formation, distinguished by the thickness, frequency and character of the sandstone bands. Dark grey shale horizons carrying necktonic and pelagic faunas occur within the succession and serve as marker bands (*Table 5.1*).

Table 5.1. CORRELATION OF THE CULM BASIN AND STANDARD SILESIAN SUCCESSIONS

Formation	Key Shales and Marker Bands	Fossil Record in Cornwall	Marker Bands in Northern England	Stage	Zone
				W	
				E	D
			'A.' cambriense	S	——
				T	
BUDE				P	C
	Warren Gutter S.	*G. depressum*	*'A' aegiranum*	H	——
	Sandy Mouth S.			A	
	Saturday's Pit S.	*Cornubon. budensis*		L	
FORMATION	Tom's Cove S.	*Rhabdod. elegans*		I	B
	Hartland Quay S.	*G. amaliae*	*An. vanderbeckei*	A	——
————	Gull Rock S.	*G. circumnodosum,*		N	
		G. cf. *coronatum*			A
		G. cf. *listeri*	*G. listeri*		
CRACKINGTON	Embury S.	*G. subcrenatum*	*G. subcrenatum*		G_2
	Deer Park S.				G_1
	Skittering Rock S.		*G. cancellatum*		
	Clovelly Court S.	*R. superbilingue*			R_2
FORMATION		*R. bilingue*			
			R. gracile	N	——
		R. reticulatum		A	R_1
		R. nodosum		M	
		R. circumplicatile		U	——
			Hod. magistrorum	R	
		H. undulatum		I	H_2
		Hd. proetus		A	——
		H. cf. *subglobosum*		N	H_1
			H. subglobosum		————
		N. nuculum			E_2
			C. cowlingense		————
					E_1
			C. leion		

A=Anthracoceras; An=Anthracoceratites; C=Cravenoceras; G=Gastrioceras; H=Homoceras; Hod=Hodsonites; Hd=Hudsonoceras; N=Nuculoceras; R=Reticuloceras; Cornubon=Cornuboniscus; Rhabdod=Rhabdoderma.

Crackington Formation

The Crackington Formation characteristically consists of closely interbedded, laterally continuous, thin (less than 0.5 m) sandstones with thicker mudrock layers. The total thickness of the formation in the coast sections south of the Bude Formation is estimated to be 305–1524 m (McKeown *et al.*, 1973).

The sandstones increase in importance upwards through the formation. They are usually mud-rich, fine to medium-grained, quartz sands, showing well-developed graded bedding. Packets of thicker (up to 4.6 m) sandstone beds with subordinate shale interbeds occur sporadically (Freshney *et al.*, 1972). Individual sandstones have a consistent type of sole mark. The thicker beds usually show flutes or grooves, whereas the thinner beds may have a plane, groove-marked or small-scale load-marked basal surface. Small-scale, low-angle, mud-rich ripple cross-lamination commonly occurs in the higher parts of the sandstone beds, and some show small-scale convolute bedding. Slumped horizons are rare and mostly confined to the upper part of the formation. The mudrocks are mid-grey to black in colour and usually coarsely laminated with sporadic thin beds of paler silty clay, mica and plant-rich silty layers. Thin carbonate bands and nodules are also developed, especially in the pyritous dark grey marker beds.

Goniatites are restricted to black shales and carbonate nodules. They range upwards from biozone E_2 of the Namurian into Westphalian A (*Table 5.1*). The Embury Shale marker bed outcropping at Embury Beach [SS 2150 1958] correlates with the basal Westphalian (G_2) *Gastrioceras subcrenatum* horizon, and Freshney & Taylor (1972) recorded a younger fauna representing the G_2 *Gastrioceras listeri* marine band in the Gull Rock Shale at Shag Rock [SS 2120 1900] and Gull Rock Beach [SS 2147 2010]. In north Devon the Hartland Quay Shale, taken as the base of the Bude Formation, yields the mid-Westphalian A goniatite, *Gastrioceras amaliae.* This record confirms the Westphalian A age of the uppermost Crackington Formation. The occurrence of goniatites in the cliff sections reveals complex folding and fault repetition of the succession.

Bude Formation

Although comprising a similar interbedded sequence of turbidite sandstones and shales to the Crackington Formation, the Bude Formation is distinguished by units of thicker (greater than 1 m) sandstone. Disturbed beds (slumped and slurried units) and dark grey to black, usually pyritous, marine marker beds (Key Shales of Melvin, 1986) occur sporadically. Faulting obscures the basal part of the sequence in the coastal sections in Cornwall. Inland the appearance of the first thick sandstones, which is used to define the base of the formation, may be diachronous (Freshney & Taylor, 1972). The estimated thickness of the formation is at least 600 m (King, 1966).

Despite reaching 9 m in thickness, the sandstone beds rarely exceed medium grain size, except for large intrabasinal shale inclusions and rip-up clasts. These sandstones are unusually featureless, although some beds show indistinct lamination, and others coarse bedding or channelling. The presence of thin clay layers suggests that amalgamation of thinner (about 0.5 m thick) beds has taken place. Sandstone bases against the underlying shales are sharp, but rarely planar. They show broad shallow scours, or large-scale load structures where the soft underlying clay was deformed during sand deposition. The general direction of current flow indicated by these sole markings was towards the southwest. This contrasts with the west–east flows recorded in the Crackington Formation. Freshney *et al.* (1979) showed that the modal current directions from sandstone sequences between Key Shales may vary. Possibly some change of sedimentary pattern took place at the time of these Key Shales. Small-scale, asymmetrical ripple cross-lamination occurs frequently in the upper parts of thin sandstone beds. Higgs (1983, 1984) reported well-developed symmetrical ripples and rare examples of hummocky cross-stratification.

Disturbed beds (Higgs, 1991) up to 16 m thick occur throughout the Bude Formation, but are more abundant towards the top. Slumps, which retain partially distorted sediment layers, are more common than slurried beds in which an almost completely homogenized texture occurs.

Marine fossils are widespread, but only in infrequent units of dark grey to black shales. These Key Shales are distinct from the lighter coloured, more silty background mudrocks of the formation. Goniatites (*Table 5.1*) show that the Bude Formation ranges up from Westphalian A at least into basal Westphalian C (Ramsbottom *et al.*, 1978). Other marine fossils include pelagic bivalves, fish, crustaceans, and trace fossils.

Outside the mudrocks of the Key Shales, the sporadic presence of *Planolites* burrows suggests that brackish- to fresh-water conditions prevailed throughout most of the deposition of the formation (Higgs, 1991). Movement trace fossils include fish swimming marks and king crab trails.

Interpretation

Both formations within the Culm Basin are characterized by the presence of fine-grained sediment, periodically interrupted by short-term events bringing in coarser grained material. Organic-rich mudrocks in the Crackington Formation suggest that sedimentation took place throughout in quiet, euxinic conditions, and that the sea-bed was unsuitable for benthic forms, but there is no firm evidence of water depth.

The events producing the sandstone interbeds are attributed to dilute turbidity currents of moderate energy. The lateral continuity of thin beds and fine grain size, together with a lack of channelling and amalgamation of the sandstones, all suggest off-fan, basin floor or laterally derived turbidity underflows. An upward increase in sandstone bed thickness and frequency through the formation suggest the approach of more proximal environments. Sole marks indicate that the predominant current flow was parallel to the basin axis from an area to the west and northwest. Weaker flows from the south and southwest were possibly laterally derived (Mackintosh, 1964). The coarse clasts of mudrock and diagenetic nodules all had an intrabasinal origin. Although there is evidence of tectonic uplift to the south of the basin at this time, none of the sedimentary material is known to have a southerly source. Indeed, the lack of labile grains suggests derivation of the sands from a stable cratonic area. Soft sediment disruption by slumping is rare, but Hecht (1992) suggested that the earliest slump horizon may reflect the tectonic uplift of a northern source area in late Namurian (G_1) times.

The interpretation of the Bude Formation is a matter of continuing debate. Burne (1995) and Melvin (1986, 1987) favoured a submarine fan setting, while Higgs (1987, 1991) argued that the Key Shales were the only marine episodes in the history of a non-marine 'Lake Bude'. The presence of *Planolites* burrows and numerous king crab trace fossils were used by Higgs (1987, 1991) to suggest shallow-water conditions. However, the limulid evidence is equivocal (Goldring, 1969; Goldring & Seilacher, 1971), not least because the animals may have been displaced by storms from their normal habitats (Tyler, 1988).

Burne (1995) and Melvin (1986, 1987) emphasized the importance of sole marks, graded bedding, the lateral continuity of most beds, and shallow channelling as an indication of deposition by turbidity currents on a constructional, channel-bearing, perideltaic submarine fan. They envisaged that the deposition of the Key Shales coincided with subdelta abandonment phases during subdelta switching in the source delta complex. Such events may have produced the changes in palaeocurrent direction observed in some of the Key Shales. On the other hand Higgs (1986, 1991) attributed the origin of the sandstones to turbidity currents flowing across a lake floor, and emphasized the presence of wave-generated structures. He suggested that a sill maintained brackish to fresh water, but was occasionally overtopped by seawater during eustatic high sea-levels to give the marine Key Shales.

Thus Burne and Melvin predicted a large-scale, southward-sloping submarine fan of sediment, while Higgs advocated an almost horizontal lake floor. The disturbed beds generally show southward vergence of their folds, confirming slump movement on a southward sloping surface. Hartley's (1991) interpretation that both slumps and basin-wide debris flows are represented, implied a significantly sloping floor. However, if they are seismically disturbed soft sediment (Higgs, 1987, 1991), then much gentler slopes could have been involved. The recognition by Enfield *et al.* (1985) of folds with a northerly sense of vergence in slumps towards the top of the succession, would suggest that the early thrusting had reversed the slope of the Bude Basin floor. A tectonically active southern margin of the Culm Basin in late Bude Formation times supports neither of the two preceding models of a deeper basin lying to the south, or a stable lake sill.

The development of the Culm Basin is usually interpreted as the fill of a foredeep trough forming ahead of the northward propagating Variscan deformation zone. However, the sediment filling this basin appears to be derived from a low-lying stable land area to the north. Both the Crackington and Bude formations could have originated from, and been marginal to, a delta complex similar to the contemporaneous Bideford Formation in north Devon. Although the uncertainty of regional structural relationships precludes firm palaeogeographic ties, the lack of coarse-grained, syntectonic wildflysch sediment in the trough is unusual in this scenario (Selwood & Thomas, 1986a). Large-scale dismemberment of the Variscan fold belt on major faults may have juxtaposed unrelated sequences and hidden others.

Chapter Six

Variscan Structure and Regional Metamorphism

The system of Upper Palaeozoic basins and rises that controlled depositional patterns also played an important part in influencing the outcome of subsequent deformations. Each depocentre responded independently to a deformation front that advanced northwards across the region to affect, in turn, the Gramscatho,Trevone–South Devon, and Culm basins. The structures revealed in each basin are complex, but detailed deformation chronologies are now known. These show that the Variscan orogeny was prolonged but episodic, and probably not synchronous from basin to basin. For clarity, the convention has been followed of nominating the first deformation in each structural area as D1, the second as D2, and so on. Letters are also added in alphabetical order to specify lesser events of decreasing age within the principal deformations. The identification of folds (F) generated in each deformation, and their associated cleavage (S-foliation), follow a similar convention. Overall, the dating of structural events is poorly constrained across Cornwall, deformation having proceeded northwards in an erratic fashion.The principal regional structures discussed are located on *Figure 6.1*.

INTRABASINAL SHELF

Structures developed in the Lower Devonian successions of Cornwall played an integral part of the thin-skinned, foreland fold and thrust models proposed in the early 1980s (Shackleton *et al.*, 1982; Coward & McClay, 1983; Coward & Smallwood, 1984). These authors indicated that primary deformation was dominated by north-northwest-verging folds associated with large-scale thrust faulting. They also identified a major southwards-verging backfold at the southern margin of the Lower Devonian succession, which they suggested could be traced between

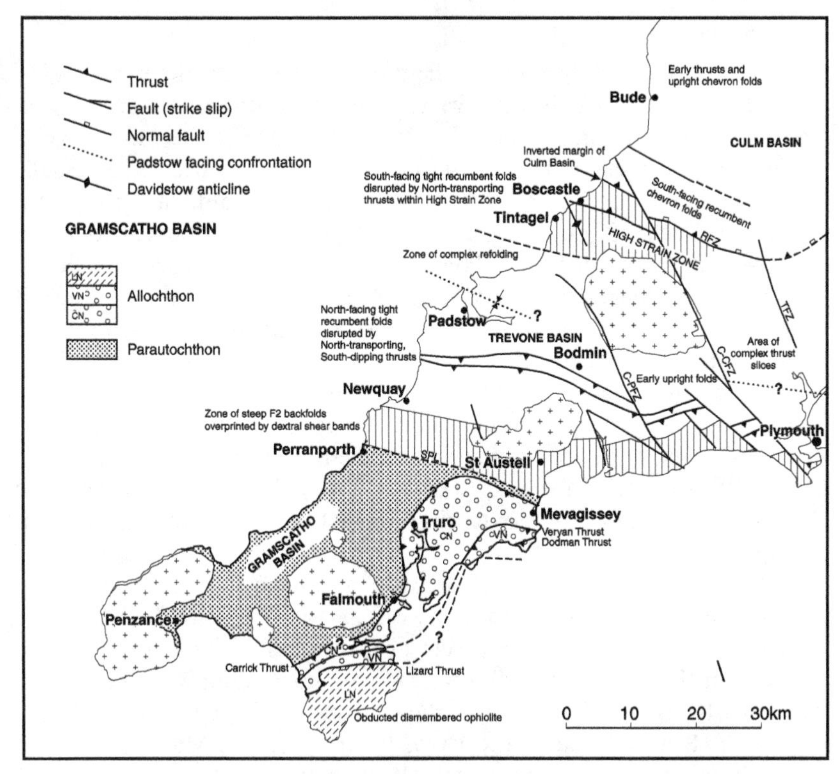

Figure 6.1. Generalized structural map: LN, Lizard Nappe; VN, Veryan Nappe; CN, Carrick Nappe; C–CFZ, Cambeak–Cawsand Fault Zone; C–PFZ, Cardinham–Portnadler Fault Zone; RFZ, Rusey Fault Zone; SPL, Start–Perranporth Line; TFZ, Tamar Fault Zone

Torbay and Perranporth. In contrast, a thick-skinned model, with elements of transpressional basin inversion, was proposed by Sanderson (1984). Later, the Start-Perranporth Line was inferred to be a zone of persistent dextral transpression (Holdsworth, 1989), and the importance of late orogenic extension was recognized (Shail & Wilkinson, 1994).

Structures Related to Convergence

Progressive Variscan deformation of the Lower Devonian succession, plus a considerable thickness of pre-Devonian basement, probably occurred above a regional décollement within the middle crust at the position of the R2 reflector of Brooks *et al.* (1984). This underlies the area at a depth of approximately 12 km. Three principal fold and cleavage generations can be recognized. The first two deformation episodes indicate a dominant top to the north-northwest sense of shear (Hobson, 1976).

D1 deformation

D1 structures occur throughout the area (Hobson & Sanderson, 1983). Primary folds (F1) exhibit a variety of attitudes, styles and scales. However, they generally trend NE–SW to ENE–WSW, face and verge to the north-northwest, and are close to tight. Fold axial surfaces are usually either recumbent, or dip gently to moderately southeast to south-southeast. Reclined folds are locally developed, and curvilinear fold axes have been described at Britain Point [SX 3515 5384] by Dearman (1964a).

Primary cleavage (S1) is approximately axial planar to the F1 folds. It shows locally developed stretching lineations defined by aligned mineral grains, clasts, and burrow fills, that usually plunge south-southeast down the dip of the cleavage (Barton *et al.*, 1993). Along the north coast, chlorite lenses within basic intrusions, and crinoid ossicles within the enclosing sediments, are often aligned within S1 cleavage parallel to the ENE–WSW-trending fold axes (Ripley, 1965). Boudin long axes trend NNW–SSE and also imply extension parallel to the fold axes (Dearman, 1964b).

D1 thrusts occur at a variety of scales and form the lithostratigraphic boundaries. For instance, to the west of the Portwrinkle Fault, the Dartmouth Group has been thrust northwestwards over the Bovisand Formation (Barton *et al.*, 1993). Elsewhere, the Staddon Grit Formation has been thrust northwards over the Trevone Basin succession in north Cornwall (British Geological Survey, 1994b), and the Middle Devonian successions in southeast Cornwall (Barton *et al.*, 1993). Cleavage and associated stretching lineations often intensify in the vicinity of thrust faults.

Deviations from this prevalent style of D1 deformation are common, particularly within the southern part of the area. South-facing F1 folds occur between the Porthnadler and Castle Dore–Coombe Hawne faults (Lane, 1966; 1970). Although various refolding models have been argued, it is also possible that primary south-facing folds developed in response to buttressing against an approximately E–W-trending basement fault block. A similar occurrence of southwards-facing F1 folds and gently northwards-dipping S1 cleavage on the west side of St Austell Bay, between Pentuan [SX 023 472] and Charlestown [SX 045 517], has also been related to primary backthrusting (Steele, 1994).

A steeply dipping S1 cleavage, associated with an E–W stretching lineation, is locally developed close to the northern margin of the Gramscatho Basin, and has been interpreted in terms of primary dextral transpression (Holdsworth, 1989). However, other studies indicate that the S1 cleavage may have been steepened by D2 backfolding (Sanderson, 1971). In the Liskeard area, the primary attitude of S1 cleavage appears to have been steeply dipping, and might be related to ramp climb during thrusting, or dextral transpression (Burton & Tanner, 1986).

Anomalous strike orientations of bedding and S1 cleavage in southeast Cornwall are associated with the Portwrinkle, Rame and Cawsand faults. Cleavage rotations of about 30° occur across the Portwrinkle Fault, and rotations of up to 75° are recorded across the Rame Fault; the sense of rotation is in a clockwise direction in the northeastern block (Barton *et al.*, 1993). Similar rotation, or the presence of lateral thrust ramps, may account for moderately eastwards-dipping S1 cleavage in the northern part of St Austell Bay.

D2 deformation

D2 structures are sporadically developed. F2 folds usually trend ENE–WSW, verge north-northwest and are open to close. Their axial surfaces dip moderately to steeply south-southeast, and they locally show an axial planar crenulation cleavage (S2) that is most commonly observed within mudstones (Hobson, 1976). A spaced cleavage is sometimes developed in sandstones (Barton *et al.*, 1993). Thrust faults also developed during D2, sometimes by reactivation of D1 thrusts

In the southern part of the area, a large-scale, south-verging F2 monoform was inferred to occur north of Perranporth by Sanderson (1971). The steep common limb of this structure was subsequently noted to contain F2 folds that are highly curvilinear about a subhorizontal E–W direction, and interpreted in terms of dextral transpression (Holdsworth, 1989). Recent work has confirmed that high D2 strain occurs on the south side of Holywell Bay, and is associated with ESE–WSW-trending phyllonite zones that contain asymmetric quartz augen indicating a dextral shear sense (Steele, 1994).

Strike-slip faults

NW–SE-trending, steeply dipping faults with dextral strike separations occur throughout the area (*Figure 4.2*). They predate final granite emplacement and appear to have a protracted movement history. Facies changes across the Portwrinkle and Cawsand faults suggest that these faults were active during the Devonian (Barton, 1994). As they are sub-parallel to the NNW–SSE mean Variscan transport trend, it is also possible that they underwent reactivation as dextral tear faults during D1/D2 thrusting (Coward & Smallwood, 1984; Barton, 1994). This view is consistent with the fact that these faults bound blocks with different deformational styles (Lane, 1970; Tanner, 1985).

D3 deformation

E–W-trending and south-verging F3 folds were identified in the coastal section to the east of the Portwrinkle Fault (Hobson, 1976), and a locally transposing, subhorizontal or gently southeast-dipping crenulation cleavage (S3) has been described around Liskeard (Barton, 1994). Low-angle extensional faults described from Pentire Point East [SW 7805 6161], Watergate Bay and St Austell Bay, may have developed in the late D3 episode (Shail & Wilkinson, 1994).

Synthesis

Stratigraphical repetition within the Lower Devonian successions has traditionally been interpreted in terms of the Dartmouth Antiform (Hobson, 1976). However, it is increasingly recognized that this concept is oversimplified, and cannot be applied with certainty in Cornwall.

D1 and D2 structures predominantly reflect overthrusting towards the north-northwest (Hobson, 1976; Shackleton *et al.*, 1982; Barton *et al.*, 1993). Anomalously orientated D1 structures usually reflect later refolding, or rotation adjacent to NW–SE-trending strike-slip faults (Hobson, 1976; Barton *et al.*, 1993). Illite crystallinity studies indicate that peak low-grade regional metamorphism, synchronous with D1, is of an intermediate pressure facies series and consistent with thrust-related tectonic burial (Warr *et al.*, 1991). Nevertheless, thin-skinned thrusting models, such as that of Shackleton *et al.* (1982) may not be entirely appropriate. If the R2 reflector of Brooks *et al.* (1984) represents a décollement within the middle crust, then a considerable thickness of pre-Devonian continental basement is likely to have been involved within regional shortening. NW–SE and E–W-trending faults within this basement not only influenced Lower Devonian sedimentation, but are also likely to have influenced the style of subsequent Variscan deformation. Transpression is likely to have been initiated in the vicinity of E–W-trending basement faults which were oblique to north-northwest-directed D1/D2 overthrusting (Sanderson, 1984), and basin inversion concepts are probably applicable (Selwood, 1990; Warr, 1993).

A major ESE–WNW-trending basement fault zone, the Start–Perranporth Line (Holdsworth, 1989), probably controlled the transition from shallow-shelf to deep-marine environments in the southern part of the area. A major backfold of D2 or D3 age may indicate possible buttressing against this basement fault zone, and structures compatible with dextral transpression are recorded on its steep limb (Holdsworth, 1989). It is also possible, although speculative, that the Staddon Ridge (Pound, 1983) was bounded on its southern side by an E–W-trending basement fault that was subsequently inverted to form the D1 thrust between the Staddon Grit Formation and the Trevone Basin (British Geological Survey, 1994b). Elsewhere, primary backfolding might occur between the Porthnadler and Castle Dore–Coombe Hawne faults, and further west in St Austell Bay.

Subhorizontal, E–W-trending stretching lineations synchronous with D1occur quite widely, and are not restricted to the region around the Start–Perranporth Line (Ripley, 1965; Shackleton *et al.*, 1982). They have been interpreted as indicating minor extension parallel to the F1 fold axes, possibly due to thrust culminations or sticking at lateral thrust tips (Coward & Smallwood, 1984). Displacement on NW–SE-trending strike-slip faults may have been synchronous with thrusting, and it appears that these faults may separate blocks with differing structural histories (Burton & Tanner, 1986).

D3 deformation represents a change from a dominant top sense of shear to the north-northwest, to a dominant top sense of shear to the south-southeast. It has been attributed to the extensional reactivation of earlier thrusts in Cornwall (Shail & Wilkinson, 1994), possibly related to granite emplacement (Barton, 1994).

GRAMSCATHO BASIN

The Gramscatho Group sediments were compared with the synorogenic Alpine flysch deposits by Hendriks (1939), who suggested that they had been deformed by northwards-directed overthrusting of a large basement nappe. The Lizard–Dodman–Start Thrust hypothesis became widely accepted, and was intimately associated with a tectonic interpretation of the Meneage 'crush breccias' (Flett, 1946). Later, these breccias were reinterpreted as being primarily of sedimentary origin (Lambert, 1965), and the concept of a single Lizard–Dodman–Start Thrust was rejected. Instead, the Perranporth–Pentewan line, along the northern margin of the Gramscatho Basin, was inferred to be the major lineament that might be correlated with the Start Boundary Fault (Dearman *et al.*, 1971; Sadler, 1974b).

Variscan Nappes

Offshore studies in Plymouth Bay, Mount's Bay and the Western Approaches have identified prominent, semi-continuous, southeasterly dipping seismic reflectors within the middle crust, which are generally interpreted as reactivated Variscan thrust faults (Day & Edwards, 1983; BIRPS & ECORS, 1986). The geometry and extent of these reflectors suggest that south Cornwall formed the relatively internal part of a Variscan orogenic wedge, similar to other collisional mountain belts (Le Gall, 1990). Onshore, it has been proposed that these thrust faults define a northern parautochthon, plus the Carrick, Veryan, Dodman and Lizard nappes (Leveridge *et al.*, 1990).

Parautochthon

The parautochthon represents the lowest structural unit in south Cornwall and includes the Porthtowan and Mylor Slate formations (Holder & Leveridge, 1986a). Its base is probably formed by the regional décollement that corresponds to the R2 reflector of Brooks *et al.* (1984). Internal thrusting also seems likely, as a set of southerly dipping seismic reflectors at a depth of 1.5–4 km (Jones, 1991) occurs just south of Newlyn East [SW 8280 5640]. Basement faults within the parautochthon along the northern margin of the Gramscatho Basin appear to have controlled dextral transpression during Variscan shortening (Holdsworth, 1989).

Carrick Nappe

The Carrick Nappe as originally proposed by Leveridge *et al.* (1984) has been redefined. Its upper parts are now incorporated into the Veryan Nappe and it only includes the Portscatho Formation (Leveridge *et al.*, 1990). The Carrick Thrust was inferred on the basis of palynological evidence that, in the absence of regional overturning, indicated the Portscatho Formation was older than the underlying Mylor Slate Formation (Turner *et al.*, 1979; Le Gall *et al.*, 1985). The upper parts of the Mylor Slate Formation were reinterpreted as an olistostrome that formed during the emergence of the Carrick Nappe (Leveridge & Holder, 1985), and sheath folds in the Portscatho Formation were related to nappe translation (Leveridge *et al.*, 1984).

The offshore projection of the Carrick Thrust trace is coincident with the projected sea-floor position of a prominent southeasterly dipping seismic reflector in Plymouth Bay and Mount's Bay (Day & Edwards, 1983; Leveridge *et al.*, 1984). It has a dip of 30–35° and can be traced to a near-horizontal reflector at a two-way travel time of 5.2 s. These data imply that the Carrick Nappe has an offshore thickness of 10–12 km and is developed above a regional décollement at a depth of 13.6–15 km (Holder & Leveridge, 1986a). Inland seismic surveys have also identified southeasterly dipping reflectors south of the Carnmenellis granite, close to the expected position of the Carrick Thrust (Jones, 1991). Seismic refraction studies imply that the Carrick Nappe at depth comprises crystalline basement (Doody & Brooks, 1986)

Nevertheless, the field evidence for a major thrust fault is often ambiguous. In Mount's Bay the contact between the Mylor Slate Formation and the Portscatho Formation is a southeasterly dipping normal fault close to Loe Bar Lodge [SW 6400 2433]. It is probable that the Carrick Thrust has been cut out by late to post-Variscan extension

(Leveridge & Holder, 1985; Shail & Wilkinson, 1994); the trace of an extensionally reactivated thrust fault may be expected to cross Loe Bar and continue along the Helford River. There are also significant variations in nappe thickness between offshore and onshore sections, and along strike in south Cornwall. These have been related to a combination of frontal and lateral footwall ramps (Leveridge *et al.*, 1984). Recent palynological studies have indicated that the lower parts of the Carrick Nappe are younger than the upper parts, and have further questioned its significance (Wilkinson & Knight, 1989).

Veryan Nappe

In Roseland, the Veryan Nappe (Leveridge *et al.*, 1990) contains the Pendower, Carne and Roseland Breccia formations. Although the contact between the Pendower Formation and the Portscatho Formation in the underlying Carrick Nappe appears conformable in a section occasionally exposed at low tide on Pendower Beach [SW 8971 3802], palaeontological evidence suggests a major tectonic break (Veryan Thrust). Conodonts recovered from the Pendower Formation range from low to uppermost Eifelian (Sadler, 1973) and are considerably older than the Frasnian miospore and acritarch assemblage in the underlying Portscatho Formation (Le Gall *et al.*, 1985; Leveridge *et al.*, 1990). Reworking of the conodonts is discounted as they preserve an upwards-younging distribution (Sadler, 1973).

Other geological evidence is also compatible with a thrust fault. Facies associations change markedly across the boundary between the two formations. Medium to thickly bedded siliciclastic sandstone–mudstone couplets that characterize the Portscatho Formation are replaced by an association of mudstones, metabasites, radiolarian cherts and calciclastic turbidites. The Mid-Ocean Ridge Basalt (MORB) type affinities of the metabasites contrast with those in the parautochthon (Floyd, 1984), and are consistent with the Veryan Nappe having been derived from a region where rifting proceeded to a greater extent. Illite crystallinity data indicate that the Veryan Nappe contains epizone grade rocks, in contrast to the anchizone grade of the underlying Carrick Nappe, and are compatible with deeper levels of thrust-related burial (Warr *et al.*, 1991).

In the Meneage, the Veryan Nappe is less well constrained and consistently indicates lower-grade anchizone metamorphism than in Roseland (Warr *et al.*, 1991). The Pendower Formation is absent, possibly as a consequence of a lateral ramp in the Veryan Thrust, or its extensional cut-out, to the west of Roseland. The provisional trace of the Veryan Thrust in Meneage is therefore marked along the northern margin of the Carne Formation (Leveridge *et al.*, 1990). The boundary between the

Carne Formation and the underlying Portscatho Formation in Mount's Bay is formed by a high-angle fault at Jangye-ryn [SW 6596 2060].

Dodman Nappe

The boundary between the Dodman Formation and the underlying Roseland Breccia Formation is faulted (Reid, 1907) and has been interpreted as a thrust (Hendriks, 1939). The Dodman Thrust has been correlated with offshore seismic reflectors and is inferred to branch off the Lizard Thrust to the east of Meneage (Leveridge *et al.*, 1984; Edwards *et al.*, 1989). The Dodman Nappe exhibits a similar metamorphic grade to the the Veryan Nappe (Warr *et al.*, 1991).

Lizard Nappe

The Lizard Complex contains oceanic lithosphere and continental basement in fault and shear zone bounded slices, which indicate a complicated pre- and syn-accretionary history prior to obduction as part of the Lizard Nappe (Roberts *et al.*, 1993; Jones, 1994). Seismic refraction, gravity, and magnetic anomaly studies indicate that the Lizard Complex is a 1 km thick sheet underlain by approximately 3 km of Devonian metasediments, and hence support a thrust model (Doody & Brooks, 1986). However, the northern boundary of the Lizard Complex is formed by high-angle faults, and the Lizard Thrust has been cut out by late to post-Variscan extension (Barnes & Andrews, 1984; Power *et al.*, 1996). Offshore seismic reflectors, inferred to be the Lizard Thrust have been mapped in Plymouth Bay (Pinet *et al.*, 1987) and the Western Approaches (Hillis & Chapman, 1992).

Normannian Nappe

The uppermost structural unit represents continental basement that originally formed the southern margin of the Gramscatho Basin, and was termed the Normannian Nappe by Holder & Leveridge (1986a). The garnetiferous granite gneiss of the Eddystone Rocks yields a K–Ar whole rock age of 375±17 Ma (Miller & Green, 1961), and has been attributed to the Normannian Nappe on the basis of unpublished data which indicate it overlies lithologies similar to the Start Schists (Holder & Leveridge, 1986b). The Giant's Rock erratic at Porthleven [SW 6236 2570] also comprises garnetiferous granite gneiss that could have been derived from a formerly extensive crystalline nappe (Goode & Taylor, 1988). The Normannian Thrust might correspond to a reflector identified by offshore seismic studies approximately 70 km southeast of the Isles of Scilly, but it cannot be clearly correlated further to the east (Evans, 1990).

Structures Related to Convergence

D1 deformation

Primary folds (F1) occur throughout the area and are asymmetrical with short overturned limbs. They are tight to isoclinal, usually verge and face north-northwest, and plunge gently east-northeast or west-southwest. Fold axial surfaces dip gently to moderately south-southeast, but are recumbent along the north coast (Rattey & Sanderson, 1984). Examples of large-scale folds occur at Nare Point [SW 8001 2512] and Black Cliff [SW 5528 3858]. The overturned limbs of these folds give rise to extensive areas of inverted strata and exhibit south-southeast-verging, but north-northwest-facing, parasitic folds.

A primary cleavage (S1) occurs throughout the area. It dips gently south-southeast and is approximately axial planar to the folds, although there is some transection of the shorter overturned limbs. S1 usually appears bedding-parallel away from fold hinges. In sandstones, S1 is a spaced cleavage defined by white mica, chlorite, and flattened and partially recrystallized detrital quartz grains. In mudstones, S1 is a slaty

Table 6.1. DEFORMATION CHRONOLOGY OF THE GRAMSCATHO BASIN

Relative Chronology		Structures & Orientations	Bulk Kinematics & Interpretation
D1		Folds verge and face NNW S1 dips gently SSE	NNW–SSE shortening during Variscan convergence and thrust nappe emplacement
D2		Folds verge NNW S2 dips moderately SSE	
D3	Distributed shear	S3 folds verge SSE; S3 dips moderately NNW	NNW–SSE late orogenic extension, possibly reactivating major thrust faults and predating granite emplacement
	Detachments	Synthetic extensional faults dip and displace to SSE	
	Brittle listric faults	Extensional faults dip SSE; slickensides plunge SSE	
Late	Brittle listric faults	Extensional faults dip NNW; slickensides plunge NNW	NNW–SSE extension coeval with granite emplacement
	Crenulation cleavage	Steep cleavage strikes NNW–SSE	ENE–WSW shortening
	Steep faults and lodes	Steep, often mineralized faults dip NNW or SSE	NNW–SSE extension and minor strike-slip
	Crosscourse faults	Steep faults strike NNW–SSE	ENE–WSW extension

cleavage. A mineral lineation, defined by elongated grains and quartz pressure shadows, is sometimes developed within S1 and plunges to the south-southeast.

There are deviations from this prevalent style of D1 deformation. At certain localities along the northern margin of the Gramscatho Basin, bedding and cleavage are subvertical, strike E–W, and are associated with subhorizontal mineral lineations. These structures have been interpreted in terms of primary dextral transpression, possibly related to oblique buttressing by a basement fault at depth, the Start-Perranporth Line (Holdsworth, 1989). In the Carrick and Veryan nappes, there is considerable variability in the orientation of F1 fold axes, and in the corresponding sense of vergence and facing, such as at Rosemullion Head [SW 796 278]. These 'oblique folds' have been attributed to sheath folding (Dearman, 1969), rotation due to differential nappe movements (Rattey & Sanderson, 1984), and translation over lateral ramps. Reorientation of earlier structures was also brought about by granite intrusion, particularly on the eastern margin of the Carnmenellis granite (Leveridge *et al.*, 1990).

Fluid inclusions within syn-D1 quartz veins, and vitrinite reflectance studies, suggest peak regional metamorphic conditions of 3.2±0.3 kbar and 320±10°C (Harvey *et al.*, 1994). These indicate a maximum burial depth of 13 km, which is consistent with D1 deformation being related to major thrust faulting and crustal thickening.

D2 deformation

In areas of coaxial interference between D1 and D2 structures, F2 folds are asymmetrical, open to tight, possess ENE–WSW-trending subhorizontal axes, and usually verge north-northwest. They commonly face north-northwest, but may face south-southeast on the short overturned limbs of F1 folds (Rattey & Sanderson, 1984). Fold axial surfaces generally dip gently to steeply south-southeast. A crenulation cleavage (S2) occurs in both the sandstones and the mudstones and is approximately axial planar to the folds, although in some areas a composite S1/S2 cleavage develops (Rattey & Sanderson, 1984). D2 strain is markedly heterogeneous throughout the area; F2 folds and S2 cleavage are often localized, or intensified, in the hanging-walls of minor D2 thrust faults. Such structures are characteristically developed at Magow Rocks [SW 5815 4227]. In zones of particularly high D2 strain, for example Little Molunan [SW 8465 3140], bedding and S1 cleavage are sometimes transposed by S2.

In areas of oblique F1 folding, non-coaxial interference occurs, and F2 folds developed on the overturned limbs of oblique F1 folds may plunge east-southeast, verge north-northeast, and face down to the south-

southwest (Dearman *et al.*, 1980). In Mevagissey Bay, S2 cleavage has a predominant moderate to steep north-northwesterly dip. Decametre-scale, upright F2 folds are indicated by changes in minor fold vergence, and may be related to continued oblique buttressing along the northern margin of the Gramscatho Basin.

Fluid inclusion and chlorite geothermometry studies of syn-D2 quartz veins indicate that late D2 metamorphic conditions were approximately 1.2 kbar fluid pressure and 270°C. This would correspond to a minimum depth of 4.5 km if fluid pressures were close to lithostatic (Harvey *et al.*, 1994). D2 deformation appears to be associated with continued imbrication of the previously deformed sequence (Wilkinson & Knight, 1989).

D3 deformation

D3 structures are best known from coastal exposures of the Mylor Slate Formation in Mount's Bay (Stone, 1966). Southeast of the Tregonning-Godolphin granite, small-scale, recumbent, ENE–WSW-trending, open to tight, F3 folds verge to the south-southeast, and commonly occur on the steeper short limbs of earlier structures, or are localized in zones of high D3 strain (Rattey & Sanderson, 1984). A subhorizontal crenulation cleavage (S3) is approximately axial planar to the folds and locally transposes earlier fabrics. In the western part of Mount's Bay, F3 folds and S3 cleavage are reoriented by large-scale, open, late D3 warping on the western margin of the Tregonning-Godolphin granite, and about the Land's End granite (Rattey & Sanderson, 1984). Further work has included extensional listric fault zones and detachment-related folds within D3 deformation (Shail & Wilkinson, 1994; Alexander & Shail, 1995). D3 deformation has generally been attributed to vertical shortening during the early stages of batholith emplacement (Rattey & Sanderson, 1984). However, D3 structures along both the north and south coast are consistent with a dominant top-to-the-southeast sense of shear (Alexander & Shail, 1995; 1996), and an alternative mechanism of pre-granite to syn-granite extension, involving partial reactivation of earlier structures, has been suggested (Leveridge *et al.*, 1990; Shail & Wilkinson, 1994).

Post-D3 deformation

High-angle, ENE–WSW-trending extensional faults, that may dip either north-northwest or south-southeast, and an ENE–WSW-trending, steeply dipping joint set, are ubiquitous. They locally host elvan dykes and a mineral assemblage that may include quartz, tourmaline, chlorite, cassiterite, chalcopyrite, pyrite, sphalerite and galena (Alexander & Shail,

1995; 1996). Formation occurred during a protracted history of granite emplacement and they have, in part, been worked as 'mainstage lodes'. A NNW–SSE-trending, subvertical crenulation cleavage, which is axial planar to open, angular folds (Rattey & Sanderson, 1984), may be contemporaneous. These structures are post-dated by a set of NNW–SSE-trending, steeply dipping, set of joints and faults, which may exhibit dextral slip and host base-metal mineralization (Dearman, 1963). Later structures include E–W and NNE–SSW-trending, steeply dipping crenulation cleavages that are associated with open folds and kink-bands (Rattey & Sanderson, 1984).

Synthesis

The sedimentary record within the Gramscatho Basin suggests that convergence between the Armorican Massif and South West England started during the Givetian, but was interrupted by Frasnian–Famennian rifting (see Chapter 3). Renewed late Famennian convergence brought about basin closure by the Tournaisian (Wilkinson & Knight, 1989). Continued NNW–SSE shortening above a regional décollement involved previously extended continental basement, deep-marine sediments, volcanic rocks, and oceanic lithosphere.

The existence of large-scale thrust-nappes is broadly compatible with onshore geological and offshore geophysical evidence, although corroboration between the two has been somewhat circular. Further detailed palynological studies are required to constrain the internal geometry of the Carrick Nappe, and the Veryan Nappe in Meneage. F1 and F2 folds generally verge north-northwest and are associated with south-southeast dipping fabrics. They are consistent with a top-to-the- north-northwest sense of overthrusting (Rattey & Sanderson, 1984). Variations in their orientation and style may be related to sheath folding and basement faults. WSW–ESE-trending basement faults along the northern margin of the basin were oblique to Variscan shortening and initiated dextral transpression in the cover sequence (Holdsworth, 1989). However, the dominant ENE–WSW-trending basement faults within the basin were orthogonal to Variscan shortening and transpression was minimal. The Carrick Thrust at depth appears to be located entirely within crystalline basement (Doody & Brooks, 1986), and may represent a reactivated ENE–WSW-trending Devonian extensional fault. Peak metamorphic conditions, synchronous with D1, imply a maximum burial of 13 km (Harvey *et al.*, 1994), and suggest moderate thrust-related overthickening of previously thinned lithosphere. Subsequent exhumation during D2 deformation is compatible with either erosion and/or syn-convergence extension (Shail & Wilkinson, 1994).

The thickened lithosphere subsequently underwent a complex history of late Variscan deformation during the Stephanian and early Permian (Shail & Wilkinson, 1994). D3 folds and faults indicate a dominant top-to-the-south-southeast sense of shear and are compatible with the extensional reactivation of major thrust faults (Alexander & Shail, 1995; 1996). Granite emplacement within the upper crust may have been assisted by ENE–WSW-trending extensional faults, and NNW–SSE-trending strike-slip faults, acting coevally above the regional detachment (Shail & Wilkinson, 1994). Late deformation included ENE–WSW shortening across the region (Rattey & Sanderson, 1984).

TREVONE BASIN

The geological structure of the Trevone Basin, north Cornwall, is complex and its interpretation has been contentious. Initially, Reid *et al.* (1910) identified a regional, west-plunging syncline, deforming a Lower–Upper Devonian sequence. Subsequently, Gauss & House (1972) defined the Trevone Succession in the southern limb of this structure, and the coeval Pentire Succession in the northern limb. The Polzeath Slate, which was recognized as the youngest formation and common to both successions, was thus placed in the core of the fold.

Later, the structure was shown to be the product of multiple deformations, and the principal regional fold became identified as the St Minver Synclinorium. Opposed northward and southward transporting deformations were recognized, which interfered with one another to give a complex zone of opposed fold facing, termed the Padstow Confrontation (*Figure 6.1*). The precise nature of this confrontation, and the timing of its origin, have been vigorously disputed. The original view (Roberts & Sanderson, 1971), which placed the northward directed deformation first, has been followed by Gauss (1973), Andrews *et al.* (1988), and Pamplin & Andrews (1988), but denied by Durning (1989), and Warr & Durning (1990) who favoured the southward directed deformation being earlier. Within the two principal deformations, local minor episodes reversed the overall sense of movement.

In these analyses the basic stratigraphy was accepted without question, and it acted as an important constraint in structural interpretation. However, the recent revision of the stratigraphy of the Trevone Basin that is reviewed in Chapter 4, is not consistent with any single large-scale fold. Rather, the Allen Valley Fault, the Polzeath, Spittal, and Trebetherick thrusts, may be used to divide the region into the distinct Pentire, St Tudy, Trebetherick, and Wadebridge structural units (*Figure 6.2*). The structures of the Trevone Basin are separated from those of the Liskeard High to the east by the Cardinham–Portnadler Fault Zone.

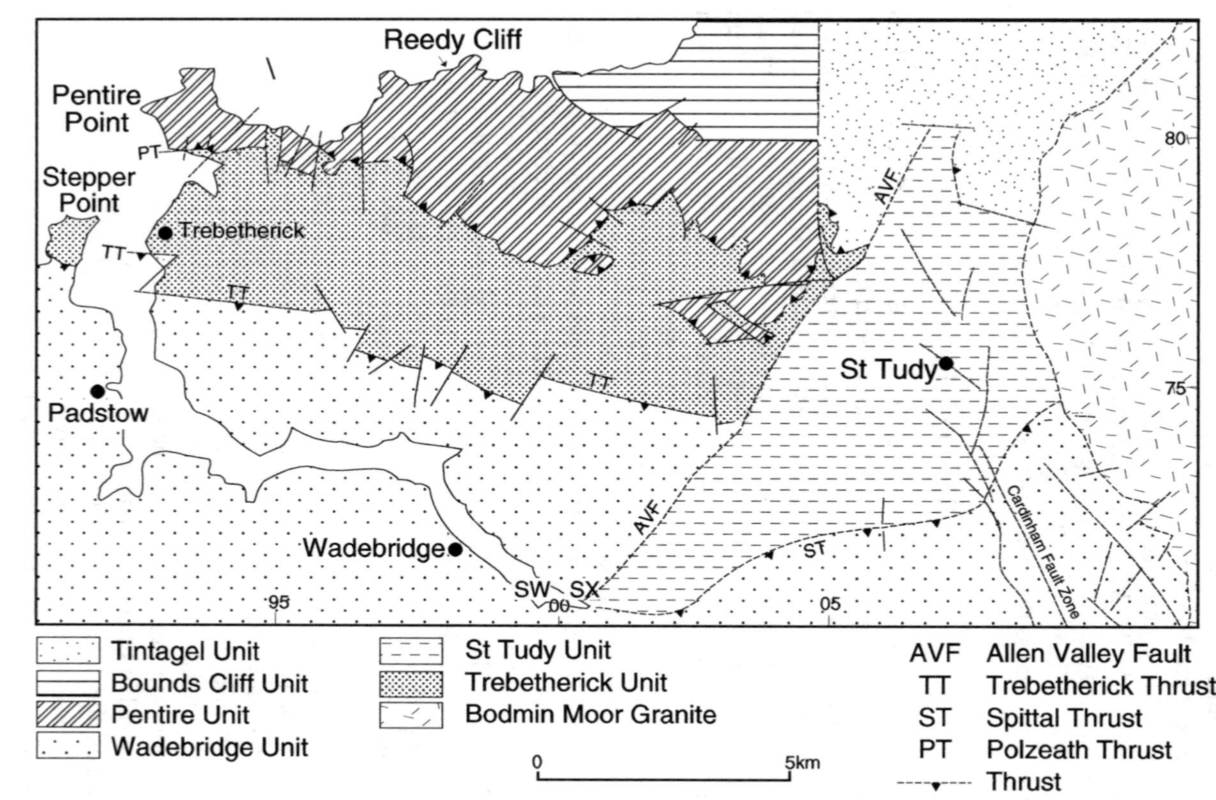

Figure 6.2. The distribution of tectonic units in the northern part of the Trevone Basin

Pentire Unit

Within the Pentire Unit the alternation of competent and incompetent horizons of different thicknesses created a multilayer complex, which influenced the style of the southward-transporting first deformation (D1) by generating buckle folds (D1a) with varying orders of magnitude. This first-order folding, controlled by the competence of the thick lavas of the Pentire Volcanic Formation, produced the present pattern of gently north-dipping formations. This regional structure is shown by the almost flat-lying lavas and dolerite sheets mapped in conformable succession, inland from the coast to the Allen Valley Fault (*Figures 6.2* & *6.3*). In the same deformation, the thick argillites enclosing the lavas developed a contrasting style of disharmonic second-order folds which are tight, and upright to south-verging. Fold closures with an axial planar cleavage (S1a) have been recorded east of The Rumps [SW 9348 8096], but such occurrences are rarely observed because of intense later deformation. Elsewhere in the argillites, folding is indicated by reversals of younging, but S1a is invariably overprinted by a later (S2a) penetrative cleavage.

The normal succession recognized in the Pentire Unit is floored by a sole thrust (Polzeath Thrust). This suggests that the presence of the thick lavas inhibited the generation of large-scale stratigraphic inversion by folding, causing D1 southward transport to be completed by thrusting and/or sliding (D1b), as these competent horizons decoupled from the enclosing mudrocks. The Polzeath Thrust (Selwood & Thomas, 1988), which emplaced the entire Pentire Unit over the Trebetherick Unit, is a prominent flat-lying structure that extends eastwards from the coast at Gravel Caverns [SX 9315 7979] to the Allen Valley Fault (*Figure 6.2*). It dips gently northwards until cut by the E–W, high-angle faulting that introduced the Bounds Cliff Succession in coastal sections (*Figure 6.2*). The inference is that the Pentire Unit is allochthonous, and is a tectonic slice derived from a root-zone lying well to the north.

Coastal sections reveal that the more obvious structures in the mudrocks of the Pentire Unit are the product of a later, northward-transporting deformation (D2). Initially, an intense, flat-lying second cleavage, which is axial planar to small-scale D2a folds, was superimposed on the earlier (D1) southwardly transported structures. Subsequently (D2b) zones of flat-lying composite structures, including multiple generations of northward-directed ductile and brittle shears, were produced. Although D2b structures are common, and obvious in the field, they are relatively unimportant in terms of total tectonic transport.

Trebetherick Unit

The Trebetherick Unit appears through a tectonic window or fenster defined by the Trebetherick and Polzeath thrusts. It reveals the lowest

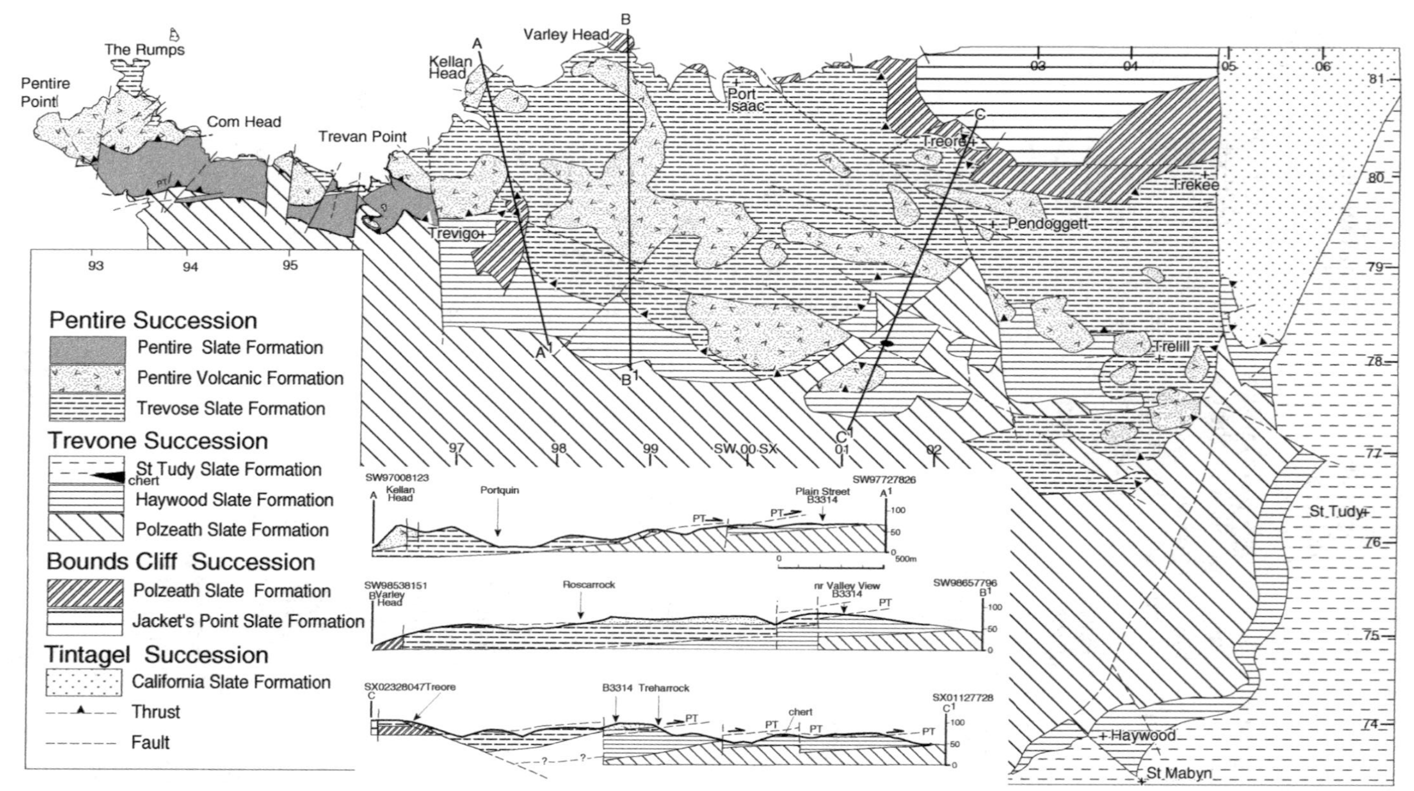

Figure 6.3. The distribution of formations in the northern part of the Trevone Basin: PT, Polzeath Thrust; B3314, road

structural level in the district, and carries the complex interference structures associated with the Padstow Confrontation Zone. The gentle dips of formational contacts in the unit (*Figure 6.3*) are not obviously consistent with the steep dips seen in coastal sections of the Polzeath Slate. As with the Pentire Succession, it appears that the presence of thick, competent, igneous bodies within the upper part of the Trevone Succession created a multilayer effect. It appears that the open, first-order regional folding was controlled by these competent bodies, while tight, second-order folds, with a steeply dipping axial planar cleavage, developed in the thick argillites of the Polzeath Slate. The sporadic occurrence of such upright folds in the Polzeath Slate has been recorded by a number of authors, including Pamplin (1990). Examples showing a steep axial planar cleavage (S1) are exposed on the south side of Hayle Bay [SW 9319 7904]. These are deformed by folds with a well-developed, flat-lying axial planar cleavage that has hitherto been ascribed to the first deformation. It appears that a previously unrecognized independent, early folding (D1a) episode is present. Spaced shears beneath the Polzeath Thrust appear to have been generated by the overriding of the Pentire Unit and are referred to D1b.

The newly interpreted early structures were reworked by a later (D2) northwardly transporting deformation, characterized by a well-developed, flat-lying cleavage that is axial planar to small- to medium-scale D2a folds. This cleavage either cuts or crenulates the first cleavage, but is observed as the only penetrative cleavage away from D1a fold hinges. The second deformation was composite and included movements in which early ductile structures were followed by brittle thrusting. The Trebetherick Thrust (*Figure 6.2*), which carries the Wadebridge Unit over the Trebetherick Unit, was interpreted as a late D2 structure with minor displacement by Durning (1989b).

St Tudy Unit

The St Tudy Unit lies in the footwall of the Spittal Thrust to the south and below the Tredorn Nappe in the north. It is arbitrarily limited to the west by the Allen Valley Fault. Hitherto, the broad north to south disposition of formations within the unit have been taken to indicate the closure of a gently west-plunging regional fold (St Minver Syncline). However, formational outcrop patterns across steeply incised valleys indicate that the beds are mainly steeply inclined, and fossil evidence (Selwood *et al.*, 1993) shows that the beds young eastwards. It appears that the Trevone Succession is deformed into an east-facing monoclinal structure (D1a). The flat limb is developed adjacent to the Allen Valley Fault, but eastwards the formations are mainly disposed in a near-vertical attitude towards the contact with the Bodmin Moor granite. This monocline is interpreted as an early compression structure, formed by the initial (D1a) reversal of movement on the basement fault zone that controlled the eastern margin of the basin.

Later, approximately N–S compression refolded the monocline into vertical to steeply eastward-plunging folds with a wavelength approaching 1 km. This deformation is responsible for the mapped arcuate pattern of formations (*Figure 6.3*). All of the structures described predate the regional penetrative cleavage, which dips uniformly at moderate angles towards the west-southwest or southwest. This cleavage is axial planar to small-scale folds (D2) which refold the steeply dipping strata. The depositional and deformational histories of the St Tudy and Trebetherick units are so similar as to indicate that they were almost certainly developed in continuity along strike. The Allen Valley Fault is a late structure showing no strike-slip movement.

Wadebridge Unit

The Wadebridge Unit is carried northward in the hanging-walls of the Trebetherick and Spittal thrusts, over the Trebetherick and St Tudy units. Within the unit the first deformation is identified as D1s, to distinguish it from the first deformation developed in the units lying to the north. D1s is dominated by a large north-northwest-facing recumbent anticline, with a subhorizontal to moderately southeast dipping axial planar slaty cleavage (Gauss, 1973). The Spittal Thrust, identified east of the Allen Valley Fault, is a D1s northward-transporting structure, carrying Trevose Slate of the eastern part of the Wadebridge Unit in its hanging-wall over footwall rocks of Lower Carboniferous age in the St Tudy Unit (Selwood *et al.*, 1993).

North-northwestward-transporting deformation was continued into the early part of D2, producing minor folds with an axial planar crenulation cleavage. The D1a and D2a folds are coaxial. Later, in D2, a change to N–S compression (Durning, 1989) produced brittle thrusts that cut up structure and carried older rocks over younger. None of these thrusts, however, is of sufficient magnitude to disrupt the sequence of formations within the Trevone Succession. The Trebetherick Thrust, the most important structure, is mapped eastwards from the north side of Daymer Bay to the Allen Valley Fault (*Figure 6.2*).

Cardinham-Portnadler Fault Zone

The northern segment of this zone (*Figure 6.2*) constitutes the most significant high-angle dislocation in the area. It forms a structurally complex area some 3 km wide, extending NNW–SSE at the western margin of the Bodmin Moor granite. Southwards, beyond the granite, it limits Trevone Basin sediments eastwards against rocks ascribed to the Liskeard High (Selwood, 1990). The zone abuts locally against the Allen Valley Fault and cuts southwards through the St Tudy and Wadebridge units. There is no evidence that the fault zone is continued northward to the coast.

Although observed displacements in the fault zone clearly post-date the D1 and D2 structures of the region, they are interpreted as representing the final expression in the sedimentary cover of movements on a deep-seated fracture that played a critical role in the origin and development of the Trevone Basin.

Structural evolution of basin

Correlations of the deformation chronologies in the Trevone Basin are shown in *Figure 6.4*. Since the deformation associated with early basin inversion (D1a) has not been previously recognized, these correlations inevitably differ from those used by other workers. Previously, the small-scale folding (D2a) and the cleavage associated with it, had been variously ascribed to D1 southward-transported structures within the Pentire Unit by Andrews *et al.* (1988, 1990) and Durning (1989); to D1 northward-transported structures in the Trebetherick Unit (Andrews *et al.*, 1988, 1990), and to D1 southward-transported structures in the Trebetherick Unit (Durning 1989).

It appears that the deformation of the Gramscatho Basin to the south led to horizontal compression of the upper crust at depth, and to reactivation of the basement thrust controlling the Trevone Basin. This initiated dip-slip reversal of the normal faults delimiting the basin, and caused regional uplift in early Namurian times. To the north, sediments filling the basin were buttressed against the footwall of its steep northern boundary fault, and upright buckle folding (D1a) was generated. Thick competent horizons within the succession became detached from the enclosing mudrocks, and first- and second-order folds developed. The same deformation reversed the movement on the normal fault defining the eastern margin of the basin. This led to the formation of the St Tudy monocline as an early (D1a) inversion compression structure. As deformation continued, upright folding advanced progressively southwards across the former basin to involve the Trevone Succession in the Trebetherick Unit. Along strike, in the St Tudy Unit, the monocline was refolded concurrently. Within thick mudrock sequences, folds tightened and cleavage developed, although in the more distally expressed folds lying to the south, this appears to have been restricted to the hinge zones.

Sustained uplift at the northern margin of the basin eventually led to D1b back-thrusting and/or sliding of the Pentire Unit southwards across the Trebetherick Unit. Such emplacement could have arisen without

Regional Structure	Structural Units			
	Pentire	St Tudy	Trebetherick	Wadebridge
Northward transporting deformation	D2b: Folding and brittle thrusting	Not recognized	D2b: Folding and brittle thrusting	D2b: Folding and brittle thrusting
	D2a: Folding and S2a cleavage	D2a: Folding and S2a cleavage	D2a: Folding and S2a cleavage	D1s: Folding, thrusting and S1s cleavage
Southward backthrusting	D1b: Thrusting	Not recognized	D1b: Footwall shears beneath Polzeath Thrust	Not recognized
Early basin inversion	D1a: Buckle folds and S1a cleavage	D1a: Buckle folds D1a: Monocline	D1a: Buckle folds and S1a cleavage	?incipient cleavage

Figure 6.4. Correlation of deformation episodes in the Trevone Basin

significant stratigraphic inversion, as a direct response to compression, or have been developed out of regional back-folding. This deformation generated southward-directed shears in the thrust footwall (Trebethick Unit), but did not affect the Wadebridge Unit.

The northward-directed deformation, which initiated basin inversion, eventually invaded the southern part of the Trevone Basin (Wadebridge Unit) where it formed the first deformation (D1s). In due course it migrated northwards, to thrust the Wadebridge Unit over structural units lying to the north, and to refold (F2a) and superimpose a second cleavage (S2a) upon them. This flat-lying cleavage crenulated S1a, but where the latter was poorly formed or absent, it developed as the principal penetrative fabric. Continued northward-directed movements were similarly expressed in all tectonic units, producing D2s structures in the Wadebridge Unit and D2b structures elsewhere. It is within this context of continuous basin inversion events that the Padstow Confrontation developed as the local expression of the overprinting of a composite southward-directed deformation by later, northward-transporting structures.

LISKEARD HIGH

This area, identified between the Cardinham–Portnadler and Cambeak–Cawsand fault zones (*Figure 4.1*), includes sediments deposited on the shelf formerly separating the Trevone and South Devon basins (Selwood, 1990). Burton & Tanner (1986) reported a structure distinct from that in the adjoining basins, with early upright, north-facing folds (D1) showing a steep slaty cleavage. In the northern part of the region these structures are refolded by major D2 folds with near-horizontal, E–W-trending axes, and a south-dipping, closely spaced crenulation cleavage.

The distinctive early structures of the Liskeard High, developed along strike from the Wadebridge Unit, correlate most closely with D1s. Since the intrusion of the Bodmin Moor granite appears to have been controlled in part by the fault zones defining the rise, the contrasting upright attitude of the early folding may have resulted from a buttressing effect produced by the rising cupola.

THE STRUCTURE OF EAST CORNWALL

The area between Dartmoor and Bodmin Moor constitutes a zone of nappes (Isaac *et al.*, 1982) characterized by a complex tectonic stratigraphy of units of diverse facies of Famennian to Namurian age (*Figure 5.1.*)

The Greystone, Petherwin and Blackdown nappes comprise relatively thin, gravity-driven fold-nappe sequences of limited areal extent and displacement. Thrust geometries are often deeply concave upwards (Turner, 1985), and propagate northwards. The structurally higher Boscastle and Tredorn nappes are of wider extent, reaching to the coast of north Cornwall (Selwood *et al.*, 1985; Selwood & Thomas, 1993). They are complex thrust-nappes showing significant northwards displacement. The internal structure of each consists of stacked thrust-sheets that repeat the stratigraphy in normal sequence. However, an extremely attenuated inverted limb of the Boscastle Nappe may be present in east Cornwall. The disposition of the major tectonic units and fault systems is shown in *Figure 6.5*.

In general, each nappe is distinguished by an association of age- or facies-related internal units. Original relationships can only be inferred by reference to simple palaeoenvironmental models (Selwood, 1990), and since there are no common marker horizons, neither internal deformation nor displacements can be estimated. However, the general sequence of thrusting is known. The Greystone, Petherwin, Tredorn and Boscastle nappes probably developed in sequence, break-back fashion, with the Tredorn and Boscastle nappes developing out of sequence, late in the deformation history. Units overlying the Crackington Formation are also out of sequence. The Blackdown Nappe and the Crackington Formation may represent the roof of a passive roof duplex modified by late, out-of-sequence thrusting (*Figure 6.6*).

Tectonic Stratigraphy

Kate Brook Unit

The Kate Brook Unit is the lowest structural unit recognized in east Cornwall, and although interpreted as autochthon, the possibility of concealed or blind thrusts cannot be ignored. The unit consists

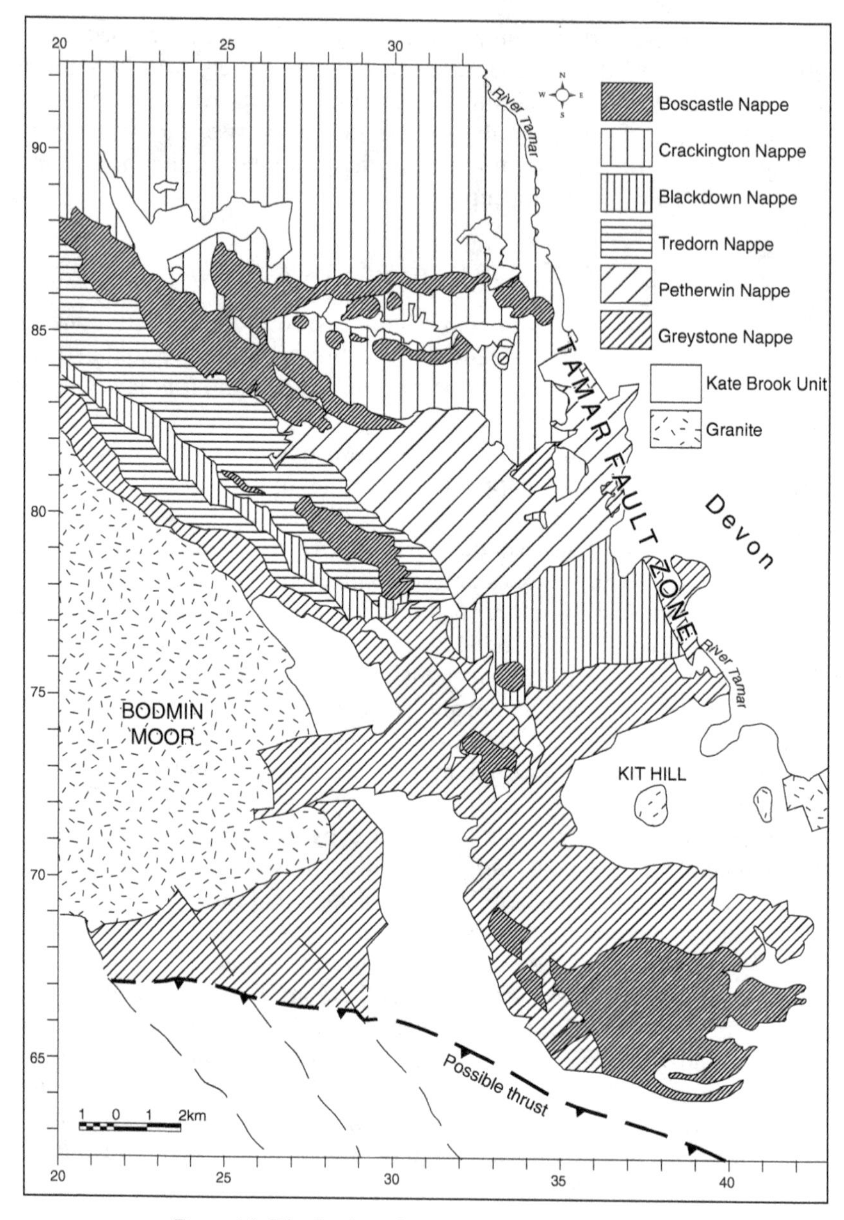

Figure 6.5. Distribution of tectonic units in east Cornwall

principally of the Famennian basinal slates of the Kate Brook Formation, with minor developments of the overlying Burraton Formation. In addition, the elongate fenster of the Liddaton and Yeolmbridge formations, north and west of Launceston (*Figure 4.7*), may be auto-

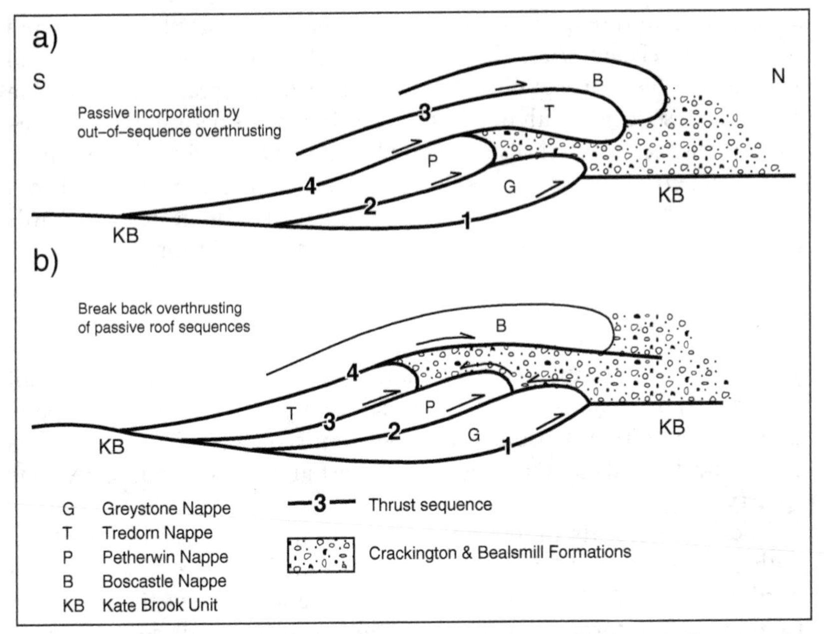

Figure 6.6. Crackington Formation relationships in east Cornwall: (a) passive incorporation by out-of-sequence overthrusting, (b) break-back overthrusting of passive roof sequence

chthonous, although the stratigraphic relationship of the two sequences is unknown. The earliest tectonic fabrics pre-date thrusting and are denoted D0. Such structures, consisting of bedding-parallel cleavages, early folds, and vein arrays, also appear in the Greystone Nappe (Turner, 1982; Isaac, 1985).

Greystone Nappe

The Greystone Nappe immediately overlies the Kate Brook Unit, but it is not thought to be present in the northwestern part of east Cornwall (*Figure 6.5*). The diverse, often inverted, successions described from west Devon terminate at the Tamar Fault Zone (Turner, 1984). Only the Brendon Formation and small fenster of Greystone Formation around Greystone Quarry [SX 3608 6775] appear in Cornwall. Continuity between the Brendon and Greystone formations is not seen, but it is assumed that a continuous Greystone Nappe is concealed beneath the Petherwin Nappe.

The lower boundary of the Greystone Nappe is the Main Thrust. This is characterized by spectacular fault rock and tectonic mélange development in west Devon, but is not well developed in Cornwall. Here, it appears as a diffuse zone of intensely deformed and intersliced green and black slates (Isaac *et al.*, 1983). Turner (1982) described the internal structure of the nappe at Greystone Quarry, where at least four thrusts in break-back sequence with northwards displacements deform the dolerite–chert sequence. These thrusts are attributed to D1 gravity sliding during emplacement (Isaac *et al.*, 1982). Small-scale structures, mapped fold structures and structure associated with D1 thrusts, indicate northwards displacement of the Greystone Nappe throughout its history.

Petherwin Nappe

The Petherwin Nappe occupies the central ground of east Cornwall (*Figure 6.5*). It comprises the Lezant Slate along with several Famennian–Lower Carboniferous rise facies association sequences. This nappe was carried northwards on the Greystone Thrust and is exposed in Greystone Quarry

The highest units of the nappe are the numerous small structural outliers or klippen of Firebeacon Chert. In the west, the nappe shows a north-facing, recumbently folded, conformable sequence of the Petherwin to Yeolmbridge formations (Stewart, 1981b). In the east, complex slices of the Stourscombe Formation outcrop on the southern margin of the nappe and overthrust Lezant Slate. The Lezant Slate, in turn, overthrusts a northern unit of generally inverted (Selwood & Stewart, 1985) Petherwin Formation. The nappe appears to consist of at least two major recumbent folds, in which the Petherwin Formation forms the attenuated lower limb of the main fold, and the Lezant Slate the core of the main fold. The upper limb appears to be extremely attenuated or missing, and is overthrust by a second recumbent fold of the Petherwin, Stourscombe and Yeolmbridge formations. These relationships are summarized in *Figure 6.7*.

Although the Petherwin Nappe is overthrust by the Crackington Formation in the north, it is unlikely to extend more than 1 km beneath this unit. Along the Tamar, the Petherwin Nappe terminates against Kate Brook Unit sequences and is not seen in any fenster to the north. An isolated klippe of the Stourscombe Formation occurs east of Launceston [SX 345 839], and an enigmatic larger klippe of the Stourscombe Formation overlies the Crackington Formation in Launceston [SX 335 844].

The emplacement of the Petherwin Nappe is the earliest D2 deformation seen in east Cornwall. Early structures in the Petherwin Nappe may be D1, but, overall, the Petherwin Nappe post-dates the Greystone Nappe, and in Greystone Quarry small-scale structures associated with the Greystone Thrust are D2 (Turner, 1982).

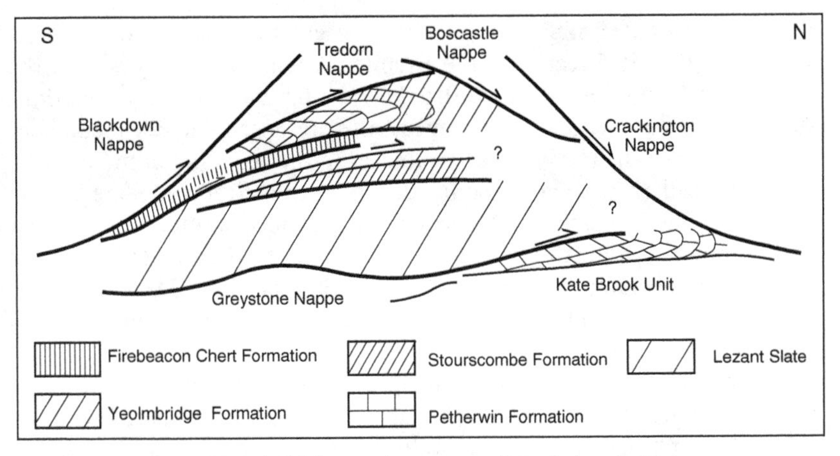

Figure 6.7. Subdivisions and sequences of the Petherwin Nappe

Blackdown Nappe

The Blackdown Nappe (*Figure 6.5*) includes large tracts of often inverted Bealsmill Formation. Only one phase of northward-verging deformation is observed, and this is correlated with late D2 deformation in other nappes. The lower boundary surface is the Blackdown Thrust, which is known to be markedly out of sequence in west Devon, but is ambiguous in its relationships in Cornwall. Around Bray Shop [SX 335 745], it oversteps both the Greystone and Petherwin Nappes, and is overlain by an isolated klippe of the Boscastle Nappe. Generally, the D2 Blackdown Thrust boundaries have been considerably modified by late D3 faults.

The main outcrop about Treburley [SX 350 777] forms a substantial klippe that abuts the Petherwin Nappe in the north, and has a thrusted contact with the Greystone Nappe in the south. Near Trebullett [SX 316 770] the Blackdown Nappe rests directly on the Kate Brook Unit. To the east, the klippe terminates against the Tamar Fault Zone. The form of the Blackdown Thrust surface is deeply concave. A second klippe runs parallel to the northeast margin of Bodmin Moor for about 12 km, and is much dissected by high-angle faults.

Tredorn Nappe

The Tredorn Nappe (Selwood & Thomas, 1993) incorporates the Tredorn Slate Formation (including Woolgarden Slates) and the Tintagel Volcanic Group. The nappe is carried on the Trekelland Thrust (Stewart, 1981b) across the Bounds Cliff Succession in coastal sections and across the Petherwin Nappe in the area between Lewannick [SX 275 807] and

Polyphant [SX 262 820]. South of this latter area, the Tredorn (Woolgarden) Slates lie directly on the Brendon Formation either within the Greystone Nappe or the Kate Brook Unit. Cleavage-bedding relationships within the Tredorn Slates suggest the presence of major recumbent folds, but, overall, the internal structure of the nappe is dominated by three stacked thrust sheets composed of either Tredorn Slate or the Carboniferous successions. On the coast, folds face south, but northeast of Bodmin Moor, north-facing has been quoted by Stewart (1981b). The sequence is overstepped by the Boscastle Nappe. Late D3 faults have significantly modified the outcrop pattern of the Tredorn Nappe, which is a late D2 structure.

Regionally, two main generations of small-scale structures are evident in both the Tredorn and Boscastle nappes. These are associated with the D2 and D3 phases of northwards displacements on thrusts (Selwood & Thomas, 1986a, 1993).

Boscastle Nappe

Northwestwards from Lewannick, the Tredorn and Petherwin nappes are progressively overstepped by the Boscastle, Buckator and Lanest Quartzite formations on the Willapark Thrust. Thrust slices of the Firebeacon Chert Formation are also intercalated with the Buckator Formation. Together, these formations constitute the Boscastle Nappe (Selwood *et al.*, 1985). It continues through to the north Cornwall coast between Boscastle [SX 099 907] and Rusey [SX 125 940], and, east of Bodmin Moor, appears as klippen overlying units of the Tredorn, Greystone and Blackdown nappes southwards to St Mellion.

The Crackington Formation derived from the Culm Basin overthrusts the Boscastle Nappe along its northern boundary, but units of the Firebeacon Chert Formation are both overthrust by, and overthrust, the Crackington Nappe. This supports the conclusion of Selwood & Thomas (1993) that the Boscastle and Tredorn nappes are the product of late D2 out-of-sequence thrusting, with subsequent D3 thrusting giving rise to further northwards displacement. It seems likely that some of the structures overlying the Crackington Formation north of Launceston are D3 klippen of the Boscastle Nappe.

The Crocadon Formation outlier around St Mellion [SX 389 655] may be a D2 klippe of the Boscastle Nappe, but this correlation is stratigraphically and structurally speculative. The Boscastle Nappe contains a distinctive shelf facies association of which the Crocadon Formation is not part. The assignment of the Crocadon Formation to a shallow-water depositional environment is not disputed, but the turbiditic and olistostromic nature of this formation is quite distinct from other units of the Boscastle Nappe.

Crackington Nappe

The Crackington Nappe constitutes the Crackington Formation, extending south of the Rusey Fault Zone between Dartmoor and Bodmin Moor. It is structurally distinct from the southward-verging structures of the Crackington Formation, derived from the Culm Basin, that overthrust the Boscastle Nappe.

The base of the Crackington Nappe is an undulating but gently northward-dipping thrust that overrides the Greystone, Petherwin and Boscastle nappes. The presence of numerous fenster of the autochthonous Kate Brook Unit suggest that none of the underlying nappes extend more than 1 km or so beneath the Crackington Formation.

Discussion

Evolution of the Nappes

Taken together, the accelerated subsidence about the Devonian–Carboniferous boundary, penecontemporaneous erosion in *typicus* Zone (*Figure 5.1*) times, intense synsedimentary deformation and olistolithic horizons, all indicate increasing tectonic instability. At this time the area lay ahead of a deformation front advancing from the south.

Since the youngest sedimentary sequences in the Greystone and Petherwin nappes are *texanus* Zone in age, nappe emplacement could not have occurred earlier than *bilineatus* Zone times. The Greystone Nappe is a gravity-driven fold-nappe, created by sliding and detachment from the advancing hinterland. D1 structures and fabrics formed at this time (Isaac *et al.*, 1982). The initial emplacement of the Petherwin Nappe, and the formation of major recumbent folds, could have occurred coevally with the Greystone Nappe. However, the Petherwin Nappe has subsequently been displaced farther northwards and, where the two nappes interact about the Tamar Fault Zone, the Greystone Thrust cuts earlier D1 thrusts within the Greystone Nappe (Turner, 1982). Attenuation of the lower limb of the Petherwin Nappe and dismemberment of the rise facies sequences, as well as intercalation of thrust sheets of the Firebeacon Chert Formation, accompanied the final emplacement of the nappe. The Petherwin Nappe is the earliest D2 structure.

The massive submarine erosion of *typicus* Zone and older sequences, followed by *anchoralis-latus* Zone slumping and the arrival of the first nappes, occurred in post-*texanus* Zone times, and probably took only a few million years. The flysch sequence of the Crocadon Formation spanned the critical time period of early nappe emplacement and ranged up into the Namurian (Whiteley, 1984). Since the youngest olistoliths in the formation are of *anchoralis-latus* Zone age, the flysch must postdate this zone. Given the incorporation of substantial olistoliths of

Famennian age into the Crocadon Formation, deposition may have been on top of, or in front of, the evolving nappes. D2 represents the main phase of thrusting and nappe formation.

Unlike the Greystone and Petherwin nappes, which are relatively local, gravity-fold structures generally only a few hundred metres thick, the Tredorn and Boscastle nappes are major thrust-nappes of regional extent. Their evolution spans D2 into early D3 (Selwood & Thomas, 1986a). These later nappes include substantial sections of stratigraphically intact sequences that were emplaced as cohesive thrust-nappes with a laterally continuous internal stratigraphy. The three-fold structural subdivision of the Tredorn Nappe in east Cornwall can be identified in a similar subdivision of sequences by Wilson (1951) on the north Cornwall coast.

The Petherwin, Tredorn and Boscastle nappes were originally emplaced break-back style. Displacement was northwards towards the foreland. In east Cornwall the Trekelland Thrust, which carries the Tredorn Nappe, has been modified and displaced by low-angle D3 faults along the northeastern margin of Bodmin Moor. The Willapark Thrust at the base of the Boscastle Nappe is out of sequence, and oversteps the Tredorn and Petherwin Nappes. This is analogous to the sequence of thrusting observed on the north Cornwall coast (Selwood *et al.*, 1985; Selwood & Thomas, 1986a). The Willapark Thrust is a late D2 structure which has been modified by D3 displacements.

Following D2 emplacement of the Tredorn and Boscastle nappes, the Namurian flysch of the Blackdown and Crackington nappes had either arrived by displacement or deposition *in situ*, in time to be involved in late D2 thrusting and D3. It is possible that the Bealsmill Formation of the Blackdown Nappe is simply a more proximal flysch facies equivalent to the Crackington Formation. Since both nappes are also contiguous, and show similar relationships to underlying nappes, it seems likely that they belong to the same tectonic unit.

Together, the Crackington and Blackdown nappes may have formed the passive roof of an early nappe duplex (*Figure 6.6*), that subsequently become involved in late D2 thrusting and early D3 low-angle normal faulting. The latter included the emplacement of the overriding Boscastle Nappe. Taking the flysch sequences as a whole, vergence has not been established, although it seems likely that D2 thrust displacements were northwards. Such a model accommodates the field evidence, and the opposing ideas of back thrusting *versus* underthrusting and southwards *versus* northwards vergence (Sanderson, 1979; Rattey & Sanderson, 1982; Shackleton *et al.*, 1982; Selwood & Thomas, 1986b).

Around Launceston, and to the west, klippen of the Firebeacon Chert Formation, Carboniferous volcanic rocks and a substantial klippe of the Stourscombe Formation, overlie the Crackington Formation on gently northwards-dipping thrusts. Similar low-angle, north-dipping faults modify otherwise subhorizontal thrust boundaries along the northeastern margin of Bodmin Moor. These faults are a combination of low-angle, early D3 thrusts and high-angle, late D3 extensional faults, which together significantly modify the outcrop patterns of the nappes. These late structures are generally attributed to uplift of the Cornubian batholith (Freshney, 1965; Selwood & Thomas, 1993).

Southern Boundary of the Nappes

The relationships of the nappe zone in the south are currently problematic. Whiteley (1983) indicated that the Kate Brook Slate Formation locally overthrusts the St Mellion klippe at its southern margin, and is itself overthrust by Frasnian purple and green slate (Gooday, 1974). However, westwards along strike, but across the Cambeak–Cawsand Fault Zone, the structural style is of major D1 upright folds and cleavage with a low-angle, south-dipping D2 cleavage overprinting the early structure (Burton & Tanner, 1986). No evidence for thrusting has been recognized. This is the area of the Liskeard High, lying between the South Devon and Trevone basins.

Seago & Chapman (1988) described the Plymouth area in detail, but extended their analysis to the western shores of the Tamar and to localities as far north as Lydford in Devon. They drew attention to numerous occurrences of south-facing structures, including some localities in the Kate Brook Unit and Greystone and Blackdown nappes, which had been interpreted as northward-thrusted units by Isaac *et al.* (1982). In fact, Isaac (1983) records both north- and south-facing structures in the Blackdown Nappe. It is not possible to resolve these conflicting interpretations here, but it is evident that a number of factors complicate the issue. Firstly, structural as well as stratigraphic relationships are probably laterally variable across NW–SE fault zones. Secondly, northwards thrusting of older south-facing structures needs to be considered because much of the evidence for northwards transport of units (Isaac *et al.*, 1982) was deduced from thrust zones and fault relationships. It is possible that north- and south-facing structural units may be interleaved, and that in some cases cleavage is not axial planar.

CULM BASIN

Within the Culm Basin, Upper Carboniferous rocks crop out between the Rusey Fault Zone (*Figure 6.1*) and the Taw estuary in north Devon, where a conformable passage into Dinantian rocks is assumed but not exposed. Initiation of the Culm Basin may have taken place during a period of crustal extension marked by Lower Carboniferous volcanism.

The tectonic setting of the basin is presently disputed. It has variously been interpreted as a foreland basin, which developed as the crust was depressed before an advancing pile of nappes (Hartley & Warr, 1990; Warr, 1993), or as a thrust-sheet top basin (Gayer & Jones, 1989). Another possibility is a pull-apart basin developed during oblique dextral convergence (Jackson, 1991; Andrews, 1993), an origin that accords well with the inverted aspect of the Rusey Fault. Whatever the setting of the basin, the contents were subsequently severely compressed by Variscan tectonism at the end of the Carboniferous. Shortening of over 50% was accomplished by both thrusting and folding. Early bed-parallel thrusts (Mapeo & Andrews, 1991) are widespread, but difficult to detect as they were subsequently incorporated into the magnificent angular E–W-trending chevron folds that now characterize all parts of the basin. Local thrusting accommodated bed-parallel slip during the folding, producing flexural-slip duplexes. *Plate 2* illustrates how the repetition of a sandstone couplet, seen above a roof thrust, is accommodated by roughly equal amounts of slip on a series of imbricate faults which anastomose into a bed-parallel floor thrust. The floor thrust eventually cuts across the overturned limb of a footwall fold-pair. This duplex could have developed during a pre-folding episode of thrusting or by flexural-slip during amplification of the fold. Early pre-folding vein arrays, a 10–20° anticlockwise obliquity between the trend of the folds and the margins of the basin, and late asymmetric folds with dextral vergence all suggest that there was an ever-present dextral strike-slip component during basin closure. The chevron folds display a regional fan-like geometry (Sanderson & Dearman, 1973), ranging from northerly inclined, north-facing folds at the northern margin of the basin, through upwards-facing folds characterizing the greater part of the basin, to south-facing recumbent folds at the southern margin in Cornwall. This arrangement is thought to have been generated during the northwards and southwards expulsion of the sediments over the basin margins during its closure. The deep structure of the basin has not been resolved.

Sometime following the late Variscan shortening there was a period of crustal extension which produced a set of widespread normal faults, predominantly dipping at moderate to steep angles to the north–northeast and south–southwest. South of Wanson Mouth [SS 194 012] it has been argued that north-dipping extensional faults dismembered the overturned limb of a south-facing overturned fold (*Figure. 6.8*). This interpretation reconciles the conflict between the structural facing and stratigraphic younging indicated by palaeontological evidence (Freshney *et al.*, 1972). It is quite probable that many of these faults developed during orogenic collapse of the regional nappe pile at the end of the Carboniferous, just prior to, or synchronous with, emplacement of the Cornubian batholith.

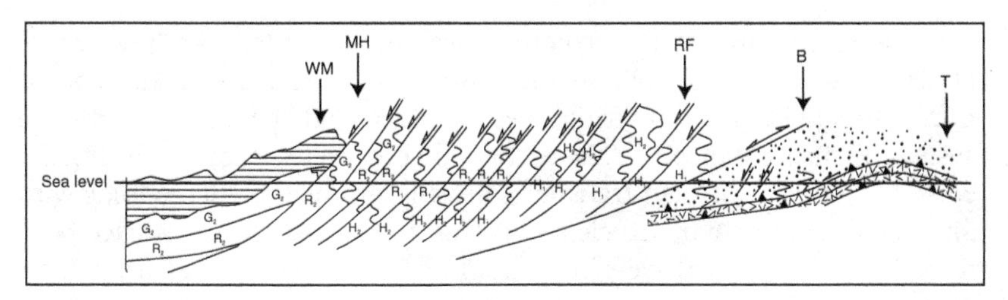

Figure 6.8. Sketch showing the structure between Bude and Tintagel. Goniatite zones: G_2, R_2, R_1, H_2, H_1 (after Freshney *et al.*, 1972; Selwood *et al.*, 1985; and Andrews *et al.*, 1988); WM, Wanson Mouth; MH, Millook Haven; RF, Rusey Fault; B, Boscastle; T, Tintagel

NW–SE-trending dextral wrench faults (Dearman, 1963) are common, the largest having displacements of the order of 1 km. Though it is apparent that they were active during the Tertiary, they are also thought to have been active in Upper Palaeozoic times, perhaps as transfer faults during rifting phases in the Devonian and Lower Carboniferous.

THE BOSCASTLE–TINTAGEL–TREGARDOCK CULMINATION

West of the Bodmin Moor granite, the Tredorn and Boscastle nappes (Selwood & Thomas, 1986b) define a structural culmination between the Rusey Fault and Jacket's Point [SX 033 831]. The antiformal structure can be compared with that of a triangle zone at a mountain front, with the Culm Basin sediments carried southwards over an advancing wedge of the Boscastle and Tredorn nappes, and the inverted Rusey Fault acting as a passive roof thrust (*Figure 6.8*).

Both units are characterized by recumbent south-facing folds dissected by northwards-transporting thrusts. This complex structure makes restoration of the original stratigraphic relationships difficult. At the southern margin of the Tredorn Nappe, the Trekelland Thrust carries the Tintagel Succession northwards over the lowermost structural unit formed of the Bounds Cliff Succession. At the northern margin of the Tredorn Nappe, the northward dipping Willapark Thrust carries the Boscastle Formation over the Tintagel Succession, which is itself repeated a number of times in thin imbricated sheets (Andrews *et al.*, 1988). The origin of the Boscastle Nappe is disputed, but may lie somewhere to the south, possibly within or north of the Gramscatho Basin.

Internally, both nappes exhibit large tectonic strains defining a high strain zone of oblique folding (*Figure 6.1*) (Sanderson & Dearman, 1973). Rotation of the fold axes away from the regional E–W trend was attributed by Sanderson (1973, 1979) to southward directed shearing during the development of large south-facing fold-nappes. Though the south-facing nature of the structures was not disputed, Andrews *et al.* (1988) showed that the high ductile strain was associated with subsequent north directed transport, which evolved to a higher-level, more brittle system of thrust transport. The early structures were reworked during the episode of north-directed transport, initially producing ductile folds with an intense axial planar S2a cleavage. This and earlier cleavages have been folded by F2b north-verging angular folds, which are frequently associated with the development of discrete thrusts.

Davidstow Antiform

The structural trend of the formations within the Tredorn nappe is locally affected by this large regional fold (*Figure 6.1*), which plunges gently northwestwards away from the northern outcrop of the Bodmin Moor granite. It is a relatively late structure, and is responsible for folding the important D2 thrusts so that they acquire a present-day extensional geometry on the northeast-dipping limb. The precise timing of the large-scale doming is uncertain. Various authors have attributed it to up-doming by the Bodmin Moor granite (Wilson, 1951) or stacking of D2 thrust sheets above the Rusey Fault acting as a lateral ramp (Andrews *et al.*, 1988; Warr, 1989).

METAMORPHISM

The Devonian and Carboniferous sedimentary and pre-orogenic igneous rocks of Cornwall have undergone regional metamorphism associated with the Variscan orogeny at temperatures of generally less than about 350°C. The regional metamorphism has been overprinted by contact metamorphism in the metamorphic aureoles surrounding the granite plutons of the region.

The low-grade metamorphic character of the rocks means that, in the field, many of the primary characteristics and mineralogy of the sedimentary or igneous rocks are still prevalent. The secondary metamorphic mineralogy is typically fine-grained and the products often invisible to the naked eye. However, one of the most distinctive features of the regional metamorphism easily seen in the field is the widespread presence of slates formed from former mudstone rocks. In these rocks the cleavage is produced by the alignment of recrystallized fine-grained phyllosilicate minerals.

Despite the fine-grained nature of the metamorphic products in Cornwall, there has been much work in recent years in documenting their characteristics, in particular by study of the pelites (mudstones) and metabasites (basaltic rocks, intrusive and volcaniclastic). Pelitic rocks are well suited to the study of low-grade metamorphism. This is because the parent rocks contained very fine-grained clay materials with a large surface area that encouraged recrystallization within the timescale of the metamorphic process. Metabasites are also useful, but in their case they contained initially a high temperature mineralogy and glassy interstitial areas that at the low pressure/temperature conditions of metamorphism were markedly out of equilibrium. Consequently, they break down relatively rapidly, allowing the growth of new products in an attempt to regain equilibrium.

Metamorphism Of Pelites

Fine-grained clastic rocks are widespread in Cornwall, and in most areas these have been metamorphosed to dark-coloured, fine-grained slaty rocks. Hand specimen and even thin section examination of these rocks reveals little of their mineralogy or metamorphic grade because of the fine-grained nature. One exception is in the Devonian and Carboniferous pelites of the Tintagel area where the rocks are somewhat coarser and a varied mineralogy has developed (Phillips, 1928; Primmer, 1985). In these rocks the slaty cleavage is formed by the alignment of muscovite and chlorite, and the rocks typically have a bright sheen, reflecting the coarseness of the phyllosilicate minerals. In addition to these minerals, biotite, chloritoid and garnet are also commonly present. Biotite occurs in lath shaped crystals up to 0.4 mm in size. The chloritoid, with some chlorite, is porphyroblastic in form, up to 0.5 mm in size (and can be seen in the rocks with the aid of a hand lens), and grows across the foliation. Garnet also occurs as porhyroblasts up to 1 mm in size. These garnets have 22–50% of the spessartine end-member (Primmer, 1985). Estimates of temperature for the maximum metamorphism in the Tintagel area based on oxygen isotope thermometry of chlorite–quartz pairs (Primmer, 1985) and from fluid inclusion studies, indicate 400–450°C.

The fine-grained slates found in most other areas of Cornwall are unsuitable for petrographic analysis; however, X-ray diffraction (XRD) methods have been applied in recent years as a means of determining the mineralogy and metamorphic grade of these rocks (Warr & Robinson, 1991). The mineralogy of such slaty rocks as determined by XRD is a simple mixture of illite and chlorite.

At very low-grade conditions of metamorphism, as in Cornwall, the determination of the illite crystallinity (IC) and *b*o value on <2 μm grain-size fractions separated from the slates has proved very useful in establishing the character of metamorphism. The <2 μm fraction is chosen to ensure that only newly formed illite is analysed, rather than any unmodified detrital clay. The IC method involves quantification of the XRD-determined 10Å peak shape by a simple measurement of the peak width (in degrees 2Θ) at half peak height (*Figure 6.10*). At diagenetic levels less than about 200°C, the 10Å peak is wide, but with increasing metamorphic grade the peak becomes increasingly narrow. Thus the actual IC value measured decreases with increasing grade. The IC scheme is used and calibrated on an international scale, and is used to recognize three broad low-grade metamorphic divisions, as shown in *Figure 6.9*. The *bo* method involves XRD measurement of the 060 reflection of illite from which the *b* unit cell size direction can be calculated. This value varies according to the substitution of (Fe+Mg) for Al in the crystal structure of the illite that occurs as a function of pressure and temperature. The measurement can thus be used as a guide to distinguish between low, medium and high pressure settings that are representative of high, medium and low geothermal gradients, respectively.

A map showing variation in metamorphic grade based on IC data (Warr *et al.*, 1991) in pelitic rocks is given in *Figure 6. 10*. This shows that diagenetic, anchizone and epizone levels of metamorphism are found in the region. The belts trend generally E–W in line with the structural pattern of the area. The lowest diagenetic grade, at less than about 200°C, is found in the Upper Carboniferous rocks of the Culm Basin in north Cornwall. The grade increases to the south into areas of anchizone

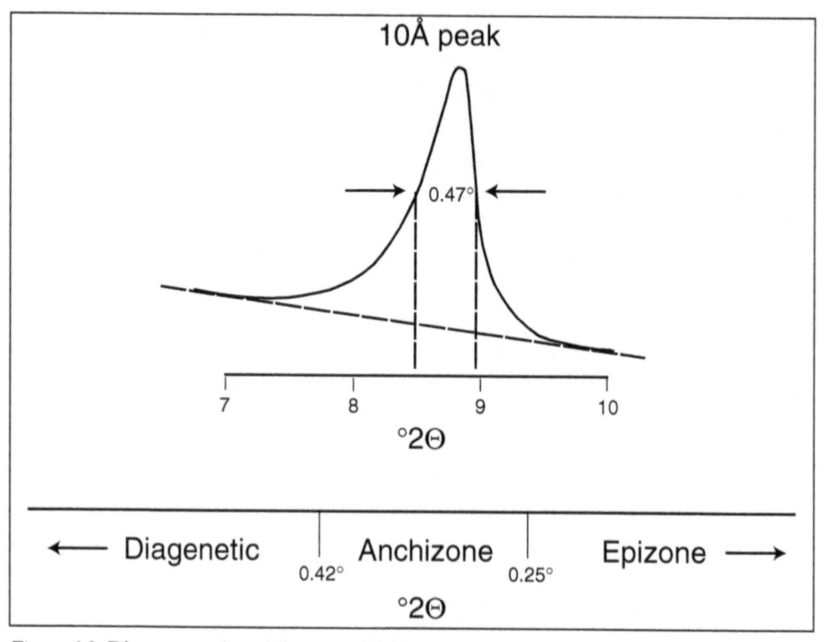

Figure 6.9. Diagram to show illite crystallinity measurement of the peak width at half-peak height of the 10Å XRD peak. Scale (in degrees 2Θ Cu Kα) shows the use of the crystallinity index value to divide between diagenetic, anchizone and epizone

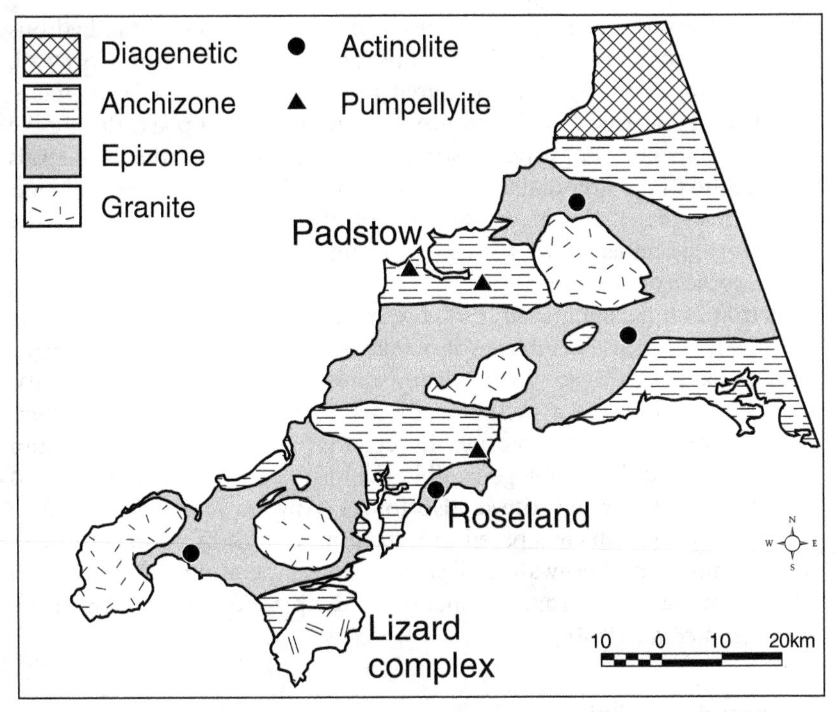

Figure 6.10. Simplified metamorphic map (after Warr *et al.*, 1991; and Robinson *et al.*, 1994). Actinolite: areas with metabasites having chlorite+epidote±actinolite assemblage. Pumpellyite: areas with metabasites having chlorite+epidote±pumpellyite±prehnite assemblage

(about 200–300°C) and epizone (greater than about 300°C) metamorphism. The Lower Carboniferous and Devonian rocks of central and southern Cornwall are all at anchizone to epizone levels. The *b*o values from the slates of Cornwall show a range 8.994–9.011Å (Warr *et al.*, 1991), which is intermediate between the low and medium pressure ranges indicative of thermal gradients more than 30°C km^{-1}.

Metamorphism Of Metabasites

Igneous rocks of basic composition are widespread in Cornwall as intrusive bodies, lavas and volcaniclastic horizons. The metamorphic minerals that developed in these rocks at low grade are hydrous mafic phyllosilicates (dominantly chlorite), various hydrous calc silicates, such as prehnite, epidote and pumpellyite, and some actinolite (Barnes & Andrews, 1981; Robinson & Read, 1981; Primmer, 1985). Also a common feature in these rocks is the alteration of the primary calcic plagioclase to albite, releasing much Ca that is used in the formation of the hydrous calc-silicate minerals. All such minerals have been recorded in Cornwall, but they are relatively rare compared to the abundance of basic rocks. This is probably due to the presence of CO_2 in the fluid phase during the metamorphic episode, which restricted the formation of these minerals and resulted in a non-diagnostic albite, chlorite and calcite assemblage. Where present in the metabasite rocks of South West England, the metamorphic minerals are typically very fine-grained and only visible by microscopic examination.

Petrographic examination of these rocks has shown two contrasting diagnostic mineral assemblages in metabasites (Primmer, 1985). In areas such as Padstow (*Figure 6.10*), where pelites have anchizone crystallinity, the metabasites have a metamorphic assemblage of albite+chlorite± pumpellyite±prehnite±epidote. In epizone areas such as Roseland (*Figure 6.10*), the assemblage is albite+chlorite±epidote±actinolite, and in the Tintagel area biotite is also found. Application of the chlorite geothermometer, which is based on substitution of Si by Al in chlorite in metabasites from Cornwall, indicates mean temperatures of 271°C and 312°C for samples from anchizone and epizone areas respectively (Robinson *et al.*, 1994).

Metamorphic Evolution

The anchizone to epizone metamorphic grades in the pelites are equated with the prehnite-pumpellyite to greenschist facies recorded in metabasites. The estimated temperature range is from about 200°C to about 350°C, although the more varied assemblages in the rocks of the Tintagel area indicate that conditions reached the biotite zone in the greenschist facies, with temperatures up to about 450°C.

The exact relationships between the metamorphic development and the Variscan orogenic processes is not well understood, although Warr *et al.* (1991) have presented an integrated model. They suggested that there is a close link between grade and stratigraphic level, especially in the northern part of the area, that is indicative of a burial control on the metamorphism. However, the widespread development of a slaty cleavage indicates some inter-relationship between deformation and metamorphism. This is particularly evident in the southern part of the region and in the Tintagel high strain zone, where metamorphic grade is not linked to stratigraphic level. A close relationship between the orogenic processes of continental thickening and metamorphism is suggested. In these regions, the pervasive foliation defined by a chlorite+muscovite assemblage is indicative of an early M1 metamorphism developed during progressive burial and increasing strain at 250–350°C and 2–4 kbar. A later M2 metamorphism, represented by the chloritoid and chlorite porhyroblasts overgrowing the M1 fabric, indicates continued heating during decompression on uplift (Primmer, 1985).

Contact Metamorphism

Little systematic work has been undertaken on the contact aureoles surrounding the granite plutons. Records of spotted slates with variable development of andalusite, cordierite and biotite, and of calc flintas with diopside and albite, are reported in various Geological Survey memoirs (Ussher *et al.*, 1909). An increase in grade as measured by IC values was also recorded by Warr & Robinson (1991) in the aureole around the Bodmin Moor granite. Metasomatic activity involving extensive fluid/rock interactions is also common in the aureoles, and often is distinguished by the development of tourmaline and axinite (Ussher *et al.*, 1909). The contact metamorphism overprints the effects of the regional metamorphism, indicating intrusion of these bodies towards the end of the orogenic event. Detailed descriptions of contact metamorphism around the Land's End and Carnmenellis granites were given by Hill & MacAlister (1908), Goode & Taylor (1988) and Leveridge *et al.* (1990).

Chapter Seven

Granites and Associated Igneous Activity

The granites of Cornwall represent part of the Cornubian batholith that extends from Dartmoor in Devon through Cornwall to the Isles of Scilly and, perhaps, even farther west to Haig Fras (95 km west-southwest of the Isles of Scilly). Its surface expressions in Cornwall are limited to the Land's End, Tregonning-Godolphin, Carnmenellis, St Austell and Bodmin Moor plutons (*Figure 7.1*). The granites are noted for their exceptionally high content of the radioactive elements K, Th and U, which account for high heat production and a steep geothermal gradient in their vicinity (Chapters 8 & 9). The economic importance of these rocks, through their association with, or hosting of, metalliferous and non-metalliferous mineral deposits, has led to a long history of study. Although ideas generated by work on the granites have shaped the development of concepts of ore formation associated with granitic rocks world-wide (Jackson *et al.*, 1989), much more work remains to be done on the nature of granite magma genesis in the province.

Geophysical evidence summarized by Floyd *et al.* (1993) indicates that from E–W the batholith varies in thickness, from 20 km east of Dartmoor to 10 km in the Isles of Scilly. It is 40–60 km wide at its base, which is marked by a southerly-dipping seismic reflector interpreted as a thrust (Brooks *et al.*, 1984), with transport from the south. That the plutons connect at depth is demonstrated by the geophysical evidence as well as the occurrence of minor cusps of granite between, and close to, the major bodies. Three-dimensional modelling of gravity data has allowed the form of the batholith to be defined; it has a total volume of the order of 68 000 km^3 (Willis-Richards & Jackson, 1989).

The major granite bodies differ in their outcrop patterns. The outcrops of the Carnmenellis and Bodmin Moor plutons are both relatively regular, and their contact metamorphic aureoles are narrow (less than 1 km). In contrast, the St Austell granite has a very irregular outcrop, particularly in its western part where the contact metamorphic aureole extends up to 7 km or more (horizontally) from the contact. The Land's End and Dartmoor

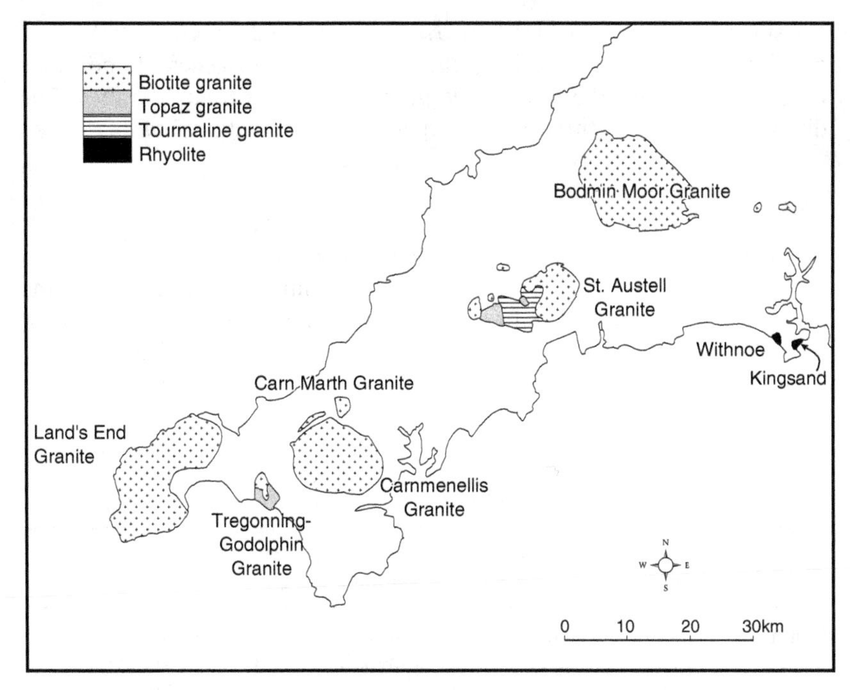

Figure 7.1. Distribution of biotite granites, tourmaline granites, topaz granites and rhyolites in Cornwall

granites are intermediate between these two extremes. It seems that both the extent of the contact metamorphic aureole and outcrop shape are controlled by the proximity of the roof to each intrusion. The St Austell granite is exposed more or less at the level of its roof, which extends at shallow depths to the northwest, while the Bodmin Moor and Carnmenellis granites have been more deeply eroded to a level where steep contacts control the outcrop pattern. In addition to natural outcrops, which include spectacular coastal sections, the Cornish granites are exposed in extensive china clay workings in the St Austell granite (and to a lesser extent the Bodmin Moor and Land's End granites), in stone quarries (particularly in the Carnmenellis granite) and underground in metal mines. The deep boreholes at Rosemanowes [SW 735 346] extend to depths of 2.6 km within the Carnmenellis granite (Richards *et al.*, 1994), but show little evidence of variation in petrographic characteristics with increasing depth.

GRANITIC ROCK TYPES

Several different varieties of granitic rock have been distinguished on the basis of petrographic and geochemical evidence, and a much simplified classification is summarized in *Table 7.1* (Floyd *et al.*, 1993). This classification places emphasis on the contrasting petrogenetic importance of biotite granites, the tourmaline granites and the topaz granites. The distribution of these three granite types is summarized in *Figure 7.1*.

Biotite Granites

In Cornwall, the biotite granites (which show variation in texture from megacrystic to non-megacrystic, and coarse to fine-grained) dominate the exposed part of the batholith (more than 90% of exposure), and extend to a depth greater than 2 km in the Rosemanowes boreholes. These granites are characteristically rich in alkali feldspar megacrysts, which

Table 7.1. PETROGRAPHIC FEATURES OF THE MAJOR GRANITE TYPES

	Megacrystic biotite granite	Tourmaline granite	Topaz granite
Grain size	medium-coarse	medium-coarse	medium
Texture	hypidiomorphic granular	aplitic-hypidiomorphic granular	hypidiomorphic granular
Phenocrysts	K-feldspar; <20 cm	K-feldspar; <8.5 cm	none
K-feldspar	euhedral–subhedral; microperthitic 32%	euhedral–subhedral; microperthitic 27%	anhedral–interstitial; microperthitic 24%
Plagioclase	euhedral–subhedral; often zoned; cores An_{25}–An_{30} rims An_8–An_{15} 22%	euhedral-subhedral; microperthitic unzoned; An_7 26%	euhedral unzoned; An_1–An_4 32%
Quartz	irregular; 34%	irregular; rounded in globular quartz varieties; 33%	irregular; some aggregates; 30%
Micas	biotite 6% muscovite 4%	Li-mica; 6%	zinnwaldite; 9%
Tourmaline	euhedral-anhedral; <70% schorl*; 1%	euhedral; 70–90% schorl*; 4%	commonly anhedral; 90% schorl*; elbaite-rich (50%); 1%
Other minerals	zircon, rutile, apatite, andalusite	rutile, apatite, topaz, fluorite	topaz 3%, apatite, other phosphates; ilmenorutile
Exley's classification	Type B	Type D	Type E

*Proportion of schorl in schorl-dravite solid solution
Based on Exley and Stone's classification given in Floyd *et al.*, 1993, with additional information from Hill, 1988; Hill & Manning, 1987; London & Manning, 1995; and Manning *et al.*, 1996

influence their appearance in the field. The size and proportions of the megacrysts vary, from about 2 cm up to approximately 20 cm, and there is a tendency for a bimodal distribution in phenocryst size. In the Carnmenellis and Bodmin Moor plutons, the biotite granite contains phenocrysts which reach an average length of 2 cm, and which rarely achieve the spectacular size (up to 20 cm) frequently observed in the Land's End, St Austell and Dartmoor plutons. This difference may correlate with proximity to the roof of the individual plutons.

The mineralogy of the biotite granites is characterized by the presence of biotite, perthitic K-feldspar megacrysts, plagioclase (An_{25-30} cores; rims An_{8-15}) and quartz (*Table 7.1*). Muscovite occurs widely, either as a discrete phase or as an alteration product of the feldspars, and tourmaline is a common accessory. Primary tourmaline is particularly abundant in the coarsely megacrystic granite. It is distinguished from tourmaline of hydrothermal origin by both its paragenesis (as coarse prismatic grains forming part of the overall magmatic texture of the rock, which lack an association with jointing or other fluid pathways within the rock) and its chemical composition (London & Manning, 1995). Compared with hydrothermal tourmaline, primary tourmaline shows weak zoning within a very limited range of substitution towards a proton deficient tourmaline end-member, and up to 70% schorl in the schorl-dravite solid solution. Other important accessory minerals include zircon, apatite, cordierite and andalusite.

Tourmaline Granites

The biotite granites are often cut by veins of leucocratic granite in which tourmaline is the dominant ferromagnesian mineral. As minor intrusions, the tourmaline granite veins have a grain size measured in millimetres; they are usually equigranular although occasional K-feldspar phenocrysts are present, and they are often associated with pegmatitic clots, segregations and selvages which contain the same mineral assemblage. These veins are generally too small to be mapped, but are widespread wherever the biotite granite outcrops.

Although the tourmaline granites differ from the biotite granites in terms of their ferromagnesian mineral species, their bulk mineralogy is similar, with K-feldspar, plagioclase, muscovite and quartz forming an equigranular matrix within which small prisms of tourmaline are evenly distributed. Accessory minerals are sparse, reflecting the lack of phases which otherwise occur as prolific inclusions in dark micas, and which tend to be less abundant in tourmaline.

Within the St Austell granite in particular, large bodies of coarse-grained tourmaline granite occur and can be mapped (*Figure 7.1*), especially within the area of china clay production (Manning *et al.*, 1996).

These tourmaline granites contain a Li-mica and correspond to the megacrystic Li-mica granites of Exley *et al.* (1983). However, they are more readily recognized in the field by their coarse tourmaline crystals (up to 8 mm) than by their micas (which deteriorate as a consequence of alteration, whereas tourmaline does not). Tourmaline granites account for no more than 5% of outcrop for the batholith overall.

The Li-mica which is present in the tourmaline granites presents a brown colour on a fresh fracture, as opposed to the lustrous black of biotite or the white of muscovite. It is essentially a zinnwaldite, but shows a more or less continuous compositional trend of decreasing Fe content towards ferroan lepidolite, according to ion microprobe analysis (Henderson *et al.*, 1989). The tourmaline granites may be porphyritic, with K-feldspar megacrysts, or aphyric. Plagioclase is strongly sodic (An_7 or less), and accessory minerals, important as inclusions in the mica, include rutile.

The tourmaline granites within the St Austell pluton show considerable textural variation, and can be subdivided and mapped as individual textural types (Manning *et al.*, 1996). Variants include a globular quartz granite, in which rounded quartz crystals (up to 1 cm) and aggregates of quartz of similar size are set in a fine-grained groundmass. Other facies are poor in tourmaline and correspond more closely to the megacrystic Li-mica granites (Exley *et al.*, 1983). Megacrysts of K-feldspar (up to 10 cm) commonly occur, and xenoliths of megacrystic granites are widespread. Contacts between individual textural variants are frequently gradational and often marked by pegmatites. Although these granite varieties are best exposed in the St Austell granite, they are also known from the Land's End and Dartmoor granites, where they have yet to be mapped in detail.

Topaz Granites

The topaz granites can be regarded as a separate granite type by virtue of their high topaz content and distinct texture. They are equigranular, with subhedral topaz reaching a modal abundance of 3%; the plagioclase is almost pure albite. The topaz granites contain a Li-mica which plots towards the Li-rich end of the compositional series shown by the Li-micas of the tourmaline granites, with an increased proportion of samples reporting lepidolite. Accessory minerals include Nb-rutile and ilmenorutile as inclusions within the Li-mica and phosphates such as amblygonite. Accessory tourmaline is sparsely found as ragged crystals rich in opaque inclusions (such as ilmenorutile), and contains up to 50% of the Li-tourmaline component elbaite (London & Manning, 1995).

The topaz granites occur as well-defined stocks in the St Austell granite, showing sharp contacts with the tourmaline granite suite (Manning & Hill, 1990), and cross-cut early wolframite-quartz veins.

They also form the Tregonning granite and display an excellent roof complex of pegmatite/aplite veins at Megilligar Rocks [SW 611 266] (Stone, 1975). It is highly likely that they occur more widely, but have not been recognized. Thus, although the mine dumps at Botallack [SW 364 335] include fragments of topaz granite, it has not been recorded *in situ* in the now flooded workings. The topaz granites are very limited in their occurrence, accounting for approximately 1% of the surface exposure of the batholith.

ALTERATION OF THE GRANITES

An important aspect of any description of the magmatic rocks which make up the bulk of the Cornish granite outcrop concerns recognition of secondary phenomena. There is pervasive alteration of feldspars and primary micas to give secondary muscovite, and recognition of primary magmatic muscovite is fraught with uncertainty. Coarse muscovite does occur as discrete grains (that is, not only as inclusions within altered feldspar), within the biotite, tourmaline and Li-mica granites, and may well be a primary magmatic mineral, although this should not be taken for granted. The separation of primary and secondary tourmaline is also problematic, but London & Manning (1995) demonstrated that the origins can be distinguished on the basis of composition and petrography.

Fluorite is an important component of many of the more exotic granite varieties, and the classification of Exley *et al.* (1983) included fluorite granite as a distinct rock type. However, although fluorite granite can be sampled as a discrete phase, it cannot be mapped as distinct bodies. Characteristically, it shows gradational contacts with topaz granite, within which it occurs as irregular patches or bodies whose geometry clearly relates to the pattern of jointing. Petrographically, fluorite is often observed to be the first phase to form as the topaz granite alters. In thin sections, topaz often shows extensive sericitization (demonstrating the mobilization of F), with fluorite preferentially associated spatially with the aggregates of sericitic mica so formed. Fluorite also occurs within the cleavage planes of altered Li-micas (which, like topaz, demonstrate loss of F as one of the first signs of alteration). Manning & Exley (1984) suggested that the fluorite granites should be considered as an alteration facies developed within the topaz granite. This view is at variance with the findings of Weidner & Martin (1987), that fluorite in chinastone has a magmatic origin.

RHYOLITE PORPHYRY DYKES (ELVANS)

The Cornish rhyolite porphyry dykes form a suite of rocks which share a characteristic style of intrusion. They often cross-cut the major coarse-grained granite varieties and some stockwork mineral vein systems, and also share the same general trend of other major lode systems. They show considerable variation in texture. The grain size of the matrix is generally less than 0.5 mm and reduces considerably at chilled margins. Phenocrysts include K-feldspar (up to 30 mm long), quartz, biotite and/or tourmaline, and flow banding is commonly developed. As illustrated by Goode (1973), the rhyolite porphyries can contain fragments of granitic rock or country rock, and may be spatially and genetically associated with intrusive breccias. A high-energy mode of emplacement, involving transport of fine rock particles by a gaseous mobile phase, was invoked by Stone (1968) and Henley (1974) as a means of accounting for the characteristic enrichment in K shown by the rhyolite porphyries, as well as their distinctive-flow banded textures. However, the rhyolite porphyries have received little attention since the mid-1970s and there are relatively few compositional data, possibly reflecting a combination of poor exposure inland and widespread weathering or hydrothermal alteration. The limited available age data are consistent with the field observations that the rhyolite porphyries were amongst the last magmatic rocks to be emplaced (Hawkes *et al.*, 1975).

KINGSAND RHYOLITES

In contrast to the considerable extent of the intrusive granitic rocks (greater than 1325 km^2) exposed in Cornwall and Devon, volcanic rocks which might be associated with the granites are restricted to the rhyolites of the Kingsand–Withnoe area of southeastern Cornwall. At Withnoe [SX 404 517], a small circular outcrop is interpreted (Cosgrove & Elliott, 1976) as an intrusive equivalent to the more extensive rhyolites exposed on the coast at Kingsand [SX 435 506]. The total outcrop of both is less than 1 km^2, but by virtue of the widespread occurrence of rhyolite pebbles within the Permo-Triassic red-beds of Devon, it is assumed that they represent the remnants of a much more extensive suprabatholithic rhyolite field (Laming, 1966).

The Kingsand rhyolites unconformably overlie the Devonian sequence, and show banding defined by alternating phyric and aphyric units (Floyd *et al.*, 1993). The consistency of this banding and a lack of internal brecciation (that would indicate a flow surface) suggests that the section at Kingsand represents part of a single flow. Petrographic descriptions (Cosgrove & Elliot, 1976) refer to the presence of glass and a groundmass of devitrified glass, within which phenocrysts include Carlsbad-twinned feldspar, biotite and quartz. Descriptions of material obtained as pebbles from Permian sediments are very similar. In addition to devitrification after eruption, the rhyolites are reddened as a consequence of oxidative weathering.

LAMPROPHYRES

The distribution of minette-type lamprophyres in South West England (Floyd *et al.* 1993, p.27) shows a regional association with the Cornubian batholith. Most of the occurrences of lamprophyres in Cornwall are in the form of steeply dipping dykes that range in width from a few metres to 15 m, but sill-like bodies are also known at Messack [SW 839 366] near St Just in Roseland and Lemail [SX 023 732]. The dykes vary in strike from NE–SW about Helford, to N–S in the Fal Estuary and Truro areas, to NE–SW in the Holywell Bay area. The lamprophyre at the north end of Holywell Bay [SW 767 599] is a particularly fine example.

The lamprophyres about Falmouth were described (Leveridge *et al.*, 1990) as reddish-brown, fine to medium-grained rocks, and classed as phlogopite-minettes or olivine-phlogopite minettes. The phenocrysts were reported to have a phlogopite core and a biotite rim, and the groundmass to be composed of feldspar, biotite and quartz, with sporadic alkali amphiboles. Other lamprophyres, such as Lemail, contain augite and hornblende (Reid *et al.*, 1910). Since all of the lamprophyres are particularly susceptible to weathering, fresh olivine is rarely seen.

The lamprophyres are relatively unaffected by Variscan deformation, and are likely to post-date the main period of orogenesis. This view is consistent with the radiometric age of 296±5 Ma given by Hawkes (1981) and Darbyshire & Shepherd (1985). Probably the lamprophyres were intruded during post-orogenic crustal extension, either just before, or at the onset of granite magmatism (Alexander & Shail, 1996). The lower-crustal anatexis which gave rise to the granites could have been stimulated initially by mantle driven melts, represented by the lamprophyres (Clark *et al.*, 1993). Unfortunately, there is no occurrence in Cornubia where the field relationship between the lamprophyres and the older granites can be demonstrated.

COMPOSITIONAL VARIATION

The compositions of the granite varieties support the petrographic classification used in this chapter. Averaged major element compositional data summarized in *Table 7.2* show that (with the exception of the tourmaline granites) the individual granite types are restricted in their compositional variation. This is emphasized when major elements are plotted against each other (Manning & Hill, 1990). The relatively large amount of variation shown by the tourmaline granites reflects the large-scale variation in their textures. In particular, this rock type variably develops large quartz crystals (up to 1 cm in the globular quartz granites) and is extensively and pervasively veined by quartz. This accounts for the high silica contents. All of the granite varieties are peraluminous, with

Table 7.2 SUMMARY OF MAJOR ELEMENT COMPOSITIONAL DATA FOR CORNISH GRANITES

	Biotite granites (non-megacrystic)		Biotite granites (megacrystic)		Tourmaline granites		Topaz granites	
	Carnmenellis Granite		St. Austell Granite		St. Austell Granite		St. Austell Granite	
	mean	sd	mean	sd	mean	sd	mean	sd
SiO_2	72.26	0.22	71.62	0.70	75.05	1.10	73.26	0.53
TiO_2	0.26	0.03	0.33	0.02	0.08	0.02	0.04	0.01
Al_2O_3	14.80	0.23	14.59	0.34	13.67	0.58	13.06	0.24
Fe_2O_3	1.80	0.15	2.55	0.28	1.36	0.52	0.83	0.24
MgO	0.36	0.11	0.50	0.12	0.15	0.05	0.08	0.03
CaO	0.69	0.10	1.06	0.16	0.54	0.17	0.49	0.15
Na_2O	3.11	0.07	3.39	0.13	2.81	0.94	3.70	0.41
K_2O	5.02	0.16	5.00	0.17	4.71	0.57	4.46	0.55
P_2O_5	0.28	0.05	0.23	0.02	0.34	0.08	0.46	0.01
Li_2O	0.082	0.016	0.07		0.15	0.09	0.42	0.10
F	0.244	0.053	0.24		0.68	0.30	1.16	0.29
B_2O_3	0.066	0.028	0.09		0.41	0.41	0.10	
Total	98.97		99.67		99.95		98.06	
O=F	0.10		0.10		0.29		0.49	
Total*	98.87		99.57		99.66		97.57	
A/CNK	1.255		1.129		1.276		1.105	
A/CNKL	1.226		1.108		1.218		0.986	
n	7		6		25		9	
Source	Charoy, 1986		Darbyshire & Shepherd, 1985		Hill, 1988		Hill, 1988	

Data for Li_2O, F and B_2O_3 for St. Austell megacrystic biotite granites taken from Harding & Hawkes, 1971
Data for B_2O_3 for topaz granites refer to two samples
O=F refers to the appropriate correction required for the presence of F as an anion instead of oxygen
*Corrected for F
A/CNK and A/CNKL are molecular ratios of Al_2O_3 to $CaO+Na_2O+K_2O$ and to $CaO+Na_2O+K_2O+Li_2O$ respectively

molecular alumina to Ca+Na+K oxide ratios (A/CNK) in excess of 1. If Li_2O (L) is included, the molecular ratio A/CNKL is reduced, to less than 1 in the case of the topaz granites.

Although not usually considered as major elements, the volatiles F, B and Li are included in *Table 7.2* because of their enrichment in the Cornish granites. Average values, which can be regarded as 'background', for Li_2O and F for the biotite granites are 0.08% and 0.24%, respectively. The topaz granites contain significantly more of each component (0.42% and 1.16%, respectively), and are also enriched in P. The tourmaline granites occupy an intermediate position for F and Li, but demonstrate the greatest B levels (reflecting their high tourmaline content). It is important to note that although the topaz granites might be regarded as volatile-rich, this description does not apply to B (Manning & Hill, 1990), which is depleted in the topaz granites relative to the biotite granites.

Trace element data contrast with those for major elements by showing considerable variation within and between granite types. The topaz granites in particular are enriched in ammonium (Hall, 1988) and alkalis such as Rb and Cs. The most valuable data for trace elements are those for Nb, Zr and Ga, which allow the different granite varieties to be distinguished. *Figure 7.2* shows a plot of Nb against Zr, which clearly discriminates the biotite granites from the tourmaline granites and the topaz granites. In addition to demonstrating the value of the Nb/Zr ratio as a discriminant, this plot also shows that the megacrystic biotite granites generally have higher Nb contents than the weakly porphyritic, non-megacrystic, biotite granites. The limited available data for rhyolite porphyry dykes (elvans) show that they generally plot with the biotite granites, with the exception of the Tremore elvan near St Austell [SX 010 649], which plots with the topaz granites. The Kingsand rhyolites and associated volcanic rocks plot with the biotite granites, supporting other evidence of their close relationship, but Floyd *et al.* (1993) note that geochemical evidence supports consanguinity with the rhyolite porphyry dykes rather than the major granites. The value of Ga as a discriminant is shown in *Figure 7.3*, which shows a triangular plot of Ga, Nb and Zr, overcoming problems of dilution affecting the samples to different extents as a consequence of variable alteration. This plot confirms the pattern shown by the Nb–Zr plot, and emphasises the distinct character of the topaz granites (unfortunately there are no Ga data for the Tremore sample).

Of the other trace elements, Rb, Sr and Ba show systematic variation which distinguishes the biotite and tourmaline granites from the topaz granites. *Figure 7.4* shows the range in Rb and Ba values, with high Rb/low Ba in the topaz granites and low Rb/high Ba in the biotite granites. Sr shows a similar relationship, but with much more scatter for the topaz granites. When Sr and Ba are plotted (*Figure 7.5*), strong linear relationships are apparent for the biotite granites (with different trends for megacrystic and non-megacrystic varieties), but the tourmaline granites and topaz granites show much scatter and enrichment in Sr, compared with the biotite granites. A triangular plot (*Figure 7.6*) emphasizes the Rb enrichment of the tourmaline and topaz granites, and their depletion in Ba and Sr.

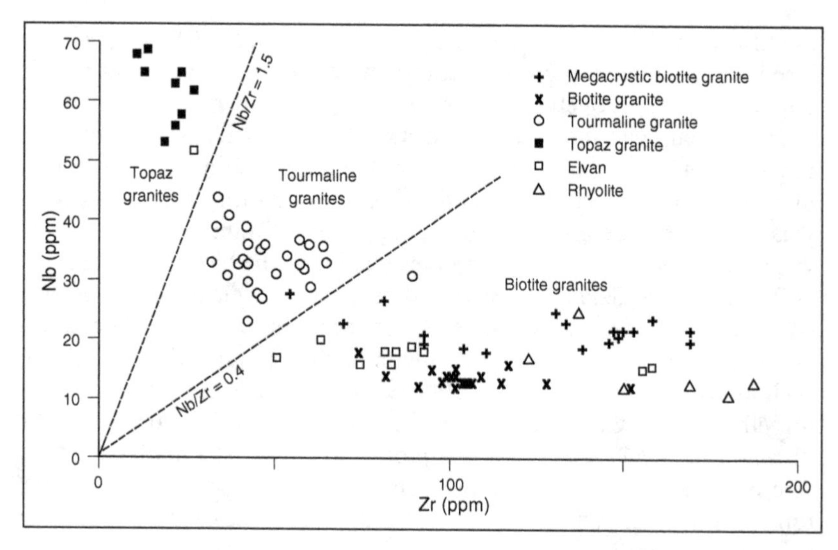

Figure 7.2. Plot of Nb vs. Zr for granitic rocks from South West England

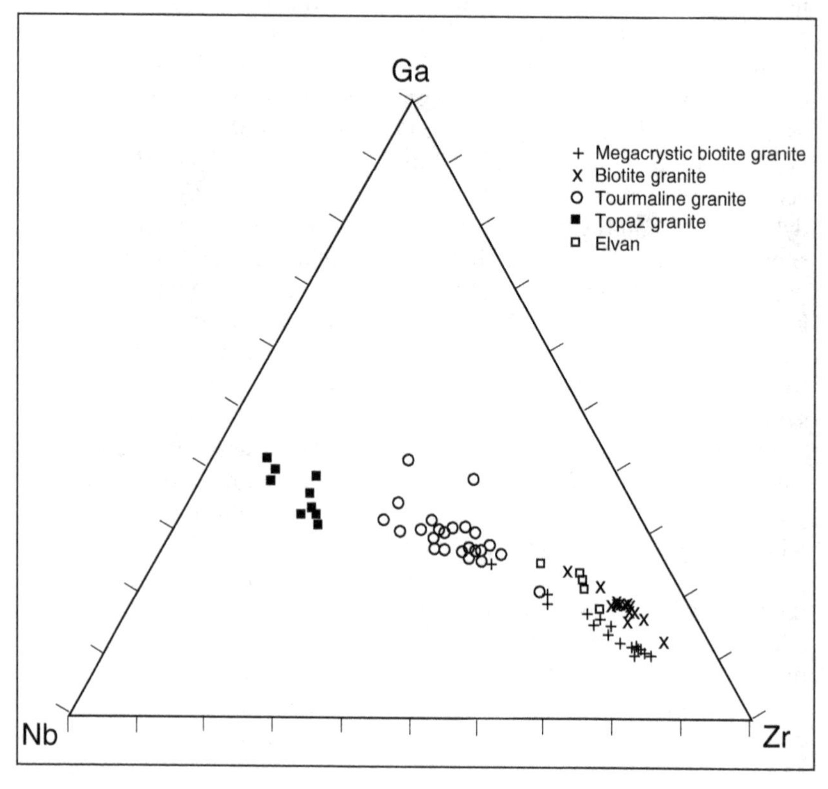

Figure 7.3. Ternary plot of Nb, Zr and Ga for granitic rocks from South West England

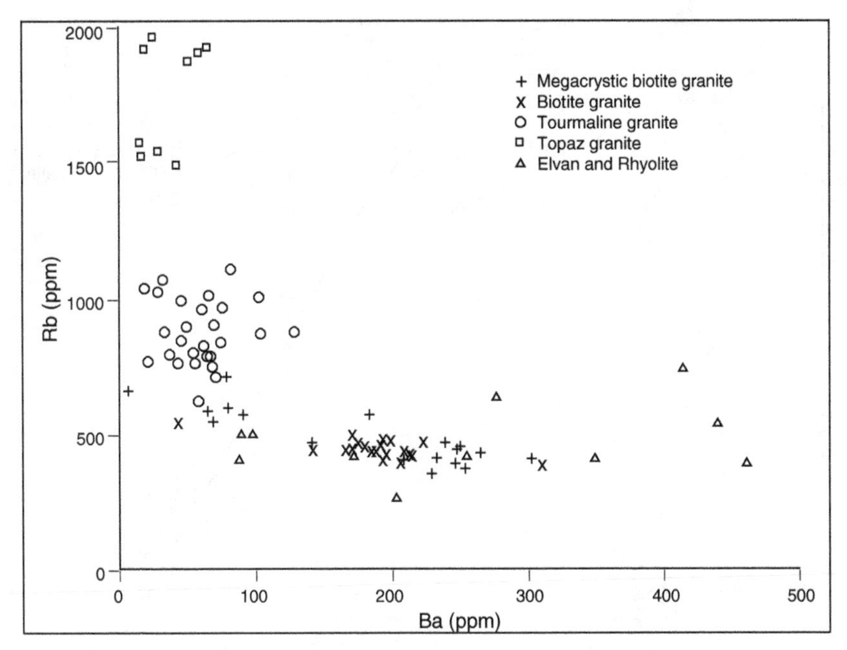

Figure 7.4. Plot of Rb vs. Ba for granitic rocks from South West England

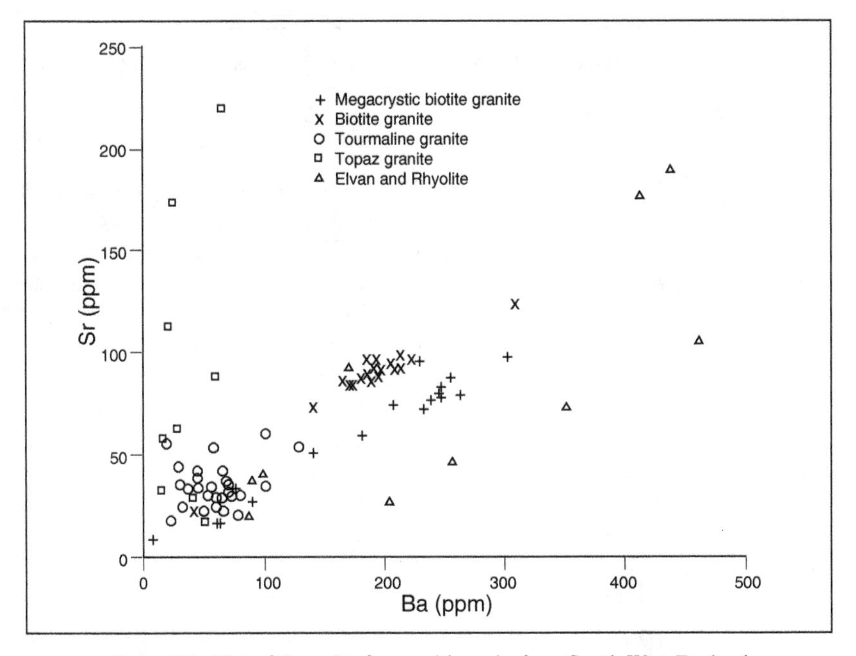

Figure 7.5. Plot of Sr vs. Ba for granitic rocks from South West England

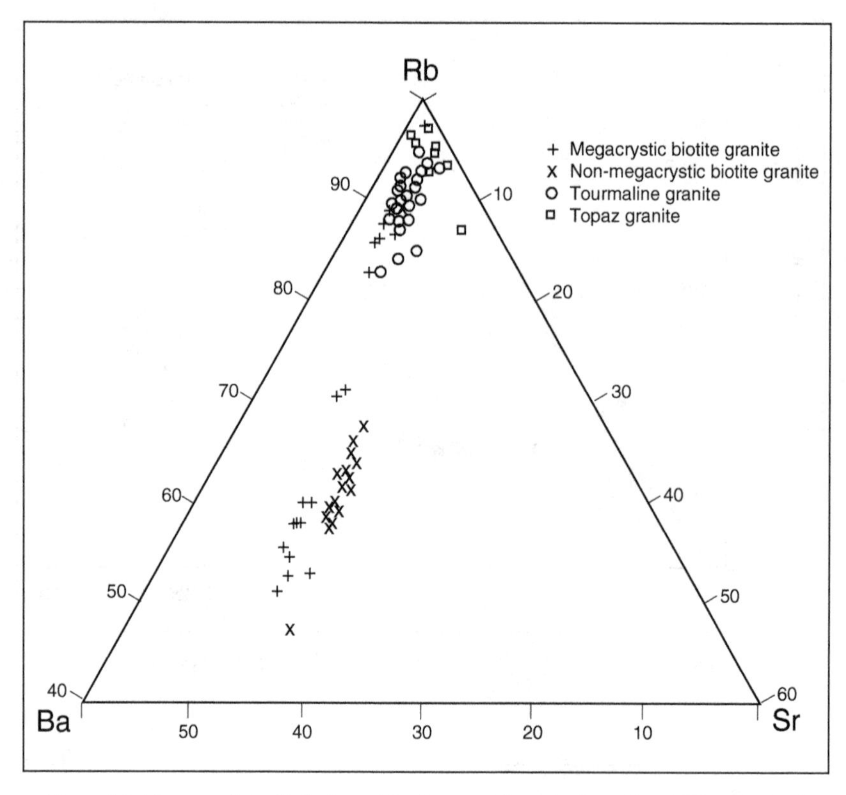

Figure 7.6. Ternary plot of Rb, Sr and Ba for granitic rocks from South West England

Data for the rare earth elements (REE) are available for samples from the Carnmenellis and Bodmin plutons (Charoy, 1986; Jefferies, 1985), and for the Tregonning Granite (Stone, 1992). The pattern of enrichment in light REE (LREE) relative to heavy REE (HREE), with a marked negative Eu anomaly, is typical of the biotite granites, the Kingsand rhyolites (Floyd *et al.*, 1993) and the rhyolite porphyry dykes. This pattern is attributed to the presence of accessory minerals such as monazite (which is the predominant host for the LREE) and xenotime (which to a lesser extent controls HREE abundance), apatite and zircon. In contrast, Eu is influenced by the major rock-forming silicates, orthoclase and plagioclase. REE abundances are greatly reduced in the tourmaline granites and topaz granites (by a factor of almost 100 for LREE), reflecting the absence or relative lack of host accessory minerals. Chondrite-normalized REE data for Li-mica, topaz and tourmaline granites typically show little enrichment of LREE over HREE (or vice versa) and modest Eu anomalies (Manning & Hill, 1990; Stone, 1992).

GEOCHRONOLOGICAL DATA

Early application of isotopic dating techniques yielded data which showed a spread of ages; this could be attributed to the uncertainties in the determinations, to resetting of isotope systems or to events taking place at different times. Only recently have major, internally consistent, studies demonstrated that individual magmatic or hydrothermal events can be recognized and dated within the broad and well-known range of ages (from about 300–250 Ma). Rb–Sr dating of the major granites and minor intrusions (Darbyshire & Shepherd, 1985; 1987) suggested that the main stage of granite magmatism was between 290 Ma and 280 Ma (*Table 7.3*). Although the Land's End data are reset and cannot indicate a magmatic age, these data indicated that the Carnmenellis and Bodmin granites, which are characterized by the least evolved chemical compositions, were earliest to form. More recent studies (Chesley *et al.*,1993; Chen *et al.*, 1993) have used U–Pb (monazite) and Ar ($^{40}Ar/^{39}Ar$ in magmatic muscovite) isotopic systems which have demonstrated considerable potential to resolve relatively closely spaced events. These studies, which concentrated on mineralization phenomena, also demonstrated diachronous crystallization of the magmatic components of the batholith (*Table 7.3*). Unfortunately, although they yielded copious data from the biotite granites and mineralization events, there are no

Table 7.3. ISOTOPIC CHRONOLOGICAL DATA FOR BIOTITE GRANITES FROM CORNWALL

	Rb-Sr age		Chen *et al.*, 1993		Chesley *et al.*, 1993	
Pluton	WR	Mineral	U-Pb	^{40}Ar-^{39}Ar	U-Pb	^{40}Ar-^{39}Ar
Biotite granite:						
Land's End	268±2*	272±2	274.5±1.4	270.1±1.0	274.8±0.5	272.9±0.8
Carnmenellis	285±19	290±2	293.7±1	289.7±0.8	293.7±0.6	290.7±0.9
St Austell	285±4		279.6±1.7	276.8±0.9	281.8±0.4	278.3±0.8
Bodmin	276±9	287±2	281.3±2.1	288.8	291.4±0.8	287.1±0.9
Topaz granite:						
Tregonning	280±4					
St Austell	274‡					
St Austell	280‡					271.0±0.4†

*Reset

†Minimum age for fluorite granite

‡Summary data from Darbyshire & Shepherd, 1994

Based on data from Darbyshire & Shepherd, 1985; 1987; 1994. Chen *et al.*, 1993 and Chesley *et al.*, 1993

U–Pb or $^{40}Ar/^{39}Ar$ data for the tourmaline or topaz granites, or for the rhyolite porphyry dykes. Overall, the geochronological data show that the dominant phase of biotite granite crystallization started at 290 Ma in the Carnmenellis and Bodmin Moor plutons, and at about 280 Ma in the St Austell, Land's End and Dartmoor plutons. Thus although magmatic activity ceased by 270 Ma, within the interval 290–270 Ma a number of separate magmatic events have been identified. Similarly, a complex sequence of hydrothermal events can be distinguished, and indirect evidence for the St Austell pluton indicates a minimum age for the rhyolite porphyry dykes of 270 Ma.

PETROGENETIC MODELS

The origin of the biotite granites (and the associated rhyolite porphyry dykes and volcanic rocks) which dominate the Cornubian batholith is generally ascribed to partial melting of lower-crustal metapelites, either of a poorly hydrated garnet granulite at 7–8 kbar at about 800°C (Stone & Exley, 1986; Stone, 1975; 1984) or a cordierite-sillimanite-spinel gneiss at 5 kbar at about 800°C (Charoy, 1986). There are some difficulties with origins postulating single source rocks, but application of new isotopic techniques has helped to develop sophisticated petrogenetic models. N isotope ratios show a range in values which demonstrate a biological source for this element (Boyd *et al.*, 1993) and confirm a metapelitic source rock. Additionally, Nd and Sm isotope systematics indicate that there was also a contribution of mantle material, although the magma was predominantly derived from lower-crustal material (Darbyshire & Shepherd, 1994). Darbyshire & Shepherd favoured a model for magma generation as a consequence of crustal underplating during continent–continent collision, using the Ivrean Zone in the Italian Alps as an example of this process. This tectonic setting has long been recognized; the diagrams which use trace element data to suggest the plate tectonic setting of granitic rocks included data from the Cornish granites to define the field for syn-collision granites.

Once generated, the magma was enriched in H_2O, Li and B as it crystallized, to give the hydrous mineral assemblage and Li-mica and tourmaline-bearing granites seen at the current exposure level. The formation of a B-enriched magma had particular consequences. Not only does the presence of B reduce magmatic liquidus and solidus temperatures, but it also increases the ability of the melt to contain H_2O. However, when the granite melt is water saturated, B partitions in favour of the aqueous phase. This leads to a catastrophic loss of both H_2O and B once water saturation is achieved. This process is believed to account for the textural variation shown by the tourmaline granites (Hill & Manning, 1987). Tourmaline breccia pipes, such as those of Wheal

Remfry [SX 924 576] (Allman-Ward *et al.*, 1979), also formed as a consequence of the irregular quenching effect associated with water loss. The involvement of an aqueous phase during crystallization of the tourmaline granites is consistent with the scatter shown by their plots on the Ba–Sr diagram. These do not coincide with the regular trends typical of magmatic crystal-liquid trace element fractionation patterns.

Although the biotite granites and the varied tourmaline granites can be regarded as a cogenetic suite, the position of the topaz granites is much less clear. They do not appear to be comagmatic with the biotite granite suite, which shows no evidence of F enrichment during differentiation. Their B content is also lower than in the biotite granite suite. They do not lie on clearly defined magmatic fractionation trends, and so alternative hypotheses involving a second stage of lower-crustal melting after emplacement of the biotite granites have been invoked (Stone, 1992; Manning & Hill, 1990). The nature of the source for the topaz granite magmas is not resolved. Stone (1992) favoured partial melting of solid biotite granite which had undergone metasomatic enrichment in trace alkalis and F, thereby reducing its melting temperature. Alternatively, Manning & Hill (1990) suggested that residual S-type granite source rocks were involved in a second episode of partial melting, with no metasomatic input, drawing analogies with some characteristics of A-type granitoids (for example, the topaz rhyolites). This model depends on the residual material in the source containing biotite, enriched in F as a consequence of the initial fusion which gave rise to the biotite granite magma. Crystal fractionation and diapiric rise to very high levels within the batholith would be enhanced by the reduction in melt viscosity caused by the presence of F.

Chapter Eight

Mineralization

The great variety of Cornwall's metal deposits, and the considerable production of Sn, Cu, Pb, Zn, Fe, Ag and As in historical times, will be described in Chapter 14. As a district hosting granite-related polymetallic vein mineralization, the South West England province has been extensively studied by geologists and is known throughout the world. Consideration of the origins of the metal deposits began in the late 18th century, stimulated by extensive contemporary mining information and by a widespread interest in mineral specimens. Thus Pryce, in his *Mineralogia Cornubiensis* (1778), was able to acknowledge the role of subterranean waters in ore genesis, although he scarcely alluded to their origins. The succeeding years saw much progress towards the elucidation of the stratigraphy and structure of South West England, and by 1839 the first Geological Survey maps of Cornwall were published, together with an explanatory memoir (De la Beche, 1839) giving details of the geology and the metalliferous deposits. De la Beche provided a detailed discussion of hydrothermal mineralization, including the structure and origins of the veins and their chronology. He also speculated on the relative contributions of fluids of direct granitic origin, and of later circulating fluids from external (essentially meteoric) sources. This debate has continued to the present day.

Important contributions to the occurrence, mineralogy and structure of Cornish ore deposits were made by Henwood (1843) and Collins (1912). Among the 19th-century workers on metallogenesis, the name of Henry Sorby must be mentioned, for it was he who demonstrated, through the study of microscopical fluid-filled cavities in quartz and other crystalline minerals, the origin of mineral veins by precipitation from hot, saline, aqueous fluids (Sorby, 1858). In the first half of the present century, against a background of declining mining activity, considerable advances were made in the study of metallogenesis. Thus Dewey (1925) and Dines (1934) considered the zoning of Cornish metal deposits and attempted to systematize the mineralogy and relative temperatures of ore formation. Dines (1956) drew together the great volume of data on South West England metal deposits that had been collected by the Mining Records Office and the Geological Survey, from De la Beche onwards.

Since the end of World War II, metalliferous mineralization in Cornwall has been the subject of almost continuous academic study. Aspects of this work have included detailed mineralogical, geochemical and structural studies, deep geophysical surveys, a major investigation of crustal heat flow, isotopic dating of the granites and vein minerals, and extensive studies on the nature of the mineralizing fluids. However, because of continued mine closures and decreasing access to underground workings, the studies have been confined to a small number of localities. This has resulted in a slight bias towards certain types of ores. The more recent literature does not always do justice to the wider range of Cornish deposits. Of the more modern general reviews of mineralization in South West England, the most informative are those of Hosking (1964), Dunham *et al.* (1978), Jackson *et al.* (1989), Willis-Richards & Jackson (1989), Bromley (1989) and Alderton (1993).

NATURE OF DEPOSITS

Most of Cornwall's metalliferous deposits are of hydrothermal origin, that is, resulting from the movement of hot aqueous fluids in the Earth's crust and the deposition of minerals by precipitation from them. Hydrothermal fluids can deposit minerals in open fractures to form fillings as veins, or can permeate rocks to form orebodies by selective replacement of the host. It is common for a combination of these processes to have operated, so that a vein is enclosed within a zone of altered rock that may include ore minerals. Normally, veins are variable in composition, with a mixture of ore (economically valuable) and gangue (waste) minerals. The minerals of economic value in Cornwall are cassiterite (the principal ore of Sn), wolframite (the principal ore of W) and the base metal sulphides (chalcopyrite, arsenopyrite, sphalerite and galena; the principal ores of Cu, As, Zn and Pb, respectively). The gangue minerals are generally quartz, tourmaline, chlorite, fluorite and either siderite or calcite.

Mineral veins

Mineral veins, known as *lodes* in Cornwall *(Figure 8.1)*, have produced the bulk of the Sn, Cu, As and W extracted. They can have a simple sheet-like form of regular geometry, but more commonly they pinch and swell, showing considerable variation in width and composition from place to place, both laterally and vertically. Workable veins usually have an outcrop length of at least 200 m and many have been traced over 1 km.

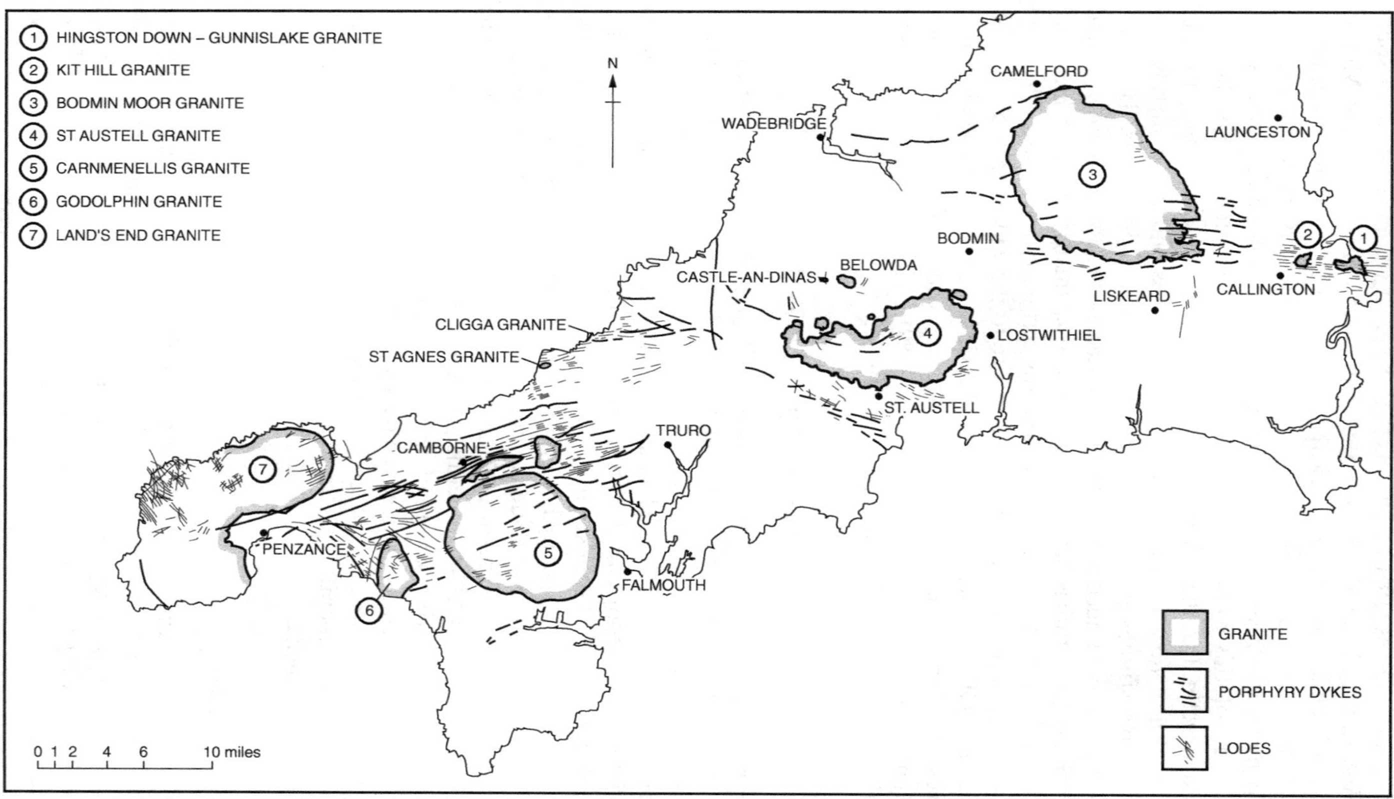

Figure 8.1. Map showing the principal vein deposits of South West England in relation to the granite plutons (after Hosking, 1964)

In exceptional cases they may be more extensive, such as the Main Lode of Devon Great Consols [SX 425 735] which was worked over a distance of nearly 4 km in Devon. Vein widths vary from 0.2–20 m or more, but most are between 1 m and 2 m. In Cornwall, some mines in the most intensely mineralized districts were worked to considerable depths; the deepest, Dolcoath Mine [SW 661 405], was exploited to a depth of more than 1000 m (*Figure 8.2*). Most workings, however, were much shallower. Veins are commonly vertical or steeply inclined, dipping at 60° or more, although flatter ones do occur, as illustrated by the Great Flat Lode [SW 690 402] near Redruth, and many veins in the St Agnes district. Two main structural trends are apparent: an early set of veins trending mostly between WNW–ESE and ENE–WSW (known generally as 'E–W' or 'normal') which carry Sn–Cu–As–W minerals, and a later set trending roughly N–S, (known as 'cross-courses') characterized by Pb–Zn–Ag ores and spar minerals. In many instances an ore deposit is made up from a number of more or less regular bodies that together form a mineralized zone. Within such zones, the individual veins may either be substantial or developed as a complex of narrow mineralized cracks. Where several phases of fracturing have occurred, the resulting veins may be built up from the sequential deposition of distinct mineral assemblages, in some cases giving rise to polymetallic orebodies of considerable complexity.

It is a feature of Cornwall's geology that a number of local words are in use for various features of the mineral deposits. As noted above, *'lode'* commonly substitutes for vein, but may also be used to define a mineralized zone, as described below. Individual lodes may be qualified by the terms *'caunter'*, denoting a vein intersecting another of normal trend at an acute angle, and also as noted above, *'cross-course'* for a vein roughly at right angles to the regional E–W trend. Other local terms which may be encountered are *'elvan'* for a quartz-porphyry dyke, and *'fluccan'* for a late-stage clay-filled fracture.

Breccias, sheeted vein complexes, stockworks and carbonas

Movement of the vein walls during mineralization, combined with hydraulic fracturing of the wall rocks due to rapid fluid decompression, commonly results in broken fragments of host rock being incorporated in the veins and forming part of the filling. These 'breccias' can account for a considerable part of the total fracture filling, and it is not uncommon for economic lodes to consist of brecciated wall rock cemented by a matrix of ore and gangue minerals. Particularly rich patches of ore may occur at the intersection of fractures of differing trend. These sometimes have the geometry of a pipe, when they are known as *'carbonas'*. Such orebodies are particularly associated with Sn deposits in the granites of west Cornwall.

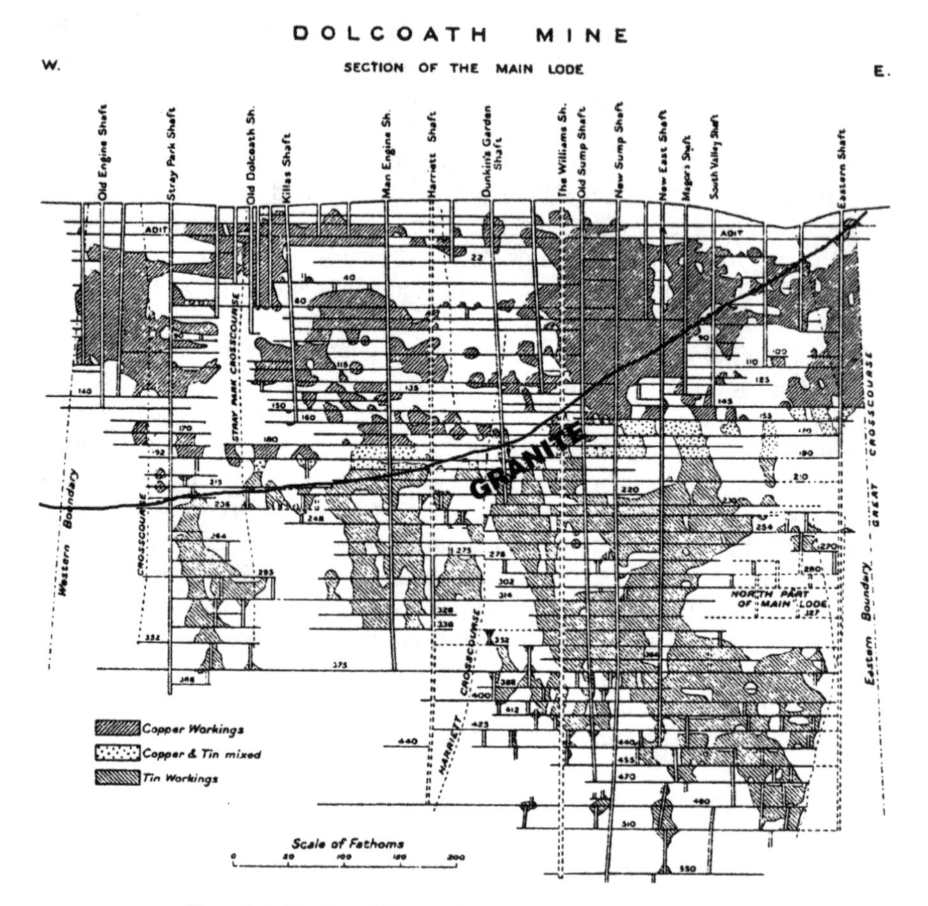

Figure 8.2. Section of Dolcoath Main lode (after Dines, 1956)

'*Sheeted vein complexes*' (*Figure 8.3*) comprise a multiplicity of veins of more or less parallel trend, while the term '*stockwork*' is applied to a mineralized zone of intersecting or ramifying veinlets. Within a stockwork, it is usual for one set of fractures to be preferentially developed and to contain wider or richer veins. The physical form of a stockwork is rarely fully resolved, shape and size being determined by mining considerations. Typically, stockworks and sheeted vein complexes are associated with small granite stocks peripheral to the main granite masses, or the margins of larger granite intrusions.

Wallrock alteration

One of the characteristic features of vein mineralization is the development of alteration in the adjacent wall rock. This may toughen or soften the host rock, thus affecting the cost of mining the vein or the

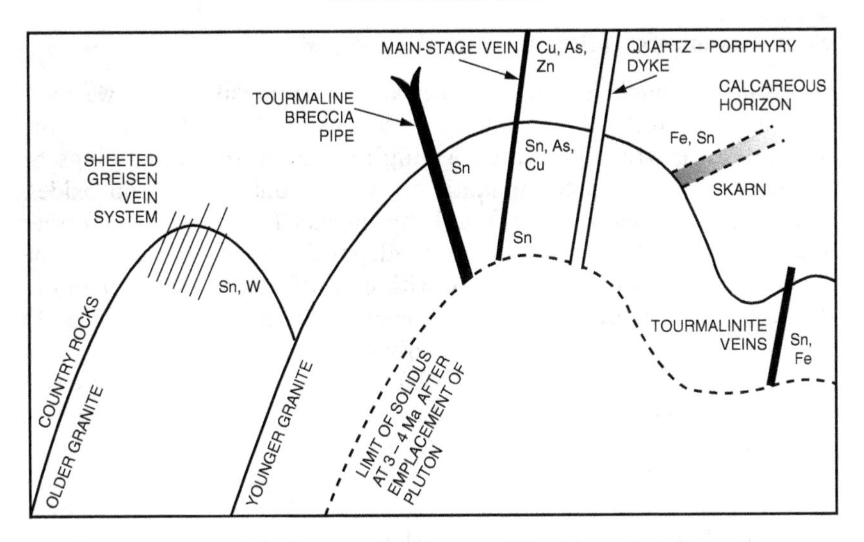

Figure 8.3. Cartoon of main types of Cornish ore deposit

stability of open workings. The zone of alteration can extend for a few centimetres to many metres from the vein walls. New minerals formed during wall rock alteration include tourmaline, feldspar and white mica (muscovite and sericite) for Sn and W veins, chlorite for sulphide veins, and clay minerals for the lower temperature mineral veins, including the prominent N–S-trending cross-courses. Where wall rock alteration is especially intense and is dominated by white or green ('gilbertite') mica, the veins are described as *greisen-bordered.*

Replacement orebodies

Replacement orebodies may be developed as irregular lenses or pods, or may form within a body of host rock such as a sill or dyke of igneous rock, or a bed within a sedimentary sequence. Direct connection with mineral veins cannot always be demonstrated, but their origin is, nevertheless, a consequence of fracture-controlled fluid flow, albeit along microcracks. Examples of replacement ores in Cornwall include locally rich developments of cassiterite in elvan dykes and bodies of magnetite in greenstone (basic igneous rocks). In certain cases, beds of carbonate-rich rock have been altered by thermal metamorphism (due to granite emplacement) to assemblages of calc-silicate minerals, which are subsequently invaded by hydrothermal fluids and partly impregnated by ore minerals. Orebodies of this hybrid type are known as *'skarns'*.

Secondary mineral enrichments and placer deposits

Near-surface, sulphide-rich orebodies are commonly altered by weathering and the effects of percolating groundwaters. Close to surface, the sulphides are strongly leached, giving rise to weathered cappings or *'gossans'* consisting predominantly of residual quartz and iron oxides. These were aptly termed *'iron hats'* by the miners. Below this zone, further leaching occurs, but at deeper levels redeposition of metals leads to the formation of secondary minerals, with considerable enrichment in ore grade. Such local pockets of high-grade ores, particularly those of Cu, Pb and Ag, were greatly sought by the early miners.

Another group of secondary ores in Cornwall derive from the erosion of hydrothermal veins, the transport of the resulting debris, and its incorporation in superficial deposits. They have an important place in Cornwall's mining history in that the earliest mining activity exploited Sn-bearing fluvial sediments. Most of these *'placer'* or *'stream'* Sn deposits are alluvial and river terrace gravels laid down during establishment of the Quaternary drainage network. However, other forms of surface detritus, such as eluvium and head, could also be included. The economic importance of placer Sn deposits is related to the inert nature of cassiterite in the weathering environment and to its very high specific gravity (6.8–7.1) which facilitates hydraulic separation. Wolframite, the principal ore of W, has also been recorded in alluvial deposits, but has only locally proved commercially viable. Deposits from northeast Bodmin Moor [SX180 823] were prospected at times of high W prices. Traces of Au are widespread in the superficial deposits of the county (Camm, 1995), but the treatment of such occurrences by the miners as 'perks' means that evidence of quantity produced is unreliable.

Metallogenetic framework

A considerable body of evidence suggests that hydrous fluids, separating from the cooling granite magma were responsible for certain types of mineralization. There is also evidence to suggest that significant mineralization is linked to the circulation of external fluids (shallow groundwaters, often referred to in the literature as meteoric waters, and deep sedimentary brines), heated and driven by the thermal energy of the granites. As noted in Chapter 7, the granites are enriched in radioactive elements, the decay of which has maintained an anomalously high heat flow (see Chapter 9) and circulation of relatively low temperature waters over and around the granites ever since their emplacement. A present-day manifestation of this process are the thermal springs (up to 35°C) encountered in the deeper workings of the South Crofty [SW 666 412] and Wheal Jane [SW 722 426] mines (Edmunds *et al.*, 1984). The interaction and relative contribution of fluids in time and place is responsible for the special character and complexity of metalliferous mineralization in South West England.

Early workers in Cornwall had commented on the distribution of the various metals in and around the granites. In the early years of the present century, Dewey (1925) and Dines (1934) suggested models in which fluids migrated outwards from 'emanative centres' in the granites to produce lateral and vertical zoning of Sn, As, Cu, Zn, Pb and Fe ores. It was suggested that this zonation reflected the temperature of formation of the ores, from Sn at the highest end (hypothermal), through Cu, Zn and Pb (mesothermal) to geographically distal deposits of Fe at lower (epithermal) temperatures. Numerous examples have been cited to illustrate aspects of this model. Vertical zoning is evident in certain of the larger Sn–Cu structures, such as the Main Lode of Dolcoath Mine, and district-scale lateral zoning, for example, in the Caradon-Phoenix mines of southeast Bodmin Moor.

However, in some districts of Cornwall are examples of polymetallic mineralization that contradict these zonal models. In the Kit Hill–Gunnislake district, for example, certain E–W trending veins carry W, Sn, As, Cu and Pb together in complex polymetallic structures. Such veins are said to be 'telescoped'. Elsewhere, there is evidence for the superimposition of contrasting vein systems. For example, mineral exploration boreholes drilled to the west of the Carnmenellis Granite (Beer *et al.*, 1975) intersected a complex Cu–Pb–Zn vein some 400 m beneath known Sn mineralization.

Understanding of the genesis, distribution and timing of Cornwall's metal deposits has increased greatly with the application of modern analytical methods. Among the most significant of these techniques is the study of the chemical nature and physical behaviour of the substances contained in tiny cavities in vein quartz and other optically transmissive mineral species. Study of these 'fluid inclusions', which may contain combinations of liquid, vapour and solid phases, by microthermometric and chemical analysis has enabled estimates to be made of the chemical nature of ore-forming fluids and of the temperature at which the host mineral was deposited. Sufficient fluid inclusion studies have now taken place to permit the characterization of the various Cornish ore deposits.

The types of ore deposits listed in *Table 8.1* are individually represented at a number of classical localities in Cornwall, but the divisions are not always clear cut and there is considerable overlap, particularly in the case of early tourmaline mineralization and mainstage veins. In overview, it is particular combinations of the various types within a local structural setting that afford each mining district a particular character of ore deposit.

Table 8.1. THE RANGE OF CORNISH ORE DEPOSITS, IN GENERALIZED CHRONOLOGICAL ORDER

	Syngenetic and epigenetic metal concentrations
Pre-granite	Pre-granite veins
	Skarn deposits
	Pegmatites
Granite-related	Greisen-bordered veins
	Tourmaline–quartz orebodies
	Main-stage polymetallic veins
Post-granite	Cross-courses

PRE-GRANITE MINERALIZATION

Stratiform metal concentrations

The source of metals in the Cornubian hydrothermal veins has been a subject of debate. The consensus of modern opinion (Alderton, 1993) suggests that the Devonian and Carboniferous greenstones made a considerable contribution to Cu, As and Sb ores, and that Pb and Zn veins probably derived from leaching of sedimentary rocks. In the Camborne–Redruth mining district the most productive Sn–Cu area is adjacent to, and partly coincides with, extensive outcrops of greenstone. Exploration levels in the Levant Mine [SW 368 345] of the St Just district intersected bodies of metabasite, originally tuff, with disseminated sulphide mineralization including Cu.

It has been recognized that specific metal-enriched horizons exist within the Devonian and Carboniferous strata of South West England (Scrivener *et al.*, 1989). In particular, these occur in Middle Devonian volcanic rocks and Lower Carboniferous black shales and cherts with basic tuffs and spilitic lavas.

Examples of primary stratiform mineralization have been described from Devon, but are less common in Cornwall, presumably due to the remobilization effects of intense hydrothermal activity around the granite masses. In east Cornwall, a group of small mines around Launceston, and extending towards Chillaton [SX 429 812] in west Devon, worked bedded Mn ores from Lower Carboniferous shales and cherts. Exploration drilling near Egloskerry (Hazleton & Gawlinski, 1982), east Cornwall, in a sequence of Lower Carboniferous shales and siltstones, with basic tuff and lava, intersected stratiform Pb–Zn mineralization hosted in beds of grey siltstone. This mineralization was considered to be uneconomic and has not been pursued.

Pre-granite vein deposits

Recognition of pre-granite veins has been possible on structural evidence, coupled with detailed fluid inclusion studies. In the Wadebridge district of north Cornwall are numerous small workings for Sb in scattered veins of roughly N–S trend. The minerals worked from these deposits include jamesonite and bournonite (originally named endellionite), with some galena, sphalerite, chalcopyrite and pyrite, in a gangue of quartz and siderite. Some of the veins, notably those at Treore Mine [SX 020 800], near Port Isaac, and at Bounds Cliff [SX 023 814], carry minor arsenopyrite and traces of Au, in addition to the Pb–Sb minerals.

While some authors (Hosking, 1964; Edwards, 1976) consider this mineralization to be typical of the low-temperature cross-course veins found elsewhere in Cornwall, fluid inclusion studies (Clayton *et al.*, 1990) have shown that the Sb veins were formed from fluids at higher temperatures, with T_H in the range 280–315°C. These inclusion fluids are of low salinity, based on NaCl, with abundant non-aqueous volatiles (CO_2 and CH_4). Such brines are typical of low-grade regional metamorphic fluids. These data, together with the structural setting of the veins in NNE–SSW shear zones, suggest that the Sb mineralization of north Cornwall was effected during the late stages of Variscan deformation.

In neighbouring Devon, it has recently been demonstrated (Hein *et al.*, 1995) that CO_2-rich fluids, similar to those described above, were responsible for the iron ore veins (siderite with secondary Fe oxide minerals) of Exmoor. This suggests that some of the Cornish siderite/Fe oxide veins were of similar origin, but at present this has not been proved.

At the northern end of Perran Bay one of the county's most enigmatic mineral deposits, the Great Perran Iron Lode [SW 772 426], crops out (*Figure 8.4*). From Penhale it extends southeastwards for at least 10 km and varies in width from 1–30 m. Typically it comprises brecciated slates cemented by siderite, minor quartz and masses of black sphalerite. Opinion as to the age of the black sphalerite is controversial, although it would appear that it predates the siderite. Surface oxidation of the siderite to depths of 60 m has produced oxides and hydrated oxides of Fe (hematite, goethite and limonite). This lode was extensively worked for iron ore during the19th century, but ceased when cheaper sources became available towards the end of that century. The lode is cut in places by later N–S cross-course veins containing pyrite and argentiferous galena. On the basis of comparison with Exmoor veins in Devon, it could be that the Perran Iron Lode represents a mineralization event predating granite intrusion (Henley, 1971).

Any discussion of Cornish mineralization would be incomplete without reference to the rather special Cu mineralization developed near

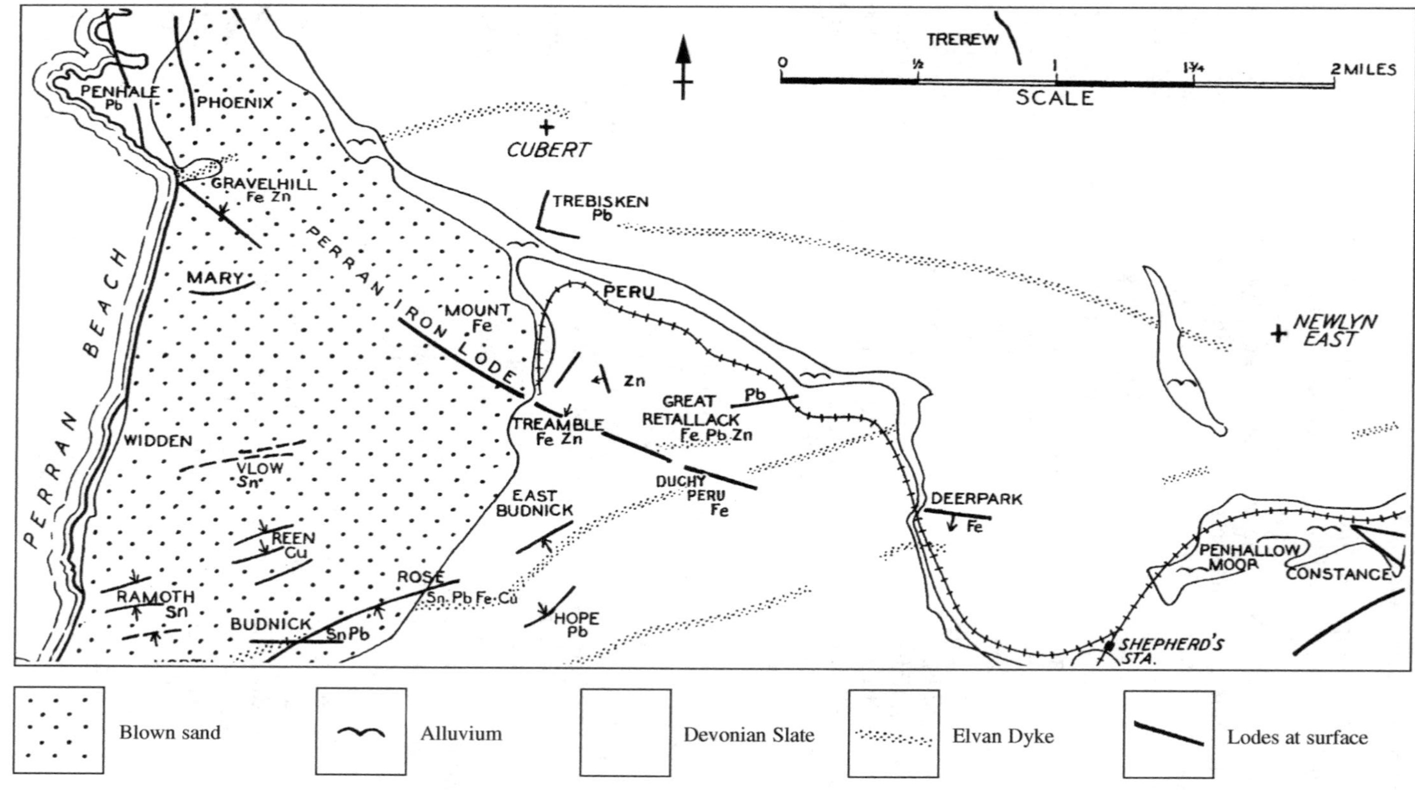

Figure 8.4. Map of the Perran Iron lode and contiguous deposits (after Dines, 1956)

Mullion in the Lizard district. This is unusual in that the Cu occurs principally in its native elemental form as disseminations and irregular masses throughout serpentinite. No other rocks of the Lizard metamorphic complex are hosts to the mineralization, so it is likely that the Cu ores are related to the complex, pre-granite, hydrothermal history of the serpentinites. Sadly, virtually nothing remains of the old 18th-century workings, and one must refer to museum collections to appreciate the richness of these rare and very localized deposits.

GRANITE-RELATED MINERALIZATION

Many ore deposits in Cornwall are related to the emplacement and cooling of the Cornubian granite batholith. It is suggested that two main processes were involved: firstly, the separation of hydrous saline fluids directly from crystallizing granite magma, and secondly, the later induction and circulation of external crustal fluids, partly of meteoric origin. Apart from the pegmatites and skarn deposits considered below, the first of these processes gave rise directly to two distinct styles of mineralization:

Type 1. Veins of quartz with cassiterite and/or wolframite; minor lollingite and base metal sulphides may be present. The veins are enclosed by selvedges altered to a mass of white mica and quartz, commonly with topaz. This type of alteration is known as 'greisen'.

Type 2. Bodies of quartz and tourmaline with cassiterite, commonly also with haematite. Sulphides are rare. Wall rock alteration involved replacement of the host by tourmaline and secondary feldspar, and is commonly marked by intense reddening.

In most districts early mineralization of types 1 and 2 above, was followed by larger-scale fluid movements that resulted in complex polymetallic 'mainstage' ore deposits. Within and close to the granite masses these are dominated by cassiterite–chalcopyrite–arsenopyrite assemblages with gangue minerals of quartz, tourmaline and chlorite. In more distal deposits, tourmaline and cassiterite decline and give way to chalcopyrite–sphalerite–galena assemblages with quartz and chlorite. Pyrite and fluorite are common in many localities.

Skarn deposits

Calc-silicate mineral assemblages produced by the thermal metamorphism of carbonate-rich sedimentary and volcanic rocks, are known from a number of localities in Cornwall. In the eastern part of the county they have been termed '*calc flintas*'. At certain localities close to granite contacts, bodies of calc-silicates have been mineralized, and there is general agreement (Jackson *et al.*, 1982) that these represent the earliest migration of hydrothermal fluids from the cooling granites. While these skarn bodies are of considerable interest in terms of their mineralogy and genesis, they are of minor economic importance.

At Grylls Bunny [SW 364 333], near Botallack, bodies of garnet–magnetite rock have been invaded by B-rich fluids to produce a series of *'floors'*, flat-lying sheets of quartz–tourmaline rock with some cassiterite and chlorite (Jackson, 1974). Magdalen Mine [SW 765 377] near Ponsanooth, in the metamorphic aureole of the Carnmenellis granite, exploited a number of steeply dipping, narrow bodies of magnetite–hornblende–cassiterite rock within altered greenstone. Much of the cassiterite is finely divided, and little is visible in hand specimen. Beneficiation of the ore proved difficult, even with the application in the present century of magnetic separation. The mine closed in 1929.

Pegmatites

Pegmatites, in the sense of late magmatic bodies of granitic rock rich in very coarse-grained feldspar and mica, are of widespread occurrence in the Cornubian batholith. Most comprise assemblages of K-feldspar with quartz and tourmaline, although some examples are rich in the Na feldspar albite. The great majority of these pegmatites are entirely wanting in economic minerals. Some examples, which include cassiterite and/or wolframite, are considered to be the earliest expressions of hydrothermal mineralization, and represent the tranformation from magmatic to hydrothermal processes. Others, such as the so-called Levant Carbona (Jackson, 1975), are granitic bodies that have been subjected to the invasion of later hydrothermal fluids.

Tourmaline mineralization

The presence of Fe-rich tourmaline ('schorl') is a widespread and distinctive feature of the Cornubian granites. It occurs in grains and in aggregates as a minor constituent of most varieties of the main granite plutons, and is particularly abundant in late-stage aplites and pegmatites. Larger occurrences occupy extensive parts of the granite host rock and, in some areas, of the envelope rocks as well. A well-known example of the former is Roche Rock [SW 991 596] and of the latter, the altered Meadfoot Group of the metamorphic aureole to the north of the St Austell granite.

Tourmaline veining is very common in and around the granites. The veins are commonly accompanied by wall rock alteration involving replacement of the primary granite by tourmaline and secondary feldspar. It is common for secondary feldspar developed in the tourmaline vein selvedges of the biotite granites to show marked red or pink coloration. Biotite in the proximal reddened granite wall rocks is commonly replaced by tourmaline or chlorite, while minor mineral phases present may include anatase, brookite and secondary monazite (Scrivener, 1982). More distal and lower-temperature hydrothermal alteration may be of primary feldspar to sericite (muscovite) or chlorite.

At a number of localities, early quartz–tourmaline vein complexes carry cassiterite, and though these may locally be rich, overall they are generally of low grade and limited extent. Such deposits are notable for the absence of sulphide minerals and for the presence of haematite-impregnated selvedges. A Cornish example is at Ding Dong Mine [SW 437 346], in the centre of the Land's End granite, near Morvah, where a complex of narrow, irregular veins containing quartz, tourmaline and cassiterite are enclosed by reddened and tourmalinized granite.

Early tourmaline veining is commonly accompanied by brecciation of the host rock, so that inclusions of granite or envelope rock are enclosed by a tourmaline–quartz matrix. In some examples, breccia is the dominant lithology, with the mineralized body taking the form of a vein of limited lateral extent or of a pipe. In many examples the matrix is very finely crystalline, with textures suggesting rapid crystallization from the parent fluid. Such breccias commonly contain a fraction of rounded clasts, and lack the distinctive wall rock alteration described above, presumably because the hydrothermal fluid–wall rock interaction was of short duration. Detailed study of one tourmaline breccia pipe at Wheal Remfry [SW 923 577] in the St Austell granite (Alman-Ward *et al.*, 1982), has demonstrated that it resulted from the explosive failure of a system in which B-rich fluids were separating from late-stage, highly evolved granitic magma within the St Austell pluton. Cassiterite is a very minor constituent of the Wheal Remfry breccia, and this is a common feature of such intrusive tourmaline breccias elsewhere in the county.

Greisen mineralization

Alteration of granite to a greisen assemblage of quartz, white mica and, in places, topaz, is a common feature, particularly of the marginal and apical parts of the smaller granite plutons. It is also associated with some granite–porphyry intrusions. Greisen alteration commonly encloses veins of quartz with variable amounts of cassiterite and/or wolframite, the Fe arsenide löllingite and minor sulphides. At several localities greisen-bordered veins form sheeted complexes in granite bodies, which may comprise substantial tonnages of low-grade ore. Well-known examples occur at Bostraze [SW 384 318] in the Land's End granite, at St Michael's Mount (Moore & Moore, 1979) in a small granite stock, and at Cligga

Head [SW 730 537] (Moore & Jackson, 1977]. Greisen-bordered sheeted-vein systems are also found in the envelope rocks, usually close to the granite–host rock contact. A good example is the large Redmoor [SX 369 710] system at Callington.

In contrast with the tourmaline vein complexes described above the greisen- bordered sheeted-vein systems usually show no brecciation, and appear to be the result of hydraulic fracturing, with controlled egress of hydrothermal fluids. Typical greisen complexes (*Figure 8.5*) consist of sets of more or less subparallel quartz veins bordered by quartz–mica selvedges Where these are closely spaced, the outcrop has a characteristic striped appearance (as at Cligga Head). The development of sheeted, greisen-vein complexes is commonly accompanied by pervasive argillic (kaolinite) or sericitic alteration. One of the best examples to be seen at present is in Goonbarrow china clay pit [SX 008 583]. The vein complexes may be very extensive, the most intensely mineralised ground at Cligga, for example, has an outcrop measuring some 200 m×300 m, with subsurface indications to at least 120 m.

Vein fillings are predominantly of quartz, which may be massive or vughy, together with cassiterite and/or wolframite. These major ore minerals are commonly coarsely crystalline, and it is usual for cassiterite to form very dark, lustrous crystals. Löllingite may be present in minor amounts, together with sulphides, notably arsenopyrite, stannite, and chalcopyrite. The sulphides are commonly of later formation than cassiterite and wolframite, sometimes occurring in cross-cutting fractures. In places, the greisen selvedges contain low-grade disseminated cassiterite and wolframite.

Despite containing very large tonnages of low-grade ores, only a small proportion of the production of Sn and W from South West England has come from sheeted-vein systems. However, the possibility of mining from open-pit operation has rendered such bodies of interest to prospecting companies in the recent past.

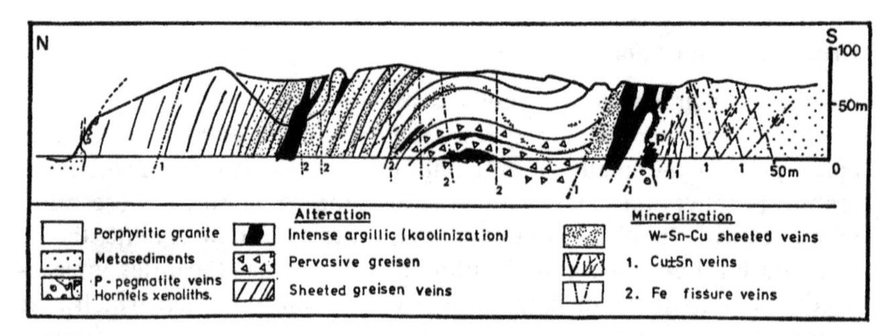

Figure 8.5. Section of the Cligga Head granite and related mineralization (after Moore & Jackson, 1977)

Metallogenesis of the early systems

The earliest recognized mineral deposits are the skarns, tourmaline-cassiterite, and greisen-bordered veins that developed adjacent to, and in the roof and walls of, the granite plutons. The vein deposits have mineral assemblages including cassiterite, topaz, tourmaline and wolframite, which are diagnostic of high temperatures (>400°C). Their isotope and chemical characteristics are indicative of magmatic hydrothermal fluids (that is, the metal-enriched fluids that separate from the granite magma during crystallization). The solutions varied greatly in salinity (5–50 wt% salts) and in the case of W veins were highly charged with CO_2 (Shepherd *et al.*, 1985). From a study of unmineralized quartz veins exposed in the cliffs around Porthleven, Wilkinson (1990) established that granite intrusion was accompanied by thermal dehydration of the host rocks with the production of CO_2-rich metamorphic aqueous fluids. Given the striking similarity of these fluids to those of the W lodes, it is possible that locally there was significant mixing between metalliferous magmatic fluids and metamorphic fluids. Indeed, CO_2 enrichment is a characteristic feature of granite-related W deposits worldwide, and is particularly evident where granites intrude black slates or greywackes. The induction of such CO_2-rich fluids may account for the difference in style between tourmaline-rich and greisen-hosted early vein mineralization. Alternatively, the suggestion of Shepherd *et al.* (1985) that this is due to separation of a parent magmatic fluid into a denser fraction (giving rise to tourmaline veins) and a lighter, CO_2-rich fraction (responsible for greisening), may be correct.

Main-stage mineralization

The most important economic concentrations of Sn, Cu and As occur in hydrothermal vein systems that are generally somewhat later than the tourmaline- and greisen-dominated systems described above. These veins are often collectively referred to the 'main stage'. Chronologically, their first development was directly related to the early cooling history of the granite plutons, but it is clear from evidence, including mineral textures, fluid inclusion studies and isotope geochemistry, that in many cases the main-stage veins have a protracted history of vein formation with multiple episodes of fracturing, brecciation and mineralization. Generally, there is a broad evolution of vein infilling, starting with an early assemblage of cassiterite, wolframite and arsenopyrite, with tourmaline and quartz as the dominant gangue minerals, to later assemblages dominated by Cu–Zn–Fe–As–Sn–Pb sulphides, with quartz, chlorite and fluorite as the principal gangue minerals.

It must be emphasized that, throughout the province, the main-stage veins develop from greisen and tourmalinite mineralization. In some cases, for example at Geevor Mine in the Land's End granite, the veins are predominantly of quartz–tourmaline–cassiterite, with later infillings of quartz and base metal sulphides. The complex mineralization at the former Wheal Jane Mine involved early greisen replacement of quartz–porphyry dykes, with later cassiterite–tourmaline–chlorite veining and subsequent sphalerite–chalcopyrite development.

In some areas the main-stage veins show a marked lateral zonation, with granite-proximal Sn and W veins giving place to base metal sulphides in the metamorphic aureole. This is evident in the Caradon–Phoenix district of southern Bodmin Moor and around the granite cupola of the Perranporth district. In contrast, the veins around the Kit Hill and Gunnislake granite cupolas of east Cornwall show complex polymetallic development within single fracture systems. At Old Gunnislake Mine, the workings developed a vein system with early wolframite and quartz, later cassiterite–tourmaline, and a final core filling of chlorite–chalcopyrite–sphalerite–fluorite. The style of vein mineralization, in which an early tourmalinite/greisen stage gave way to a later development of quartz–haematite or chlorite-base metal sulphide–fluorite, is very typical of the Cornish metallogenetic province .

Fluid temperatures for main stage veins show a wide range (200–450°C), although there is a progressive decrease with time from 250–450°C for the early Sn–W phase, through 200–350°C for the Cu–Fe–As–Sn sulphides, to 200– 250°C for the late Pb–Zn sulphides. Fluid salinities were also variable (<5–50 wt% dissolved salts). During the early cassiterite phase, salinities averaged more than 20%, decreasing to <5–20% for Cu–Fe–As–Sn sulphides, before increasing again to 20–25% during the final Pb–Zn stage. Unlike the early tourmaline veins and greisen-bordered systems, the main-stage veins were deposited from mixed magmatic–non-magmatic fluids. The latter originated outside the granite, and were probably deeply circulating groundwaters that were drawn into the hot, fractured granite where they mixed with residual magmatic fluids released from deeper pockets of magma. The trend towards relatively cool, saline fluids of non-magmatic origin heralded a marked rotation in the regional stress field and the generation of regional fractures with a N–S orientation that were to dominate subsequent mineralization in South West England.

POST-GRANITE MINERALIZATION

For Cornwall, post-granite mineralization in the strictest sense refers to the N–S-trending polymetallic (Pb–Ag–Zn–fluorite) cross-course veins. By the late Permian, (260 Ma), magmatism had finally waned, and subsurface temperatures were controlled by the radiogenic heat output of the granites. To the north and south, Cornubia was flanked by deep rift basins containing thick sedimentary deposits eroded from the adjacent elevated land masses. It was a period of crustal extension, and the N–S regional fractures provided excellent conduits for the circulation of metal-rich fluids and the formation of mineral veins. These carry Pb–Zn–Ag sulphides in a quartz–fluorite–siderite gangue and, like the older granite veins which they sharply cut, show a complex history of fracturing, brecciation and mineralization. Fluid temperatures ranged from 80–180°C with the cooler fluids associated with the deposition of the later gangue minerals. Unlike earlier mineralizing events, the ore fluids maintained a constant salinity (19–27 wt% dissolved salts) and were distinctively Ca-rich. Though other geological fluids may show Ca-enrichment, the combination of high salinity, low temperature, and Ca/Na ratios close to unity, implies a strong affinity with present-day oil field brines and to the mineralizing fluids responsible for the formation of sediment-hosted (Mississippi Valley Type) Zn–Pb deposits. Moreover, the isotope evidence indicates that the cross-course fluids contained a dominant meteoric water component (such as a highly modified groundwater or seawater). It seems probable, therefore, that the N–S-trending mineral veins owe their origin to metalliferous brines derived from sedimentary basins. One interesting feature of post-granite mineralization is that the veins are best developed outside the granites. This may simply reflect preferential fracturing of the weaker sedimentary rocks, but it also raises an important genetic question. Were the metals derived from the Permo-Triassic sediments in the rift basins or were they leached from the older Palaeozoic rocks by fugitive brines? Such shortfalls in our understanding of metallogenesis are not unique to Cornwall. Considerable debate flourishs amongst ore geologists as to the source of metals, and it is possible that answers to some of these questions may come from further research on Cornish mineral deposits.

CHRONOLOGY OF MINERALIZATION

Precise radiometric ages have now been obtained for the five major granites (Land's End, Carmenellis, St Austell, Bodmin and Dartmoor), which demonstrate that emplacement took place over an interval of approximately 30 Ma, from the Permian/Carboniferous system boundary throughout the early Permian. The oldest granite is the Carnmenellis (293 Ma) and the youngest the Land's End (275 Ma). As the sequence of granite-related mineralization described above (*Figure 8.6*) is directly linked to magma emplacement, it follows that each granite has its own suite of mineral deposits within distinct temporal frameworks. Thus the skarns and greisen-bordered veins associated with the Carnmenellis

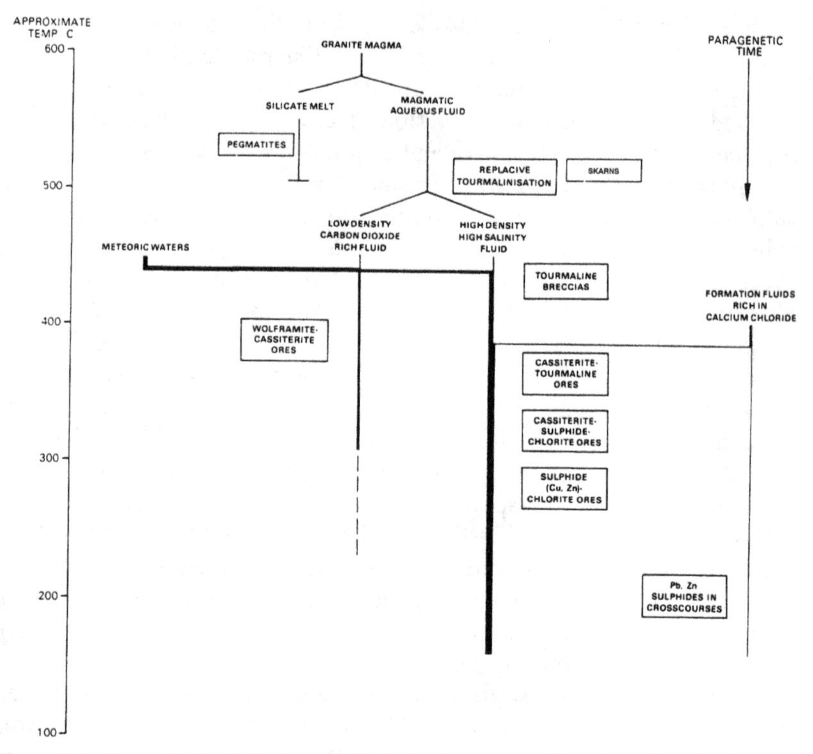

Figure 8.6. General scheme for metallogenic development and its chronology in Cornwall (after Leveridge *et al.*, 1990)

granite may be older than those hosted by the Land's End granite. Furthermore, mineralization may be associated with unexposed deep granite emplacement events, which may either not be exposed at surface, or be represented by late minor intrusive bodies such as aplite sills or elvan dykes.

Several methods have been used to date the mineral deposits (Darbyshire & Shepherd, 1985; Chen *et al.*, 1993 and Chesley *et al.*, 1993), and the evidence supports the concept of several early magmatic hydrothermal events concommitant with granite intrusion. However, ages for the main-stage mineral veins are equivocal, and indicate that mineralization may post-date the host granites by between 5–30 Ma (that is long after their initial cooling to 300°C). Fluorites from the South Crofty [SW 669 413] and Wheal Jane mines [SW 783 434], for example, yielded ages of about 259 Ma and 266 Ma, respectively (Chesley *et al.*, 1991), implying a considerable hiatus between emplacement of the Carnmenellis granite and mesothermal mineralization. The best explanation for this discrepancy would appear to be regional-scale hydrothermal circulation of fluids through the granites, sustained by magmatic activity deeper within the batholith over an extended period of time.

Until recently, the timing of post-granite N–S cross-course mineralization was highly speculative, and some researchers invoked a Tertiary event. However, radiometric ages for Pb–Ag–Zn veins in the Tamar valley district of east Cornwall, based on Rb–Sr isotopic analyses of inclusion fluids and fluorite, have demonstrated a mid–late Triassic age of 236 Ma (Scrivener *et al.*, 1994). This corresponds closely with a period of intense faulting and the development of rift basins across northern Europe containing thick red-bed sequences. These ages are also consistent with the relatively late development of fluorite in the E–W-trending mineral veins prior to the rotation of the regional stress field and the opening up of N–S-trending fault systems.

SOUTH CROFTY MINE

The last working mine in Cornwall, South Crofty Mine, was closed in 1998. It produced annually about 1900 tonnes of Sn metal in concentrates, from some ore grading about 1.4 per cent Sn. The workings included the former setts of South Crofty, Dolcoath, East Pool, Tincroft and Roskear mines (*Figure 8.7*). By the end, activity was centred about the ground between the Cook's Kitchen shaft [SW 665 407] and the New Roskear shaft [SW 656 410]. The mine lies to the north of the Carn Brea granite outcrop, with its surface workings on metasediments and metabasites of probable late Devonian age. The Carn Brea granite is encountered about 270 m below surface, and all of the present workings, on a swarm of ENE–WSW-trending veins, are enclosed by granite. The granite is cut by later quartz porphyry dykes ('elvans'), and includes bodies of aplite and pegmatite. It is a feature of the veins in this district, one of the most richly mineralized in the world, that near-surface deposits of Cu give way to Sn with some minor W at depth, and that the mineral assemblage is to some extent controlled by the granite/killas contact.

The workings of South Crofty are remarkable in that they demonstrate most of the features of the complex history of granite-related metallogenesis in South West England. A simplified sequence is as follows:

(1) Subhorizontal 'floors' of quartz and wolframite, with greisened margins, also some wolframite-bearing veins of pegmatitic aspect.
(2) Fe-rich tourmaline (schorl) veinlets, carrying some cassiterite and mostly forming steeply dipping swarms.
(3) The main economic Sn-bearing veins in steeply dipping fractures trending ENE–WSW. The dominant mineralogy is of very finely crystalline tourmaline ('blue peach'), with aggregates and discrete grains of fine-grained cassiterite. Brecciation is a common feature of these veins.

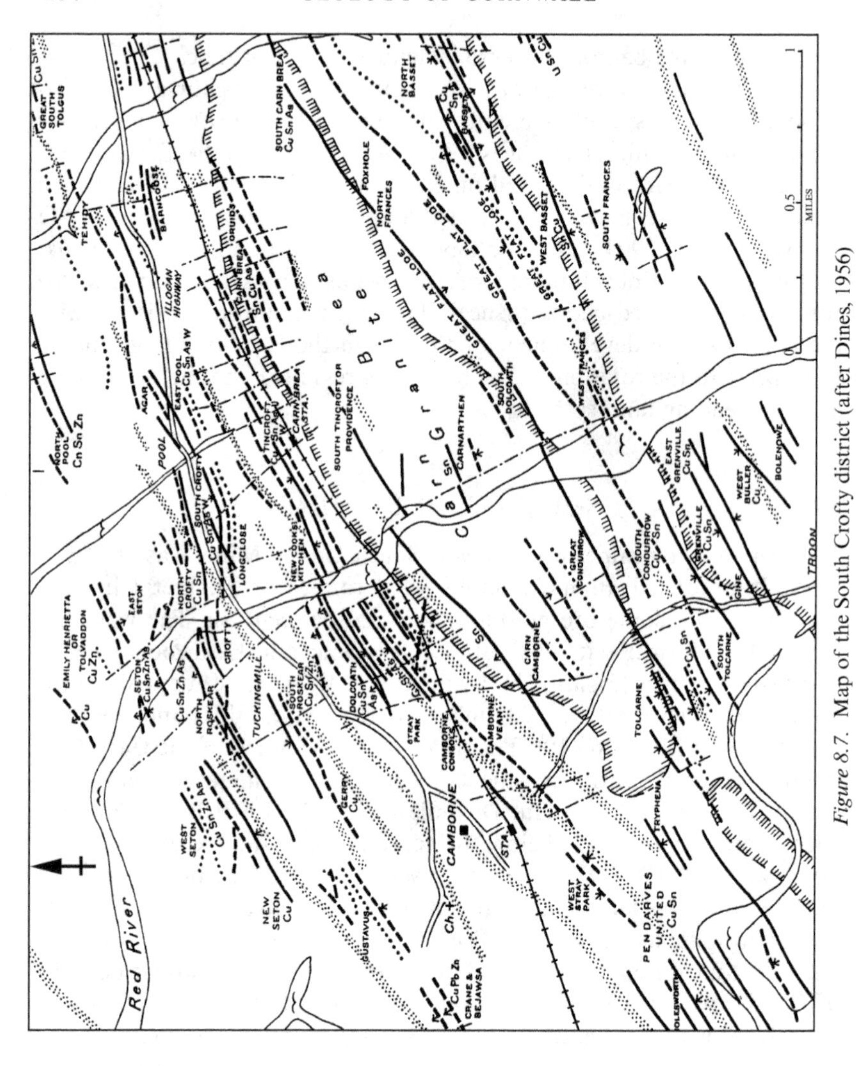

Figure 8.7. Map of the South Crofty district (after Dines, 1956)

(4) The blue peach veins are overprinted in places by chlorite ('green peach'); there is some remobilization and recystallization of cassiterite, which in this assemblage commonly occurs as coarsely crystalline grains and aggregates.
(5) Steeply dipping, WNW–ESE-trending fractures are infilled with assemblages of fluorite, quartz, earthy chlorite and hematite. These fractures may form individual veins ('caunter lodes') or form cross-cutting zones within the earlier peach lodes. Cassiterite in these structures is scarce or absent, and the caunter lodes may carry Cu–Pb–Zn–Bi minerals.
(6) Chalcedony and clay-filled wrench faults of NW–SE to NNW–SSE trend, associated with extensive kaolinization. The principal example of this type is a major fracture complex, called The Great Cross-course, which has a diplacement of more than 100 m, and physically divides the mine workings into two distinct compartments.
(7) Extensional fractures of roughly N–S trend, typically infilled with comb-layered quartz and fluorite These veins may carry minor amounts of low-temperature sulphide minerals, and are not commonly associated with kaolinitic alteration.

The phenomena listed above span the entire range described in this chapter and serve to demonstrate the complexity of post-granite emplacement events in this intensely mineralized part of the Cornubian orefield. *Figure 8.6* attempts to place the South Crofty mineralization in a genetic and temporal framework.

Techniques for dating the granites are more advanced than those for the mineral deposits. No doubt future studies will help to unravel the multistage, complex history of mineralization that can be observed throughout Cornwall, and allow chronological correlation across the peninsula.

Chapter Nine

Modelling the Mineralization Framework

The Cornubian batholith is composed of High Heat Producing (HHP) granites. The heat accompanying the radioactive decay of isotopes of U, Th and K has had a significant influence on the thermal equilibrium to which the granite has cooled. This heat generated from within the granite has also been an important contributory factor in promoting the circulation of ore-forming fluids in and around the granite.

The Hot Dry Rock Geothermal Project carried out at Rosemanowes quarry [SW 735 346], near Penryn, by the Camborne School of Mines, looked at the possible utilization of high temperatures found at relatively shallow depths through the generation of electricity by circulating water through an engineered reservoir in the granite. Other aspects of the thermo-mechanical behaviour of the granites which relate to mechanisms of metalliferous mineralization, are reported below.

CATACLASTIC FLOW AND LODE FORMATION

Lodes, barren veins and granite porphyries (elvans) in a given mining area often appear to be developed within fractures having a common origin, as near-parallel, steeply dipping, normal or strike-slip faults, with offsets up to a few decametres. The lode fill may be brecciated by brittle deformation and subsequently cemented, or it may show grain-size reduction and silicification as in 'capel'-type lode material. This latter suggests cataclastic flow of quartz. In this, the small grain size allows pressure solution and redistribution of quartz at grain boundaries at rates sufficient to allow quasi-continuous deformation without macroscopic rock breakage. Feldspars in such fine-grained rocks generally remain within the brittle field and show signs of mechanical disruption when examined under the microscope. Most lodes show evidence of many episodes of movement, in contrast to the sheeted-vein systems which often appear as single-generation infills of open fractures with little relative displacement of the walls. Variation in lode width may be caused by the relative movement of the opposing sides of the lode/fault where there are changes in strike and/or dip. The highest ore values are often found in the steepest parts of the lode. According to Collins (1912), enrichment of the ore is also found where a steep lode intersects a shallower lode, such as at Vottle shaft, Wheal Kitty [SW 725 512], or where a lode is cut by lodes of opposite dip such as at East Pool [SW 671 416]. Within the host rock the lithology can affect the productivity of a lode. A good example occurs at West Chiverton [SW 793 508].

Apart from intersections, variation in lode productivity has been related to changes in lode aperture caused by movement on an irregular fault plane. Deformation of the lode filling accompanies such movement, and may either be completely within the brittle field (lode breccias) or as quasi-continuous cataclastic flow producing fine-grained banded rock. Low minimum effective rock stresses would promote brittle failure, while high effective stresses would allow the accumulation of high shear stresses and so promote cataclasis. Cyclic changes in fluid pressure with time within the lodes could allow change from one form of deformation to the other and back again.

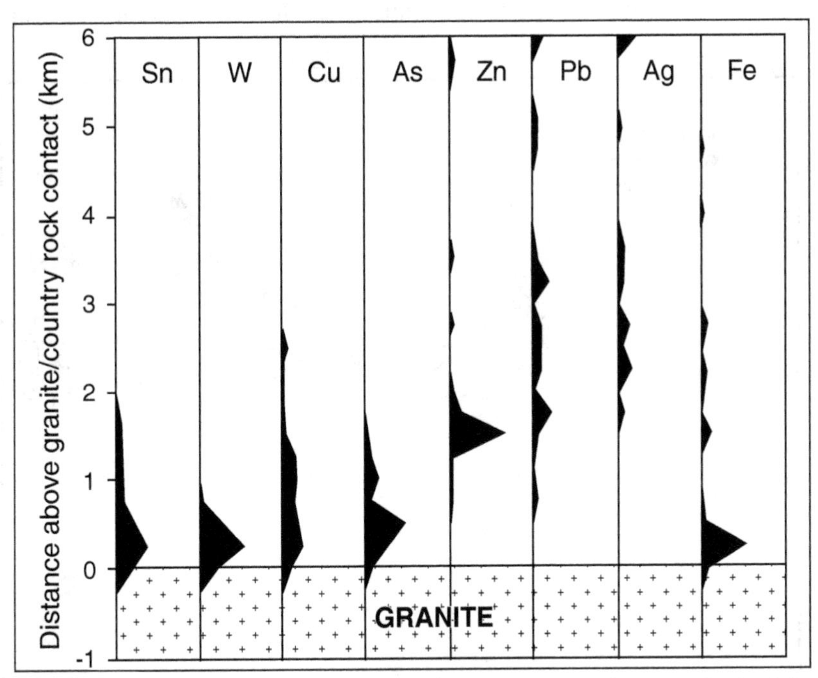

Figure 9.1. Mine productivity in relation to the granite–country rock contact. Productivity on a scale 0–1 for each element is defined as the productivity per km² for 250 m 'depth to granite' intervals as a fraction of the total productivity. All production from within granite is assumed to have been from within 250 m of the contact

Historically, a relationship has been recognized between lode mineralization and the granite–country rock contact. *Figure 9.1* shows strong contact control for Sn and W, with little production from rocks more than 1 km above the granite. Production of As also appears to be contact-related, while Cu has been mined from the granite margin to about 2 km above the granite–country rock contact. Unfortunately, primary Zn productivity data is dominated by a single mine; estimates for production from the reworking of mine dumps are not reliable. Pb and Ag are found in quantity only far above the contact zone. Much Fe was mined close to the granite contact, but deposits are found at a variety of positions in relation to the contact.

HEAT FLOW

The genesis of the Cornubian ore field appears to be intimately related to the high U content and great thicknesses of the granite, and the consequent high heat flow values found in the region. Conduction appears to be the dominant present day heat transfer mechanism, but this can be perturbed by convection seen as hot spring activity in deep mines. Away from the influence of the granite batholith, values close to 60 mW/m^2 are found (*Figure 9.2*), while on or close to the granite outcrop

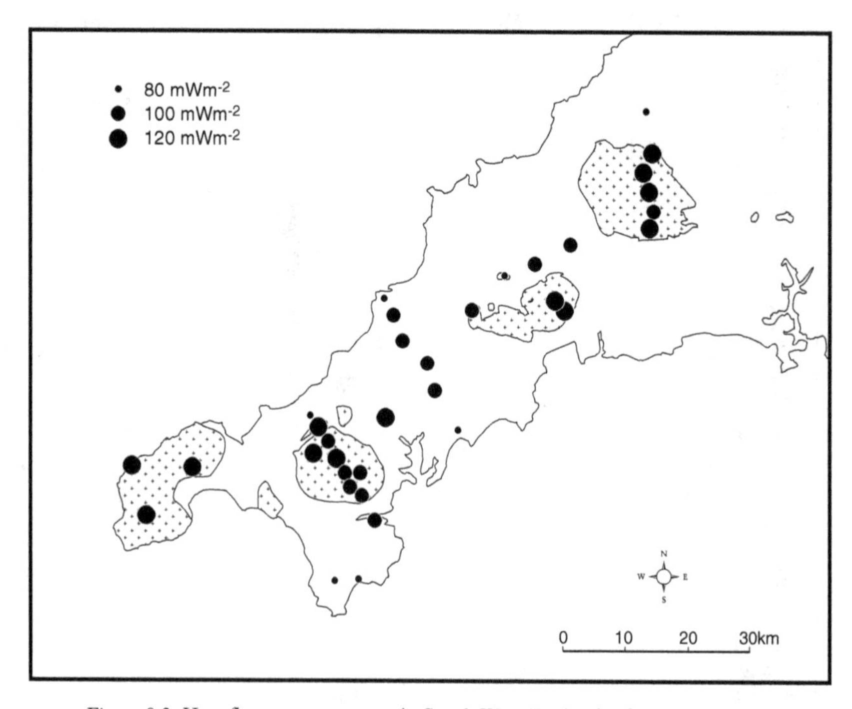

Figure 9.2. Heat flow measurements in South West England (after Lucas, 1993)

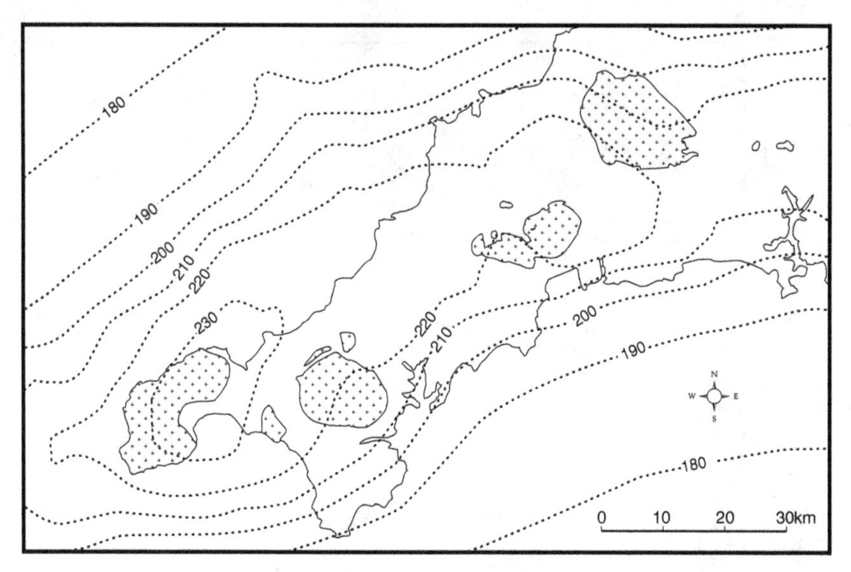

Figure 9.3. Model isotherms °C at a depth of 6 km under South West England for the present day (after Willis-Richards, 1990)

120 mW/m^2 is typical. *Figure 9.3* shows model temperatures for the present day at a depth of 6 km beneath the surface (Willis-Richards, 1990).

Although the undisturbed underground temperatures may determine the sense of any natural convection, the most important control on fluid circulation through the crust is the existence and maintenance of adequate permeability through the formation and repeated opening of fractures. In the context of the lodes which formed along these fractures in Cornwall, Jackson *et al.* (1989) and Willis-Richards (1990, 1993) examined the interaction between visco-elastic creep of warm rock in response to thermal expansion or contraction, and to tectonic stresses.

The contact zone of a granite emplaced beneath a few kilometres of host rock was found to accumulate tensile horizontal stress deviators as the granite batholith cooled over a time span of a few million years. Faults will dilate and slide most easily within this zone, and hydraulic fracturing from pressurised fluids may preferentially propagate along it (*Figure 9.4*).

The thermal structure of the lithosphere controls its strain-rate response to tectonic stresses, again through the temperature sensitivity of visco-elastic creep of silicates. Willis-Richards (1993) presented a theoretical map of the present-day instantaneous lithospheric strain rate response of South West England to an applied remote stress (*Figure 9.5*). This suggested that the hotter parts of the lithosphere can deform up to 40

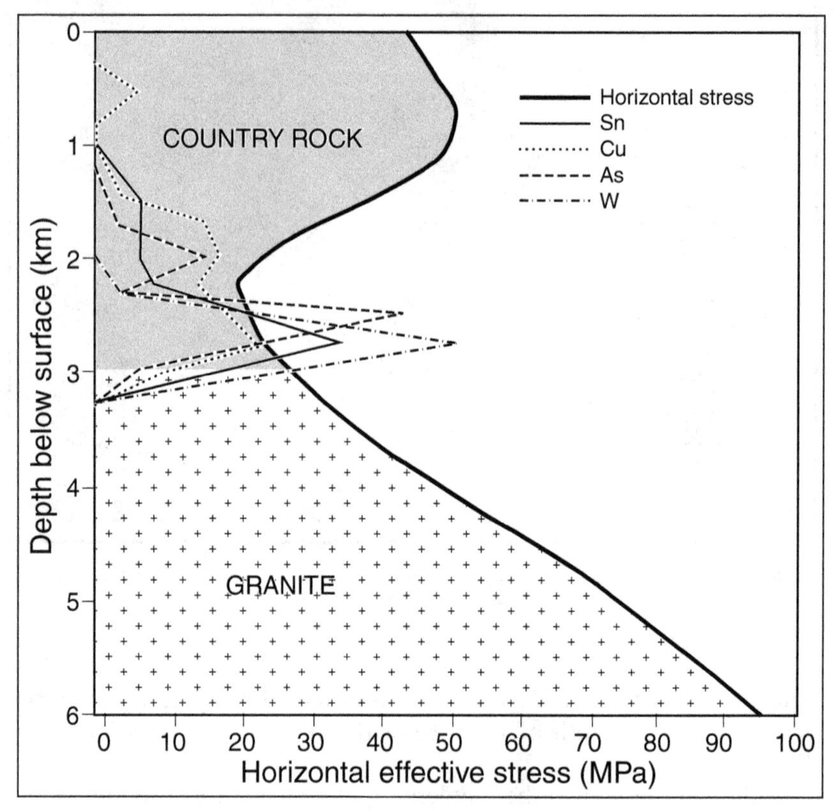

Figure 9.4. Production of Sn, Cu, As and W compared with thermo-visco-elastic horizontal stresses due to cooling and erosion imposed on lithostatic initial conditions.(Emplacement depth 3.5 km, 15 ma cooling to steady state with 500 m erosion; after Willis-Richards, 1990)

times faster than the colder regions. This model links, in a simplified way, the localization of the strain rate to thermal structure determined by the steady-state radiogenic heat production of the granite batholith. During cooling of the batholith the relative strengths of hotter and colder parts of the lithosphere are likely to have been even more exaggerated. The persistence through time of this thermal control on lithospheric deformation, realized through fault movement in the upper brittle portions of the lithosphere, helps explain why both Sn and Pb mineralization show a similar relationship to the present day thermal structure.

CONVECTION MODELLING

Sams & Thomas-Betts (1989) produced models of natural convective flow for the Cornubian granite batholith and surrounding regions, both for the present day and with 3 km of cover rock restored (*Figure 9.6*). Their

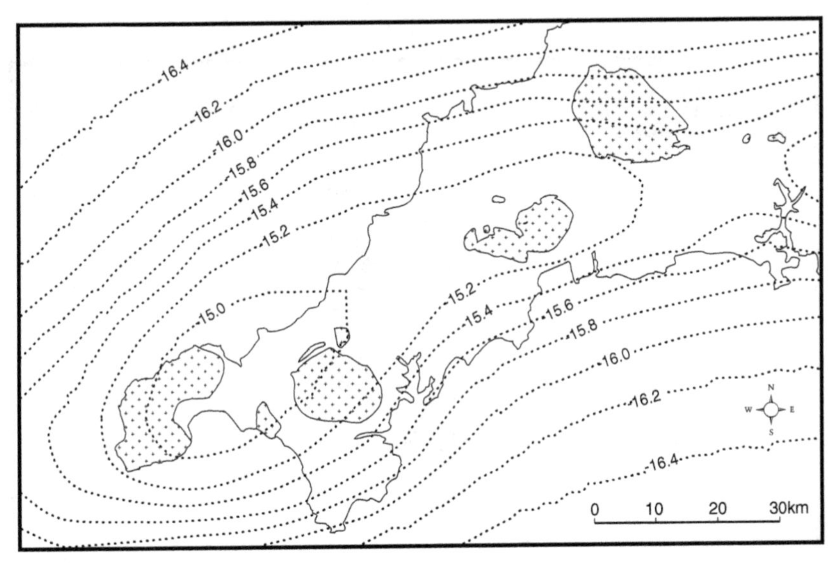

Figure 9.5. Model of the present-day lithospheric strain-rate response to an applied tectonic force of 10^{12} N m^{-1} Units: Log_{10} strain per second (after Willis-Richards, 1993)

models incorporated permability decreasing exponentially with depth, but no lateral variation in permeability. Thus, the resultant flow patterns depend only on the geometry of the heat generation anomaly caused by the U content of the granite batholith. With 3 km of cover rock the major upflows through the present-day surface correspond well to Sn and Cu mineralized areas. This model of convective circulation permits the Cu, Zn and S content of many of the lodes to have been leached from the country rocks. When the cover rock is removed, the pattern of circulation changes; areas of upflow move away from the granite outcrop to regions of highest near-surface temperature gradient, while the granite outcrops are loci for gentle downward movement. The upward flows show some correlation with regions of epithermal (cross-course) mineralization; the downward flow over the granite might be linked to the kaolinization process.

METASOMATIC ZONING OF GREISEN-BORDERED VEINS

At a much more detailed level, Schneider (1993) constructed a numerical thermochemical model of the microclinization (deeper levels) and greisen alteration (shallower levels) seen in the Hemerdon sheeted-vein Sn/W deposit in Devon [SX 257 584]. His model and conclusions may apply

a)

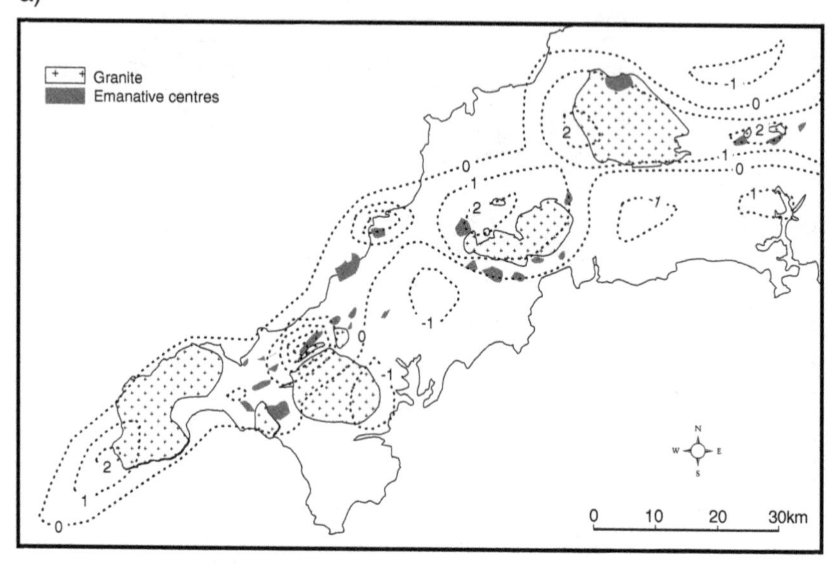

b)

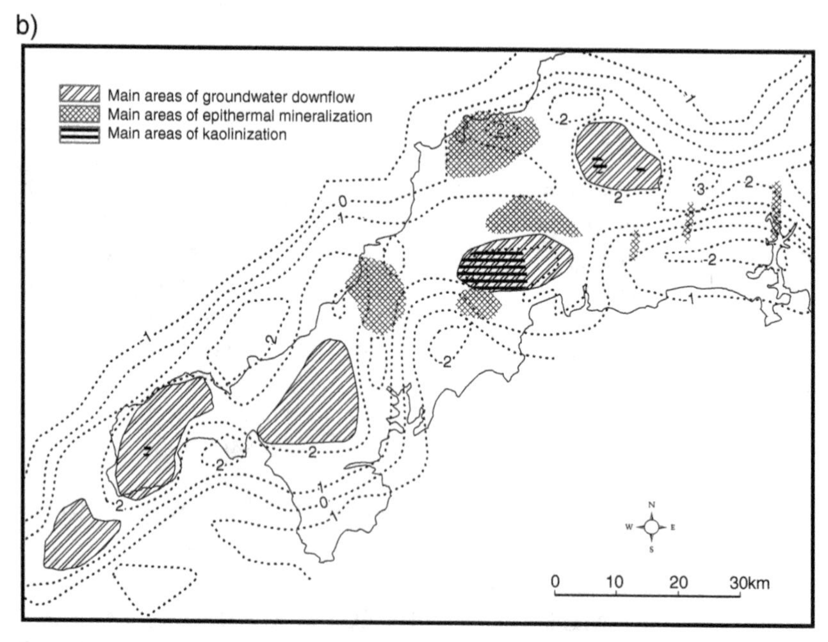

Figure 9.6. Numerical model of natural convection around the Cornubian granite batholith (after Sams & Thomas-Betts, 1988). (a): present-day surface with 3 km of cover rock, contour intervals 2×10^{-8}kg m^{-1} s^{-1} (b): present-day surface, contour intervals 2×10^{-9}kg m^{-1} s^{-1}

equally to similar deposits in Cornwall. If they do, greisen-altered granite may be divided into six zones that reflect distance from a fracture, and differences between the amount of fluid percolating into the rock and that simply moving along the fracture.

In order to explain these changes, Schneider's model examined percolation of fluid through the solid granite laterally from a single vein carrying an H_2O, KCl, HCl, $CaCl_2$ fluid of a fixed composition inferred to have been in equilibrium with granitic magma. Temperatures along the vein were taken to decrease from above 600°C to less than 400°C, while the rock temperature was assumed to remain constant away from the vein over the relatively short distances of interest. Chemical mass balance equations governing the reactions of quartz, feldspar and micas with fluids moving through the rock away from the vein then allowed calculation of the disposition of metasomatic alteration facies with temperature and relative distance from the vein. These results show good agreement with field observations, orthoclase feldspar being seen at higher temperatures and the replacement of feldspar by muscovite below about 400°C.

FRACTURE PROPAGATION IN ENDO-GRANITIC SHEETED-VEIN SYSTEMS

In the past, there has been some debate on the origin of the fractures in relation to greisen-bordered sheeted-vein systems. Hosking (1964) was of the opinion that the fractures in the Cligga Head sheeted-vein system were thermal contraction joints, but most later authors have emphasized the role of fracturing by high fluid pressures (Moore & Jackson, 1977).

Inspired by an Indian quarrying method in which a fire gradually moved across the floor of a quarry cracking off a granite slab about 130 mm thick, Marsh (1982) derived an approximate analytic expression for the stress effects of a periodic thermal stress applied at a free rock surface. He found that as the thickness of the heated zone increases so does a tensile buckling stress normal to the free surface; this is used as a model for the igneous stoping process. Similar stresses will arise normal to the surface of an open crack into which hot fluid is injected. If hot fluid passes rapidly along a fracture in granite with a temperature difference in the order of 200°C between the fluid and the rock, then normal tensile stresses adequate to cause fracture of a slab of the order of 1 m thick can be generated. The thickness of the spalled slab increases with the wavelength of the thermal anomaly and with decrease in the temperature difference between the fluid and rock. If this mechanism was influential, then the expected decrease in temperature difference and increase in thermal perturbation wavelength along the fracture would lead to upward increase in the fracture spacing; this is mirrored in the upwardly widening fracture pattern seen at Cligga Head.

If this mechanism is accepted, then the critical control on the formation of subparallel sheeted-vein systems would appear to be the close physical proximity of hydrous fluid at high temperatures and a relatively cool enclosing granite which has perhaps cracked during cooling. These conditions are most likely in apophyses from, or apical culminations to, magma bodies, but can also be generated during multiphase intrusion.

Chapter Ten

China Clay

One of the latest processes to have affected the granites of South West England is kaolinization. This has led to the formation of world-class deposits of kaolin, locally known as china clay. The related product chinastone is a weakly altered to unaltered low-Fe granite which is used as a fluxing material in the ceramics industry, and is still exploited in a small way in the western part of the St Austell granite.

About 150 million tonnes of china clay have been produced in total from Devon and Cornwall. Most has come from the western part of the St Austell granite, and the industry has had a major environmental impact on the landscape of mid-Cornwall. The superb exposures in the china clay pits have allowed significant research into the granites, their post-magmatic alteration processes, and their mineralization.

HISTORICAL DEVELOPMENT OF CORNISH CHINA CLAY PRODUCTION

The significance of the china clay deposits of Devon and Cornwall was first recognized by William Cookworthy in 1746. He subsequently built a kiln in Plymouth that produced the first true hard paste porcelain to be made in Britain (Bristow & Exley, 1994). By the mid-1840s, production had reached 30 000 tonnes per annum (tpa) and thereafter the industry grew steadily (*Figure 10.1*) to reach an output of nearly 500 000 tpa by 1889. The use of china clay in the paper industry was already an important market by the middle of the 19th century and much of the growth was associated with this application.

From 1889 to 1950 the industry changed remarkably little; production averaged around 600–700 000 tpa, with pronounced lows in the two world wars and in the depression years of the 1930s. After 1950, investment in a new generation of plants began, prompted by the amalgamation of many of the smaller companies to form English China Clays. Rapidly increasing demand for china clays for coating paper led to a second period of

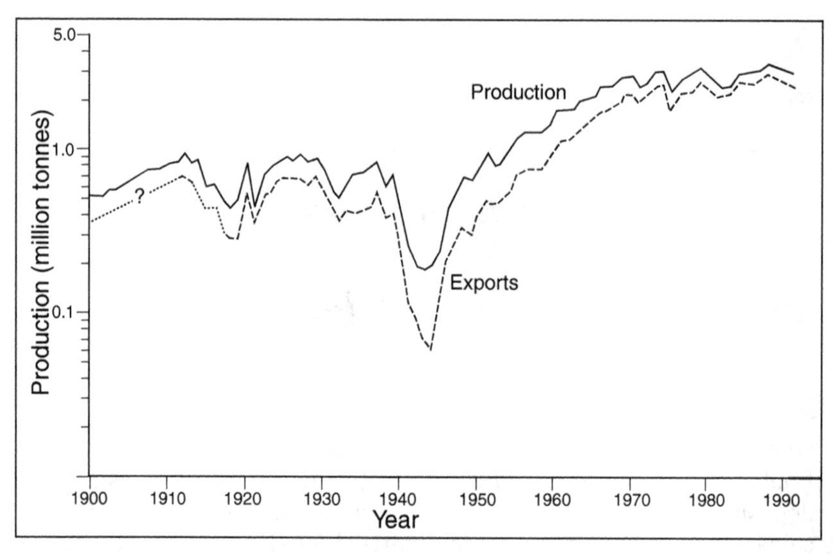

Figure 10.1. UK china clay production and exports 1900–1992

rapid expansion from 1950 to 1970, taking production up to over 2.5 million tpa. Further details of the contemporary industry will be found in Chapter 15.

THE FORMATION OF CHINA CLAY

Deposits of kaolinite-rich material can be laid down as sediments, as the ball clays of Devon were, or they can be formed by the alteration *in situ* of rocks such as granites by hydrothermal and/or weathering processes. Many commercial kaolin deposits owe their origin to more than one process, for instance a bed of sedimentary clay may be laid down and then have its mineralogy and chemistry transformed by later weathering.

The Cornish china clay deposits were formed by the kaolinization of granites. Kaolinization is essentially a process of hydrolysis, accompanied by the removal of alkalis and silica:

$$\underset{\text{feldspar}}{2NaAlSi_3O_8} + \underset{\text{water}}{3H_2O} = \underset{\text{kaolinite}}{Al_2Si_2O_5(OH)_4} + \underset{\text{silica}}{4SiO_2} + \underset{\text{alkali}}{2NaOH}$$

The most easily altered component of the granite is sodic plagioclase. Other minerals such as K-feldspar, mica and topaz were altered less readily, and often through intermediate stages. Some of the kaolinite was also derived by alteration of smectite and illite formed during early stages of argillic alteration (*Figure 10.2*).

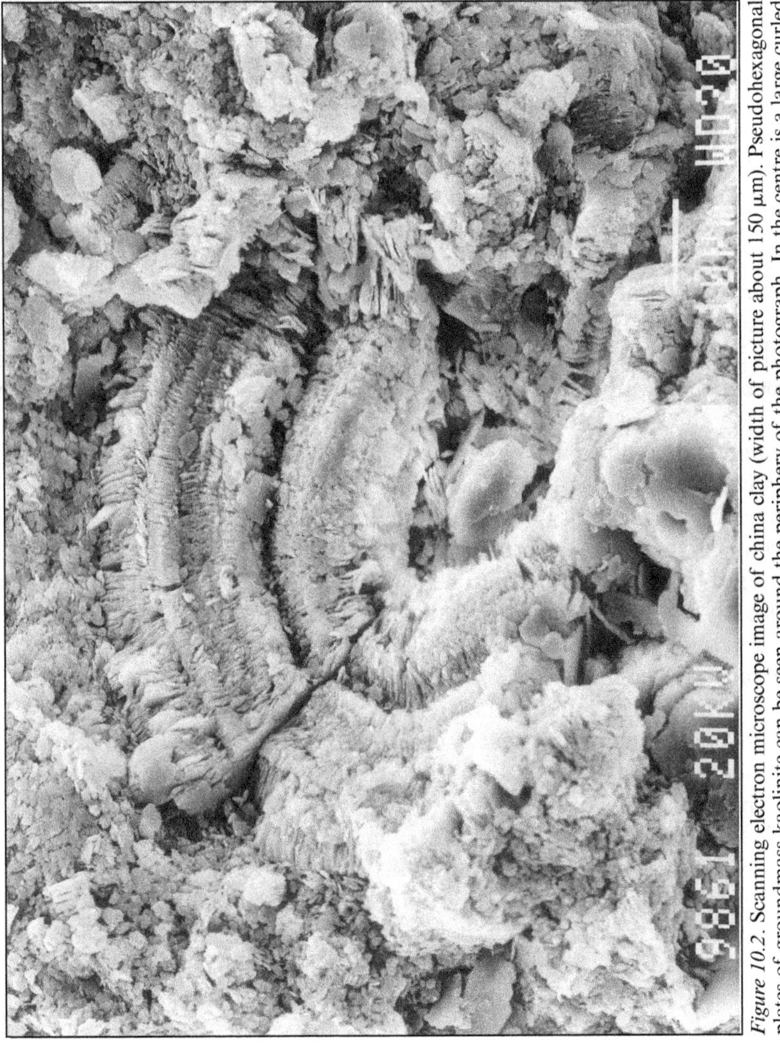

Figure 10.2. Scanning electron microscope image of china clay (width of picture about 150 μm). Pseudohexagonal plates of groundmass kaolinite can be seen around the periphery of the photograph. In the centre is a large curled stack of kaolinite. Photograph courtesy of ECC International

There has been considerable debate over whether the process of kaolinization was due to hydrothermal action or weathering. However, recent research has shown that it probably involved both processes (Bristow & Exley, 1994). The stages for the St Austell granite are set out in *Table 10.1* and *Table 10.2*.

One of the most crucial influences on the quality of the china clays is the composition of the granites (*Table 10.2*, stages I–IIIa). The lithian-mica granites found in the western part of the St Austell pluton provided a low-Fe, high Na-feldspar parent for the formation of china clay, although china clay can be formed from biotite granite, as the deposits on Bodmin Moor show.

The striking quartz–tourmaline veins seen in nearly every china clay pit are sometimes Sn and/or W bearing, and were formed in Stage IIIb, together with associated tourmalinization and greisenization. The presence of B in these high-temperature hydrothermal fluids meant that any Fe liberated by the deferruginization of biotite would have combined with Si and B to form tourmaline, which remained unaltered in the later process of kaolinization. This is an important contributory factor in making the Fe content of Cornish china clays lower than most other commercial kaolins from elsewhere in the world.

Chemical changes, microfracturing, the development of grain-boundary permeability and phyllic alteration during Stage III probably increased the overall permeability of the granite, and made it more susceptible to later kaolinization. Kaolinite is unlikely to have formed at this early stage, as it would not have been stable in the high-temperature, moderately saline conditions indicated by the fluid inclusions (see Chapter 8). The final phase of granitic intrusion is represented by felsitic elvan dykes (Stage IIIc), which are seen in many china clay pits.

Once the magmatic heat from these elvan dykes had dissipated, radioactive decay of U, Th and K continued to provide a source of internal heat. By Stage IV, which probably took place in the late Permian and early Triassic, the cover of overlying rocks had probably been reduced by erosion to around 1 km or less, and the intrusions had completely crystallized. Mineralization now took on a different character, with cross-course veins carrying epithermal mineralization such as Fe, Pb and Zn. Evidently the stress direction had changed, and these cross-course veins are usually oriented N–S or NW–SE. They are often associated with a small dextral strike-slip fault component.

Stage IV fluids lacked B, so the Fe produced by continuing deferruginization of biotite could no longer form tourmaline. In the St Austell granite, Fe-bearing fluids moved outwards to form a ring of Fe lodes around the granite. These lodes can be traced into the granite where, in the china clay pits, they are seen to form sub-vertical N–S or NW–SE quartz veins, often stained red with Fe oxides.

Table 10.1. ALTERATION PROCESSES OF THE ST AUSTELL GRANITE

Stage	Process	Age Ma	Depth km	Temperature °C	Fluid salinity	Heat source	Min stress
I	Emplacement of biotite granite forming main batholith.	290-285	?3	500-600		Magmatic	E–W
II	1st phase of post-magmatic alteration and mneralization.	285–275	2–3	?200–500	Moderate	Magmatic	E -W then N–S
IIIA	Emplacement of evolved Li-rich and biotite granites in western part of St Austell granite.	275–270	2–3	500–600		Magmatic	N–S
IIIB	1st part of 2nd phase of post-magmatic alteration and mineralization.	275–270	?2	380–450	Moderate	Magmatic, with some radiogenic	N–S or NW–SE
IIIC	Emplacement of felsitic elvan dykes.	275–270	?2	500–600	Moderate	Magmatic	N–S
IV	1st phase of argillic alteration and NW–SE or N–S quartz haematite veins and faulting.	*c.* 240	?1–2	250–350	Moderate to high	Radiogenic.	E–W
		POSSIBLE QUIESCENT PERIOD					
V	2nd phase of argillic alteration. Main period of kaolinization. Deep Mesozoic supergene alteration.	180 to present	0.2–1.5	30–200	Low	Radiogenic	E–W or N–S
VI	Mesozoic and early Tertiary chemical weathering.	180 to present	0–0.5	20–50	Low	High surface temperature	Vertical

Radiometric dates from Darbyshire & Shepherd, 1985; 1987

Table 10.2. CONSEQUENCES OF ALTERATION PROCESSES ON THE ST AUSTELL GRANITE

	Main changes in mineralogy			
Stage	Feldspar	Quartz	Mica	Mineralization
I				
II	Limited greisenizaion alongside veins			Sn, W
IIIa				
IIIb	*Greisenization*: conversion to quartz, mica and topaz by F-rich fluids, and mica (gilbertite). *Tourmalinization*: replacment by tournaline.	Repeatedly fractured. Fractures annealed by fresh growths of quartz.	Some recrystallization. Biotite looses Fe, which is taken up by tourmaline growth.	Sn, W, Cu
IIIc				Sn, W, Cu
IV	*Na-feldspar*: altered to smectite/ illite, a little kaolinite. *K-feldspar*: altered to illite, and possibly smectite.	Free Si, released by argillation, forms overgrowths on quartz and new Fe-stained non-tourmaline bearing lodes (NW–SE or N–S).	Much Fe liberated from biotite, which is carried out of the granite to form Fe lodes. Some mica hydrated to gilbertite.	Fe, U, Pb, Zn
		POSSIBLE QUIESCENT PERIOD		
V	*Na-feldspar*: altered readily to kaolinite.*K-feldspar*: less readily altered to kaolinite.*Smectite*: altered readily to kaolinite.	Free Si, released by argillation forms overgrowths on quartz and some minor quartz veins.	Some Fe liberated from biotite, is not carried out of granite and colours matrix. In areas of intense kaolinization mica/illite altered to kaolinite.	Fe, U (minor)
VI	Altered kaolinite is b-axis disordered in Eocene–Oligocene weathering.	Some solution of Si from quartz grains.	Some Fe, liberated from biotite, not carried out of granite and colours matrix. In areas of intense kaolinization mica/illite altered to kaolinite.	

Because there is clear field evidence for an association between kaolinized zones and Stage IV veins, it is possible that Stage IV included the first phase of argillic alteration. However, fluid inclusions (Jackson *et al.*, 1989) indicate that the Stage IV mineralizing solutions had temperatures of 350°C and were moderately or highly saline. Probably these saline groundwaters had developed under the hot arid conditions of the Permian and Triassic. In conditions like these, kaolinite is not stable (Bristow & Exley, 1994), and a mixture of smectite and illite is the most likely product of argillic alteration.

Throughout most of the Mesozoic, Cornwall appears to have been part of a large offshore island. With the onset of wetter climatic conditions in the late Triassic, the saline waters of Stage IV were flushed out and replaced by fresh meteoric water, ushering in the main period of kaolin formation (Stages V and VI).

Convective circulation systems driven by radiogenic heat now became established. Fehn (1985) and Sams & Thomas-Betts (1988) showed that radiogenically driven convective circulation of meteoric water was likely to have been one of the major enabling mechanisms for kaolinization in South West England, a process which is probably still active. Durrance *et al.* (1982) demonstrated that areas of high and low heat-flow exist on Dartmoor, and by relating this to ^{222}Rn concentrations, were able to map contemporary convection cells in which down-flow correlates strongly with kaolinized areas, and up-flow with fresh granite. Computer modelling of convection flows by Sams & Thomas-Betts (1988), based on the present land surface, confirmed this discovery. Circulation on a considerable scale was possible only because of the enhanced permeability, which had been created by the high-temperature hydrothermal events of Stage III and by repeated movements along older lines of weakness, notably the cross-courses. The distribution of zones of kaolinitic alteration was clearly influenced by these zones of enhanced permeability.

Although china clay deposits are usually funnel-shaped, with the stem extending down to several hundred metres (*Figure 10.3*), troughs, inclined slabs and irregular bodies also occur (Bristow & Exley, 1994). It is not uncommon to find unkaolinized granite overlying kaolinized granite, particularly where there is a granite which is less susceptible to kaolinization overlying a granite which is more susceptible. Examples of this situation are seen in Gunheath [SX 003 568] and Goonbarrow [SX 009 584] china clay pits.

During the Mesozoic, the climate throughout northern Europe was warm and humid, caused by a combination of lower palaeolatitude and warmer global temperatures. As a result, exceptionally deep chemical weathering profiles, with the development of thick kaolinite-bearing pallid zones, are known from many places across northern Europe, as far as Ukraine and the Urals. In the Palaeogene, deep chemical weathering

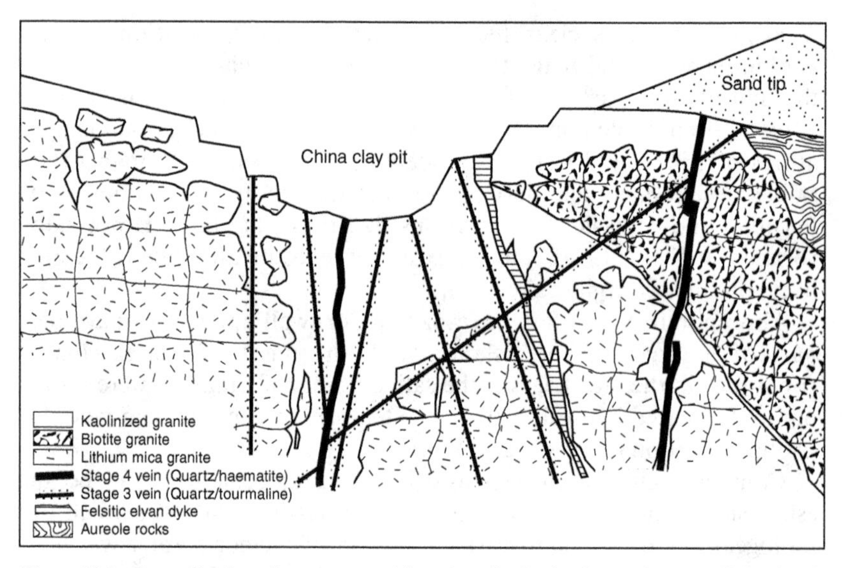

Figure 10.3. Generalized section across a china clay pit. Greisening and tourmalinization is indicated by stippling

profiles are also known to have developed in several parts of the British Isles, Belgium, Germany and France (Bristow, 1968; 1990). Weathering in the Mesozoic in Europe seems to have generally produced well-ordered kaolinite, while the shallower weathering of the Palaeogene seems to have usually produced a pallid zone characterized by b-axis disordered kaolinite. It is from this latter material that the ball clay-bearing sediments were derived (Bristow & Robson, 1994).

The intense weathering in the Mesozoic and Palaeogene could have contributed significantly to the kaolinization of the granite (Stage VI). Isotopic studies (Sheppard, 1977) of O and H have suggested that the kaolinite was formed by fresh water of meteoric origin. Water containing organic acids is known to accelerate kaolinization, so the role of surface waters derived from an organic-rich swampy or forest environment could have been significant. The distinction between alteration caused by cool meteoric water being drawn down into the granite on the downward limb of a convection cell, and deep supergene alteration may be almost impossible to draw. Pockets of bleached and kaolinized Devonian sediments are found in many places in Cornwall, particularly associated with the older planation surfaces (Bristow & Exley, 1994), presumably formed by Mesozoic or Tertiary weathering.

There is one further strong indication of a supergene origin for the china clay deposits. When the base of the kaolinized zone is plotted out from boreholes, the familiar pattern of funnels and troughs emerges, but the depth of these kaolinized zones is much the same over the entire area of the western part of the St Austell granite. On Bodmin Moor, much the same is seen. There is also no evidence for large kaolinized masses at depth below unkaolinized rock being formed in the deeper metalliferous mines. Kaolinization must therefore have developed in relation to a surface parallel to the present land surface. This effectively excludes the possibility of hydrothermal kaolinization at depth being the principal cause of kaolinization and points strongly to a surface-related process as the final stage in the formation of the kaolinite we see today.

MINERALOGY AND PROPERTIES OF THE KAOLINIZED GRANITE

The kaolinized granite consists of a mixture of two groups of minerals (Bristow & Exley, 1994); those formed during kaolinization and those remaining more or less unaffected by kaolinization.

Secondary minerals formed by kaolinization

Most of the kaolinite is well-ordered, euhedral or subhedral, and in the form of 1–10 μm pseudohexagonal plates in small stacks up to about 10 μm in height. They are visible around the periphery of *Figure 10.2*. Much larger curled stacks, 50-100 μm across, also occur, as can be seen in the centre of the figure. The morphology of these two varieties make it likely that the finer-grained 'groundmass' kaolinite was formed by replacement of feldspar and other minerals, while the large curled stacks were formed by a process comparable with the growth of authigenic kaolinite in sediments. Re-treatment of refining residues is now enabling these large curled stacks of kaolinite to be recovered and ground to make 'synthetic' fillers (Bristow, 1994). A third morphological variant forms irregular mantles on mica flakes, suggesting direct replacement.

The SiO_2 released by the kaolinization of feldspar was transported through the rock to the joints and veins, where it either crystallized as vein quartz or was conveyed out of the system. Kaolinization involved the removal of something like 25wt% of the granite, possibly accompanied by a slight reduction in volume. Kaolinite-filled cracks are commonly seen in the china clay pits, and may represent a slight contraction of the overall mass during alteration.

Most of the secondary mica is fine-grained, and it occurs around the edges and in the cleavage and fractures of earlier minerals, such as K-feldspar and topaz. It has been referred to variously as 'illite', 'hydromuscovite', 'sericite' and 'gilbertite'. The first three of these are colourless, but the last is usually identified by its golden yellow or green colour. All these mica varieties have less K_2O and more SiO_2 and H_2O than muscovite.

Small quantities of smectite are found in most kaolinized granites, usually concentrated in the finest particle size fraction. This may be a relic of Stage IV alteration, notably occurring near alteration 'fronts' (Exley, 1976).

Primary minerals inherited from the parent granite

Feldspar

Together, feldspars constitute approximately 50–60% of the volume of unaltered granite, the ratio being about 3:2 in favour of K-feldspar in the early biotite-bearing granites and the reverse in the later varieties. The plagioclase is generally albite and is more susceptible to kaolinization than K-feldspar. K-feldspar, on the other hand, may persist until an extreme stage of kaolinization. Intact, crystallographically perfect phenocrysts of orthoclase up to 100 mm long may sometimes be obtained from partially kaolinized granite. These are known to clay workers as 'pigs eggs', and are much prized by mineral collectors.

Quartz

This is the most stable primary mineral and is relatively unaffected by kaolinization. Originally making up 30–50% of the granite by volume, the large waste tips associated with china clay working are mainly composed of quartz.

Mica

The total amount of mica in unaltered granite ranges from 6% to 10% and includes both coloured and white varieties. Any biotite present will lose Fe during kaolinization, which will tend to stain the kaolinized granite various shades of red, yellow or brown. While much of the mica in kaolinized granite is original magmatic mica, a substantial proportion of the mica removed from the clay stream during refining is secondary and not primary mica.

Accessory Minerals

Tourmaline is the most abundant accessory amounting to up to 4% in some unaltered granites, although much of the tourmaline seen in china clay pits is associated with veins formed during the high-temperature hydrothermal mineralization of Stage IIIb. Topaz also occurs, almost entirely in the later granites as a result of the high F content of the magma. It never attains more than about 3% in the parent granite, although it can rise to as high as 30% in some greisened areas. It is often altered to secondary mica. Under extreme kaolinizing conditions it can be altered to kaolinite. Fluorite is most common in the chinastone areas, where it occurs at about 2%, but it tends to be removed during kaolinization. Goethite, limonite and haematite are often found as nodules or as a general stain through the matrix. Most of these oxide 'stains' are associated with the deferruginization of Fe-bearing micas. Nodules are surprisingly widespread and are found in every pit. In some of the deeper china clay pits, notably Lower Ninestones [SX 007 562] and the old Stannon sink [SX 133 813], similar nodules are seen in the unoxidised state to be composed of radiating marcasite. Apatite is a common accessory mineral and is a F-bearing variety. It does not, however, exceed 0.5%. Other accessory minerals frequently found in trace amounts are chlorite, zircon, monazite, cassiterite, wolframite, rutile, turquoise, wavellite, libethenite, varlamoffite, and various Sn and Cu sulphides, arsenates and phosphates.

RESOURCES

The intensity of argillic alteration in the Cornubian batholith is uneven (*Figure 10.4*), although kaolinization is known to exist in all the granite plutons, with the exception of the Isles of Scilly granite.

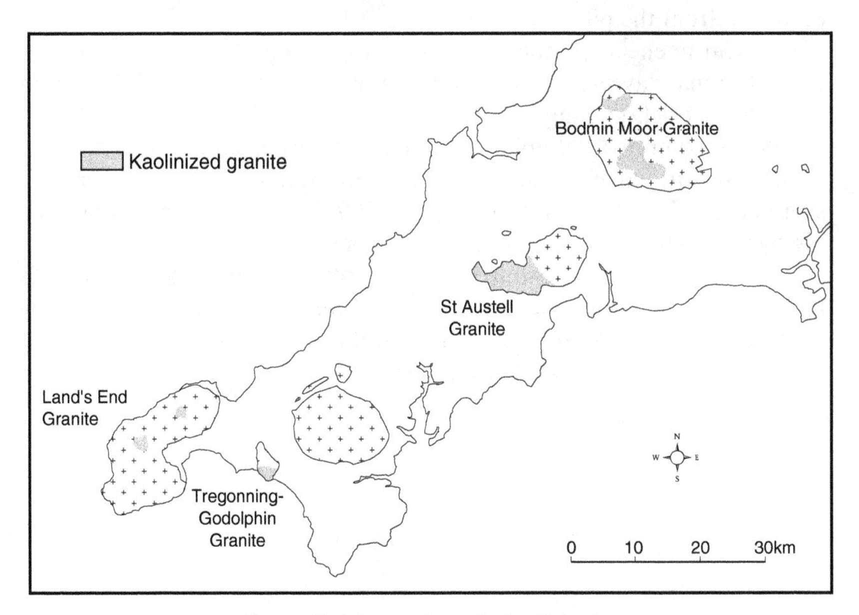

Figure 10.4. Areas of granite kaolinization

The St Austell granite has the most complex history of all the Cornubian plutons. Here, most of the important china clay deposits were formed from lithian-mica granites with low biotite contents, so yielding china clays of high brightness. China clays destined for paper-making come mainly from the central area between Nanpean, Bugle, Carclaze [SX 024 548] and Foxhole , while those used for ceramics generally come from the area west of Nanpean. However, many pits are capable of producing both paper and ceramic clays. There is some kaolinization along the southern margin of the biotite granite which forms the eastern part of the intrusion, with Bodelva pit [SX 050 550] situated where a major dextral strike-slip fault ('Great Cross Course') intersects the granite margin. In the early part of this century, china clay was extracted in a small way from Belowda Beacon [SW 967 627], a small boss of granite to the north of the main St Austell mass. Chinastone has been obtained from many quarries in the topaz and fluorite granites between Nanpean and Treviscoe [SW 947 559], it was also formerly exploited in quarries (now backfilled) between Stenalees and Hensbarrow [SX 013 569]. Stone from this quarry was used in the construction of St Paul's church, Charlestown.

The Bodmin Moor intrusion is mainly composed of biotite granite, and sporadic kaolinization occurs in a broad belt running from the north, through the west-central parts, to the southwest. Although many small pits have been worked in the past, only Stannon in the north [SX 130 810] is currently producing. About 100 000 tpa of filler and ceramic clays are produced from the pit.

The Carnmenellis pluton consists mainly of biotite granite. It has no current china clay production, although there was some exploitation in the past for brick-making.

The Tregonning-Godolphin pluton comprises a biotite granite in the north and a lithian–mica granite in the south. In the south there were two small china clay pits [SW 593 291; SW 599 293]. Cookworthy's original discovery of china clay was made on Tregonning Hill.

The Land's End intrusion is wholly biotite granite. Kaolinization is restricted to small areas, and much of the china clay is badly stained by Fe oxides. Until recently there was a small production from a narrow zone of kaolinization near Bostraze [SW 383 317].

Chapter Eleven

Offshore and Mesozoic Geology

Away from the coastline of Cornwall, Palaeozoic rocks are buried beneath a cover of Mesozoic and Tertiary strata that reaches up to 9 km thick in local sedimentary basins. These basins are closely related to the structures seen in the Palaeozoic basement, but towards the edge of the continental shelf, rocks older than the late Cretaceous are broken into a series of rotated blocks that were displaced by the initial rifting and subsequent opening of the North Atlantic Ocean.

Unfortunately, the marine area around Cornwall and the sedimentary basins have been given a large number of differing names with little evidence of consistent usage. However, in this account the nomenclature suggested by Evans (1990) has been employed and is shown in *Figure 11.1*.

BASEMENT GEOLOGY

The rocks of the area shown in *Figure 11.1* can be divided into two main units, separated by a major unconformity, that pre-date and post-date the Variscan orogeny. The post-Variscan succession embraces the Permian–Recent interval, although this chapter will only consider the Permian–Cretaceous strata. The pre-Variscan, basement succession is continuous and presently exposed in the Irish, Welsh, Cornubian and Armorican massifs that surround the area.

Cornubian Massif

This Massif comprises a structurally complex succession of Devonian and Carboniferous rocks intruded by an elongate granite batholith, and includes the Lizard Complex, a fragment of oceanic lithosphere. Offshore, the Eddystone Reef, some 50 km east of the Lizard, is an isolated outcrop of garnetiferous, granitoid gneiss surrounded by an area of mica schist and other gneisses that appear to underlie the inner part of Plymouth Bay. The Eddystone rocks are probably Precambrian in age, but were

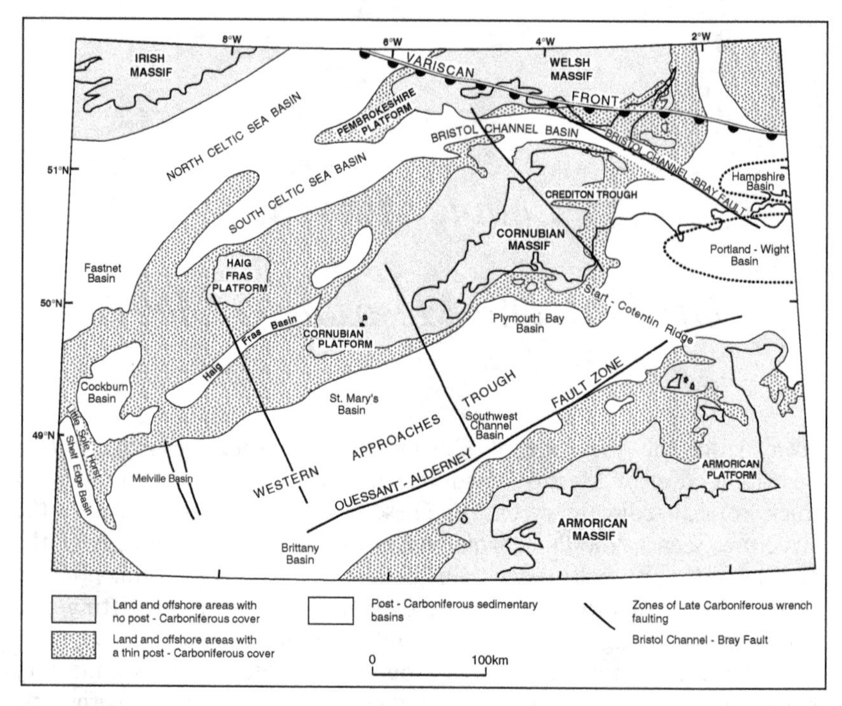

Figure 11.1. The major tectonic elements of the region offshore Cornwall (after Ziegler, 1987; Evans, 1990; Day, 1986; and Hillis, 1988)

re-metamorphosed in the Devonian (Holder & Leveridge, 1986a). The relationship of these rocks to those at Start Point and the slightly metamorphosed greywackes and slates of Dodman Point, is still not completely understood (Harvey *et al.*, 1994). The Cornubian Massif extends beyond the land area of Cornwall and forms a basement ridge, the Cornubian Platform, west of the Isles of Scilly. To the northwest of the Cornubian Platform lies the Haig Fras Platform. This is mostly formed of granite similar to that elsewhere in Cornwall. The exposed body trends NE–SW and occupies an area approximately 45 km by 15 km. The remainder of the Haig Fras Platform is composed of Devonian and Carboniferous metasediments.

Armorican Massif

The Armorican Massif occupies the area represented by Brittany, the Channel Islands and the Cherbourg Peninsula (Cotentin). Many of the granites, diorites, schists, gneisses and amphibolites of the Armorican Massif are the result of the Icartian (2700–2550 Ma), Lihouan (2000–1900 Ma) and Cadomian (690-570 Ma) orogenies. The older two events are generally called the Pentevrian, while the younger event is termed the Brioverian. Overlying these metamorphosed rocks are limited outcrops of Cambrian–Carboniferous sediments, including the famous 'Grès Armoricain' orthoquartzites of Ordovician age. The northern boundary of the Armorican Massif is the northward dipping Ouessant–Alderney Fault Zone (*Figure. 11.1*), where basement rocks are both faulted against, and overlain by, Mesozoic and Tertiary strata.

Welsh Massif

This area of Precambrian–Carboniferous strata preserves evidence of both the Caledonian and Variscan orogenies. While outside the area under consideration, the Pembrokeshire Platform extends southwestwards from South Wales and forms the northern edge of both the Bristol Channel Basin and the South Celtic Sea Basin.

Irish Massif

The south of Ireland is an area of Devonian and Carboniferous clastic rocks, intensely folded by the Variscan Orogeny. The North Celtic Sea Basin, which contains the Kinsale Head gas field near the Irish coast, is bounded on the north by the Irish Massif and on the south by the Pembrokeshire Platform.

POST-CARBONIFEROUS DEVELOPMENT

Following the late Carboniferous Variscan Orogeny and the intrusion of the granite batholith of South West England, the whole area was characterized (in the early Permian) by a rugged, arid landscape with intermontane basins containing volcanic rocks. Breccias and mudstones seen underlying the late Carboniferous/early Permian rhyolite lava at Kingsand in south-east Cornwall [SX 442 511] belong to this phase. By the late Permian, crustal rifting had led to the formation of a number of fault-bounded basins (Evans, 1990) in which exceptionally thick, continental red-bed sequences accumulated (Harvey *et al.*, 1994). In the latest Triassic, a marine incursion was followed in the Early–Middle Jurassic, by marine sedimentation. In mid-Jurassic to early Cretaceous times, renewed crustal uplift caused a major hiatus and/or erosion. The early to late Cretaceous was characterized by renewed subsidence as the North Atlantic opened. Later, Alpine tectonics (45–6 Ma) caused uplift, tilting and erosion in the area and effectively shaped the present landscape. During both of these periods of instability (late Jurassic and mid-Cenozoic) salt movements (halokinesis) created numerous salt domes and salt pillows in the Melville Basin and the South Celtic Sea Basin

(Evans, 1990). Hillis (1988) has produced a series of models for the tectonic history and subsidence of the Mesozoic to Cenozoic interval, as shown in *Figure 11.2*).

Sedimentary basins can either be formed when the lithosphere is stretched and, as a result, thinned, or developed by movement along pre-existing low-angled faults. In the area under discussion, especially to the south of Cornwall, however, the latter mechanism was dominant. It appears that a number of Variscan structures relaxed (Harvey *et al.*, 1994) during basin formation. Subsequently, areas of inversion occurred above ramps in the basement thrusts. Working on data from the Celtic Sea, Coward & Trudgill (1989) suggested that reactivation of Variscan structures caused uplift in the late Jurassic. This view has been further developed by Ruffell & Coward (1992) using seismic reflection data from the North and South Celtic Sea basins.

The various basins developed during the Mesozoic are shown in *Figure 11.1*. The early geological history of the Plymouth Bay Basin has been described by Harvey *et al.* (1994) while Chapman (1989) has given a comprehensive account of the development of the Melville Basin.

MESOZOIC HISTORY

During the Permian and Triassic a system of rifts began to fracture the northwest European area. Crustal extension along these fractures created a number of sedimentary basins around the British Isles (Hillis, 1988). All these basins are a part of the Arctic–North Atlantic rift system which included basins in the Bay of Biscay and Rockall Trough (Ziegler, 1987). The sediment fill of some of these basins began in the Triassic, while others, such as the southern part of the Melville Basin and the Plymouth Bay Basin (Harvey *et al.*, 1994), began to receive sediment in the early Permian. In these basins, the sediments are associated with early Permian volcanics that may have been related to the final phases of granite emplacement. The sediments immediately overlying the volcanics comprise sandstones, conglomerates and breccias, that are reminiscent of the Aylesbeare Group of east Devon (Laming, 1982). The lower parts of the succession were probably deposited as intermontane alluvial fans, combined with river sediments and braided streams. In all basins there is a general fining-upwards into siltstones and mudstones. Within the mid-Triassic of the Melville Basin, the Melville Halite is associated with anhydrite and carbonate. In the northern half of the basin some 1200 m of evaporites accumulated (Evans, 1990). Evaporites of this age are also known in the South Celtic Sea Basin (Naylor & Mounteney, 1975; Van Hoorn, 1987). *Figure 11.3* shows a generalized cross-section of the basins both north and south of the Cornubian Massif.

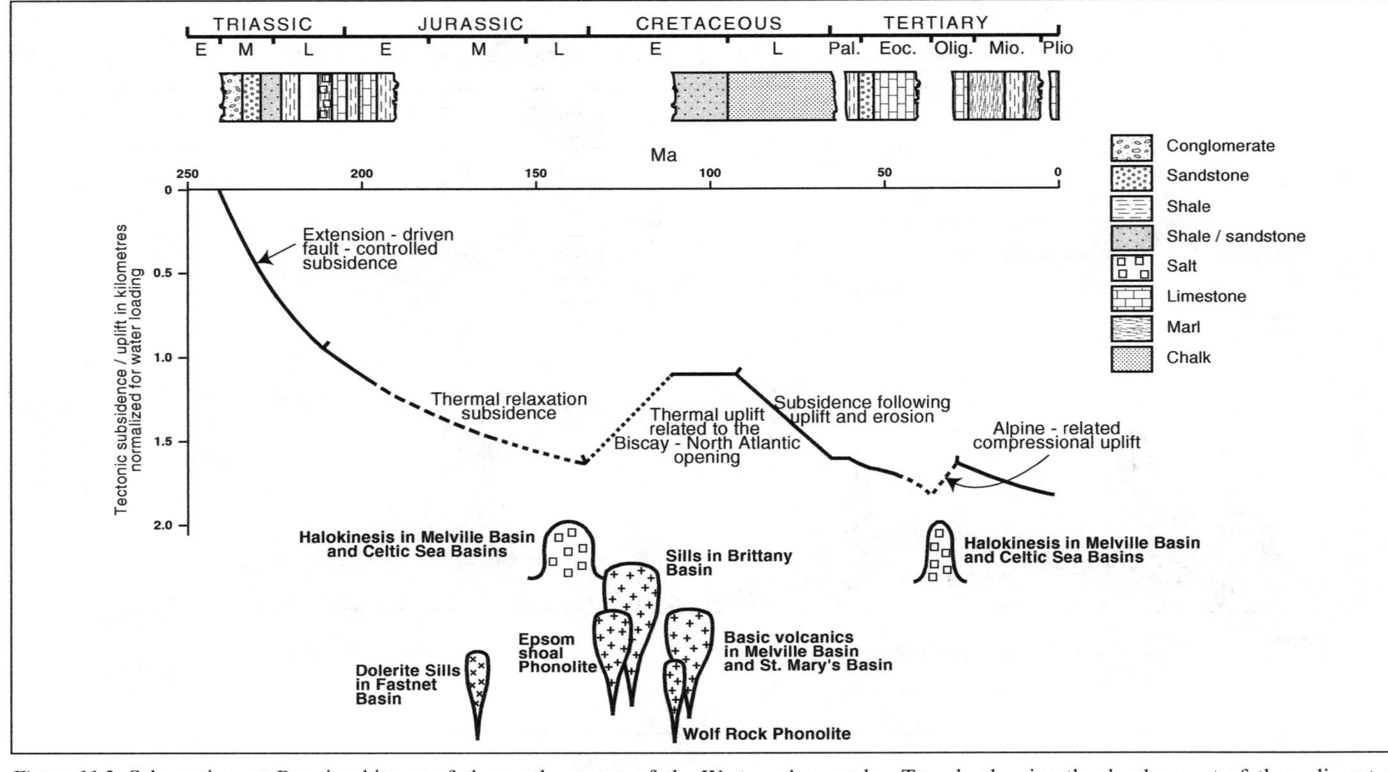

Figure 11.2. Schematic post-Permian history of the northern part of the Western Approaches Trough, showing the development of the sedimentary succession and periods of halokinesis and igneous activity (after Hillis, 1988; and Evans, 1990)

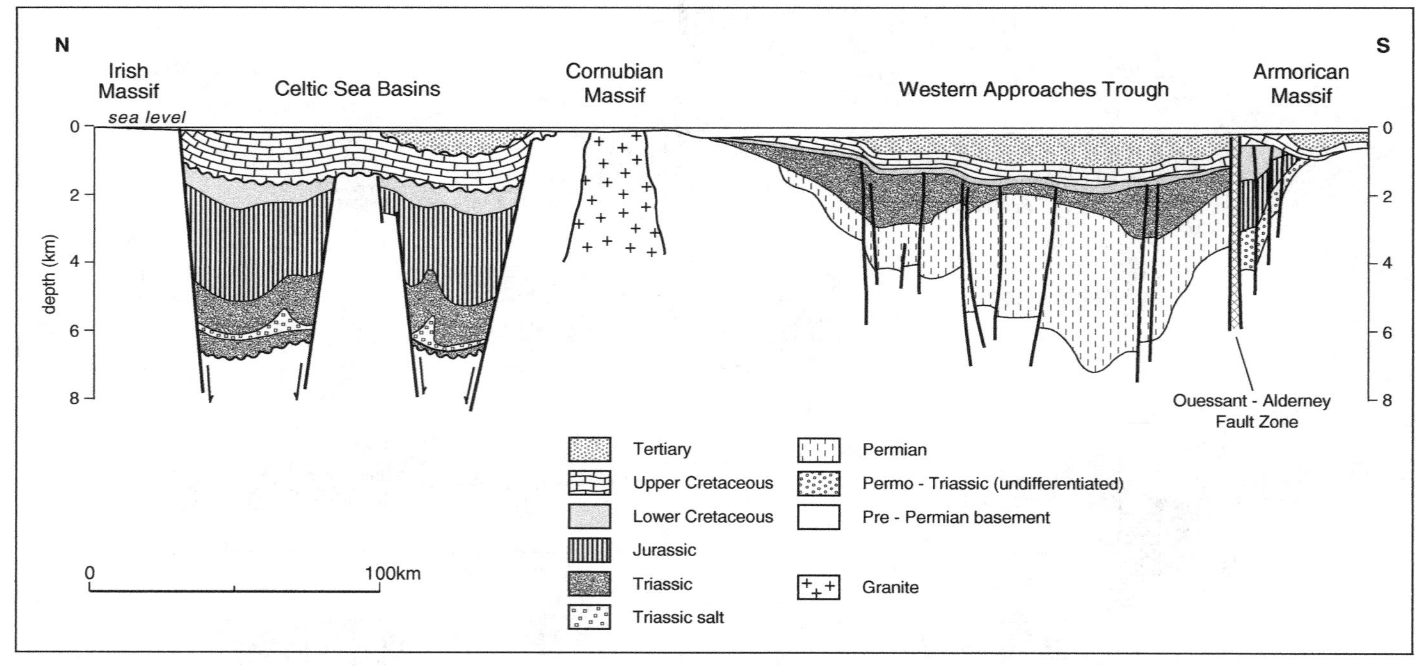

Figure 11.3. Schematic cross-section across the Celtic Sea basins, Cornwall and the Western Approaches Trough (after Hillis, 1988; and Naylor & Mounteney, 1975)

Towards the close of the Triassic, fault-controlled subsidence was replaced by thermal subsidence (*Figure 11.2*) and this, coupled with a eustatic rise in sea-level, led to a marine transgression over the Upper Triassic sediments. This transition to marine sediments is well preserved in many wells drilled in the Melville Basin, where the lowermost Jurassic sediments are grey claystones interbedded with dolomitic limestone.

Lower Jurassic sediments are preserved in the Melville, Brittany and Celtic Sea basins as well as in the eastern part of the Plymouth Bay Basin. In most areas the sediments are characteristically 'Liassic' in appearance, although those in the South Celtic Sea Basin became more carbonate-rich higher in the succession and were eventually succeeded by limestones (Millson, 1987). The Melville, Brittany and Plymouth Bay basins are characterized by a variety of dark grey mudstones, interbedded micritic limestones, fibrous calcite ('beef') and occasional carbonaceous mudstones. In places their geochemistry indicates that they have a moderate to good potential for hydrocarbon generation.

Middle Jurassic sediments are more limited in distribution as late Jurassic to early Cretaceous erosion has removed almost all the Middle and Upper Jurassic strata from the Melville, St Mary's and Plymouth Bay basins (Evans, 1990). This erosional episode was less marked in the Brittany Basin, where a more complete succession is reported. In the Celtic Sea basins (Millson, 1987) there is a more complete Middle and Upper Jurassic succession, typified by arenaceous sediments. Some of these may be marginal marine or fluvial in character.

The late Jurassic to early Cretaceous uplift (*Figure 11.2*) was not completely synchronous across the area, and was probably made up of a number of discrete pulses related to rifting in the Bay of Biscay. A considerable thickness of the Jurassic succession may have been removed, especially in the basins along the northern margin of the Western Approaches Trough.

In the Brittany and South West Channel basins, sedimentation was essentially continuous across the Jurassic/Cretaceous boundary. During the early Cretaceous a considerable thickness of shallow marine sediments were deposited. In the South West Channel Basin approximately 1,200 m of sediment is preserved. This consists of multicoloured sandy clays and medium–coarse-grained lignitic sands. There are some freshwater sediments reminiscent of the Purbeck facies of Dorset. In the North Celtic Sea Basin 'Wealden' sediments are also known (Ainsworth, 1987). Deep Sea Drilling Project (DSDP) Site 401 (*Figure 11.4*), on the continental slope, recorded reefal limestones and algal material of Jurassic to late Aptian age (Montadert *et al.*, 1979). This is a part of a reef system which pre-dated the break up of the continental margin (Masson & Roberts, 1981).

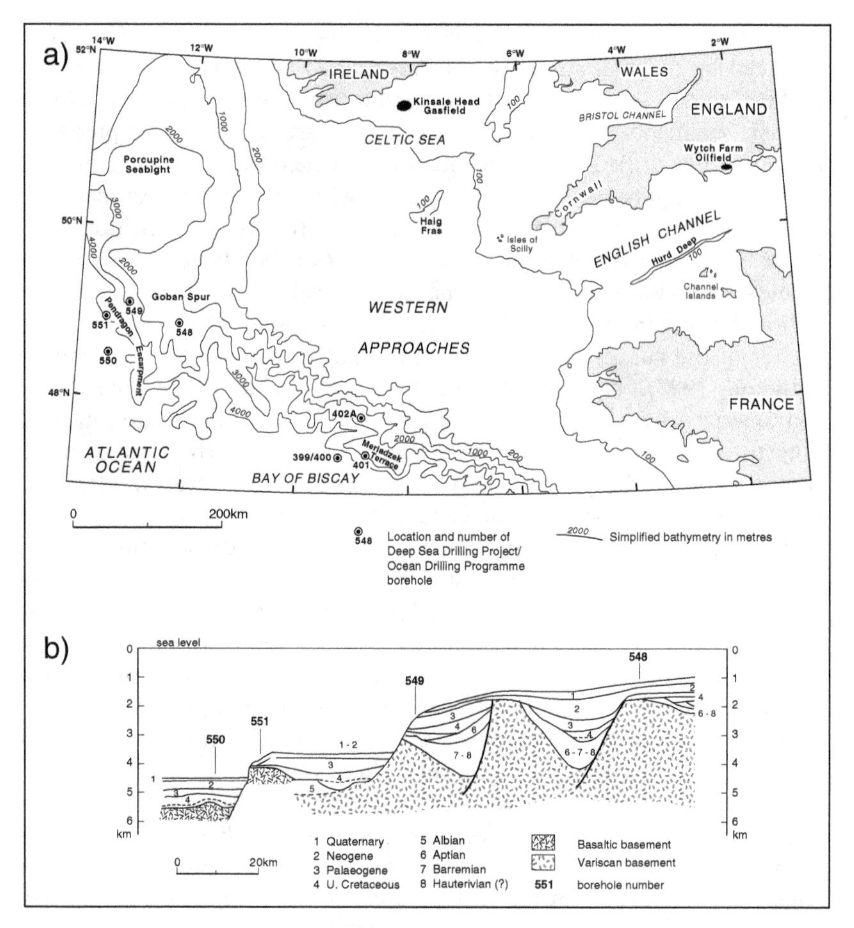

Figure 11.4. a) Outline map of the Western Approaches showing location of boreholes (399–402) drilled by DSDP Leg 48 and boreholes (548–551) drilled by IPOD Leg 80 (after Roberts *et al.*, 1977; and Evans, 1990). (b) Schematic profile across the Goban Spur showing the location of the boreholes and the geological succession (after Hart, 1985; and Hart & Duane, 1989)

Throughout the Hauterivian-Barremian (early Cretaceous) interval, either shallow-marine, brackish-water or fresh-water sedimentation continued, especially in the southern parts of the Western Approaches Trough. During this time, deposition also migrated northwards.

During the same interval a number of rapidly subsiding half-grabens formed under the present continental slope, as shown by International Programme for Ocean Drilling (IPOD) Site 549 (*Figure. 11.4*) which penetrated a succession of Barremian mudstones and claystones (Graciansky & Poag, 1985). The rifting on this margin was accompanied by isolated basaltic volcanism on the outer shelf (Evans, 1990; Chapman, 1989). Sills of possible Barremian age are also known from the Brittany Basin.

Isolated igneous intrusions (*Figure 11.2*) of comparable age are also known from the Western Approaches Trough. The Wolf Rock is a fine-grained, fluxioned, microporphyritic phonolite (Harrison *et al.*, 1977) and is reportedly 112±2 Ma (early Aptian) in age. The nearby Epson Shoal is a similar phonolite with an age of 130±6 Ma (Berriasian). Both these bodies appear to be volcanic necks and may be associated with movement along the Ouessant-Alderney Fault Zone.

The Aptian–Albian interval was characterized by transgressive deposits of mudstones, limestones, glauconitic sandstones and, finally, chalks. With continued sea-level rise during the Cenomanian (Hancock, 1989), chalk deposition gradually transgressed onto the basement massifs and eventually covered much of the Irish Massif, the Cornubian Massif (Hart, 1990) and the Cherbourg Peninsula. Much of the chalk succession comprises carbonate muds with cherts, although in places there are glauconitic limestones (Melville Basin), nodular chalks, carbonate banks and mounds (similar to those seen in Normandy) and, at the base of the succession, bioclastic limestones and glauconitic sandstones (Lott *et al.*, 1980). In many borehole successions the distinctive gamma-ray 'spike' of the Plenus Marls (uppermost Cenomanian) can be detected (Hart & Duane, 1989). In the IPOD boreholes on the Goban Spur (Hart & Ball, 1986; Hart & Duane, 1989) similar 'anoxic events' are recorded in the Santonian and the Upper Campanian. IPOD Site 550 records a series of possible Campanian to Maastrichtian debris flows and turbidites, all within a chalk succession. Hart & Duane (1989) compared this to redeposited chalks in the North Sea Basin.

Chalk sedimentation continued into the Danian over much of the continental shelf, but the sea-level fall during the Thanetian effectively terminated deposition of this facies.

Chapter Twelve

The Tertiary

Unlike Devon, no large basins containing Tertiary sediments exist in Cornwall. Apart from the small Dutson Basin near Launceston, occurrences of Tertiary rocks are restricted to isolated outliers that appear to be erosional relics of a widespread sedimentary sheet. The survival of these outliers may result from their position on or near prominent watersheds that originated in mid to early Tertiary times. There is also limited evidence of widespread Tertiary weathering conditions in Cornwall, which may be of Palaeocene age.

EOCENE

Dutson Basin

Before the Second World War, bricks for local use were made from clays worked from a small deposit [SX 344 862] northeast of Launceston. The extent of this Palaeogene deposit was mapped by Freshney *et al.* (1982) who reported (*Figure 12.1*) that these sediments occur in a fault-controlled basin near Dutson [SX 341 859]. The basin is comparable in style, if not in size, to those at Bovey Tracey and Petrockstow in Devon, and Stanley Banks, offshore in the Bristol Channel. These last three basins are aligned with the Sticklepath fault complex. The Dutson Basin lies in a NW–SE-trending strike-slip fault system (Tamar Fault Zone), parallel to the Sticklepath fault zone, which lies 45 km to the east. Such NW–SE fault systems are widespread in South West England, and the most significant movements on these took place at a late stage in the Variscan orogeny, immediately before the intrusion of the granites. Tertiary movement was demonstrated by Dearman (1963), although Turner (1984) suggested that there is little evidence for significant Tertiary movement on the strike-slip fault zones between Dartmoor and Bodmin Moor. Bristow & Robson (1994), when considering the Palaeogene basins of Devon, suggested that parts of the Sticklepath fault complex were 'locked' within the granite, and the Bovey Basin was initiated by a compensating flexure that translated the dextral movement south of the granite ridge into a dominantly vertical movement. They proposed a 'pull-push' model in which an extensional phase (transtension), related to the opening of the Atlantic in Eocene times, allowed deposition of most of the sediments, and was followed by compression (transpression). This latter phase, dated as Oligocene, would correspond to the N–S compressional regime of the later stage of the Alpine orogeny. As basins tend to form at right-stepping offsets in dextral strike-slip faults, it is conceivable that more Tertiary basins in Cornwall have been lost by erosion or await discovery.

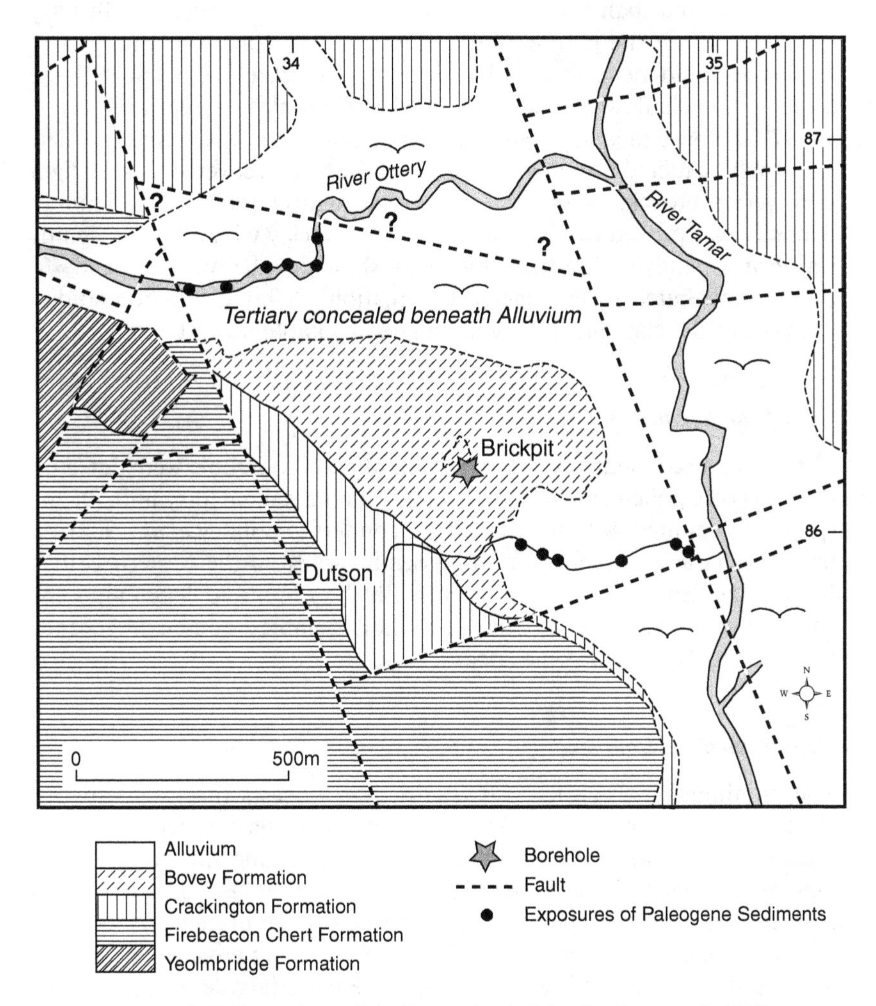

Figure 12.1. The geology of the Dutson Basin (after Freshney *et al.*, 1982)

Stratigraphy

The Dutson Basin occurs over an area of approximately 0.5 km^2, although its northern and eastern boundaries are relatively ill-defined. Freshney *et al.* (1982) reported lithological data from a series of isolated outcrops in the bed and banks of the River Ottery [SX 3371 8660–SX 3432 8681], near the supposed most northerly expression of the basin. The sediments in this area were referred to the Bovey Formation by the British Geological Survey (1994a). They consist of pale grey to brown and white plastic clays with some silts and sands. Abundant orange mottling is visible in the clays, and brownish-orange clay fragments occur in the soils of ploughed fields approaching the River Tamar to the east. A dense Mn–Fe-rich hardpan up to 0.3 m thick commonly overlies the Tertiary sediments. Since the nature of the outcrop is sporadic, no stratigraphic sequence could be deduced, but Freshney *et al.* (1982) reported the findings of a shallow borehole [SX 3444 8618] sunk to a depth of 9.35 m in 1977. Below 6 m, a grey sandy clay and clayey sand with minor rootlets and small, scattered fragments of black lignite were recovered. A soft, light grey, very sandy clay with scattered angular quartz grains up to 5 mm in diameter was proved to 2 m, overlying a brown clay with a bed of lignitic clay. Some bands within the brown clay show dips between 35° and 40° that are attributed to periglacial deformation.. X-ray diffraction studies have revealed a clay mineralogy dominated by kaolinite and illite.

Age of the Sediments

The Tertiary sediments from the Dutson brick pit borehole were dated on the basis of a pollen assemblage from the middle sedimentary unit. Of the specimens counted, 83% were ferns and conifers (Freshney *et al.*,1982), an unusual characteristic for Tertiary assemblages from the western British Isles. A late Eocene age is consistent with the Eocene to ?late Oligocene age for the basins along the Sticklepath fault complex, and an Oligocene–Miocene age for the St Agnes Beds.

Depositional Environment

The dominance of conifer and fern pollen suggests that these species must have constituted a significant part of the Tertiary vegetation about Dutson. From this, an environment of poor soils and little tree canopy may be envisaged. According to Freshney *et al.* (1982), the mean annual temperature was 15°C and the mean annual range of temperature was 16.5°C. The clay mineralogy could be explained by erosion and subsequent sedimentation of weathering-profile materials developed on the Upper Palaeozoic rocks during the Tertiary. When the mineralogical data, an assemblage of quartz, kaolinite, illite, mixed-layer minerals and smectite, is combined with the palynological data, an immature tropical and subtropical weathering environment is indicated.

It is envisaged that the area had a climate lying on the fringe of the tropical weathering zone, and a river, flowing either northwest or southeast, drained from the central upland spine of Cornwall and deposited its sediment in the Dutson Basin. As a result of the close control that the NW–SE faults had on the palaeo-drainage patterns, it is highly unlikely that this river was a tributary also depositing in the Bovey or Petrockstow basins in Devon.

OLIGOCENE AND MIOCENE

St Agnes Outlier

The sands and clays to the north and west of St Agnes Beacon (*Figure 12.2*) have been recognized as quite distinct in Cornish geology since Pryce (1778) reported them as the St Agnes Beds. They were first assigned to the Tertiary by De la Beche (1839). Their origins have been problematic, with explanations ranging from marine through fluviatile to aeolian. The marine origin held sway for many years because of the discovery of an apparently shingle-worn cliff face, a postulated sea stack and caves, in mine workings east of the Beacon in 1875. Later a fluviatile, and subsequently an aeolian and colluvial origin, has been favoured.

The St Agnes Formation outcrops over an area of 1.6 km^2 to the north and west of the Beacon. The maximum thickness is about 10 m. Sections of the deposits may be examined in the New Downs Sandpits [SW 706 509], which are also referred to as 'Dobles Sand Pits'. The sections observable depend on the state of the sand extraction operation, but many of the exposures described earlier in this century are now collapsed and overgrown.

Stratigraphy

In the vicinity of New Downs Sandpits, an upper sand horizon (Beacon Member) is separated from a lower sand (Doble Member) by a clay (New Downs Member). The grouping of these members into the St Agnes Formation has been suggested by Walsh *et al.* (1987).

The unconformity beneath the St Agnes Formation surface is occasionally visible in the lower levels of New Downs Pits and in the openworks of Trevaunance Mine [SW 7121 5128]. The surface is roughly subhorizontal and was relatively deeply weathered before deposition of the Doble Member. Reid & Scrivenor (1906) observed a bed of pebbles, often containing 'tinstone', at the base of the St Agnes deposits. In recent years,

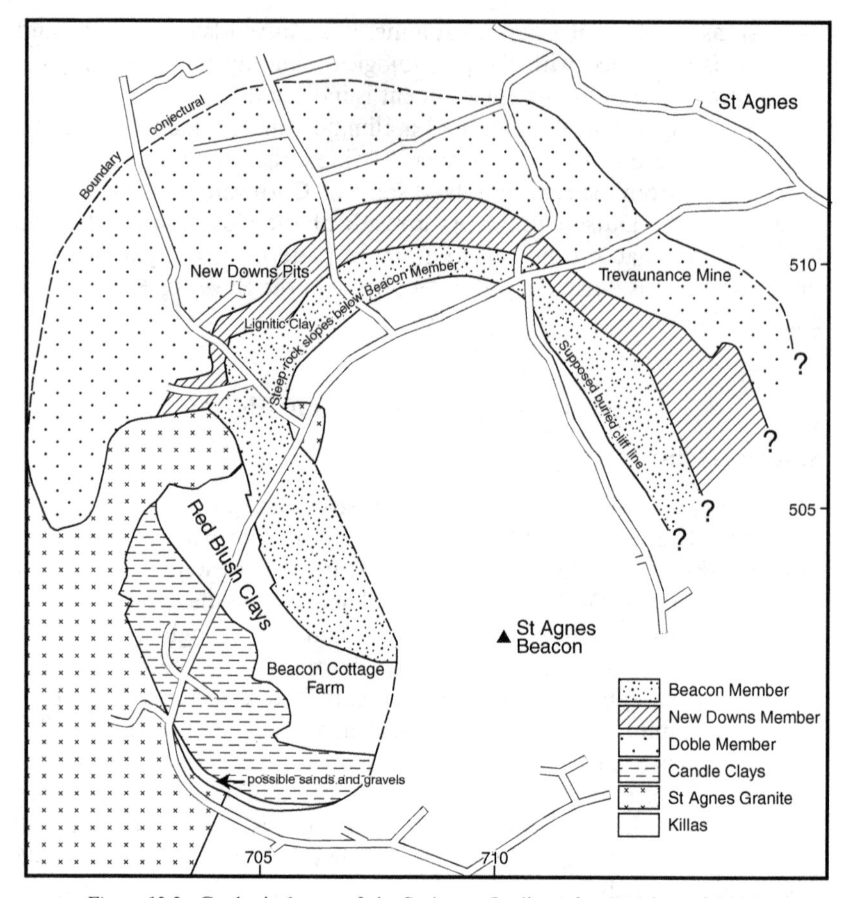

Figure 12.2. Geological map of the St Agnes Outlier (after Walsh *et al*, 1987)

where the base has been exposed in the New Downs Pits, the interface has been observed to be characterized by a non-pebbly, Fe-cemented sand, or 'iron pan'. This material varies from black, dark blue to brown or yellow. Occasionally, large, isolated angular pieces of stained Devonian killas occur in the iron pan. A distinctive feature of the pan is the occurrence of concentrations of tubular structures up to 2 m long and more than 0.3 m across. These pipe-like features have their long axes sub-horizontal and orientated NE–SW. They often have unconsolidated yellow sand in their cores.

Doble Member. These yellow, or buff, fine-grained silty sands are well sorted and about 6 m thick in their type section at Doble Pit. The colour becomes paler towards the top of the member and the contact with the overlying clay is gradational. The source of these sands is questionable since the palaeocurrent evidence exposed in New Downs Pit suggests a provenance to the northwest whereas foreset bedding evidence from New Pit, 250 m to the northeast, indicates palaeocurrent flow from the southeast. The heavy mineral assemblage is dominated by species which could be locally derived: tourmaline, haematite, cassiterite and topaz. Scanning Electron Microscope studies of sand grains from the Doble Member strongly suggest aeolian transport, but post-depositional chemical surface alteration of the grains may obscure evidence of other transport mechanisms (Walsh *et al.*, 1987). Although no conglomeratic basal layer has been recognized in the St Agnes Formation in recent years, occasionally a band of pebbles and cobbles occurs about 2 m above the base of the Doble Member. These pebbles and cobbles are dominantly locally derived and their long axes show a preferred alignment NW–SE, which tends to confirm the provenance directions mentioned above.

New Downs Member. This member, named after the New Downs Sandpits [SW 706 509] by Walsh *et al.* (1987), is typically a pale grey, almost structureless, poorly sorted, silt-dominated clay. At the type locality it is about 3.5 m thick and is predominantly micaceous, with about 20% each of kaolinite and quartz. Some sand is present in the clays and isolated pebbles of vein quartz occur sporadically towards the top. The base of this member is transitional from the Doble Member below, but the top is sharply demarcated and may represent a minor erosion surface. Some evidence for the erosive nature of the upper contact is gained from slight mottling of the upper levels of the clay. Faint stratification in the silty clay horizons dips northwards at less than 10°.

Since Hawkins (1832) there have been many reports of carbonaceous material in the New Downs Pits. In recent years dark grey and brownish lignitic clays have been intermittently exposed about 1.3 m below the overlying Beacon Member.

Beacon Member. The highest 3 m of sands in the New Downs Pits was assigned to this member by Walsh *et al.* (1987). It was named after St Agnes Beacon [SW 710 502]. The Beacon Member is a very well-sorted, cross-bedded fine-grained to medium-grained yellow and orange sand. Some weak cementation by Fe compounds occurs, as do irregular lenses of greenish silty sand. Earlier, Macfadyan (1970), had recorded up to 7.3 m of 'brown sand' (which undoubtedly would now be classified as Beacon Member sediments) in the New Downs Pit. The surface textures, and the rounded or subrounded nature of the grains, clearly indicate aeolian transport.

Age of the St Agnes Formation

The absence of animal fossils and suitable minerals for radiometric dating has caused considerable speculation about the age of the St Agnes Formation. After they were first classified as Tertiary by De la Beche in 1839, Reid (1890) assigned them to the 'Older Pliocene' times; but for most of this century they have been regarded as an equivalent of the St Erth Beds, sitting on a Pliocene marine-cut platform. However, work by Walsh *et al.* (1987), using pollen samples collected from the carbonaceous clays of the New Downs Member, have established a Neogene age, most likely Miocene. The formation is now recognized to be quite different from the St Erth Formation that is of late Pliocene age.

Depositional Environment

Due to the lack of animal fossils, the interpretation of the conditions of deposition for the St Agnes Formation has to be based upon palynological and sedimentological evidence. The Neogene pollen groups suggest a Mediterranean climate, with a typical European vegetative cover of coniferous forest, mixed woodland, and scrub. The pollen evidence also suggests that marine influences were lacking. An inland environment where winter temperatures fell below freezing is envisaged. The sand members are well sorted and show characteristics typical of aeolian sediments, whereas the clay member shows features that may be seen in modern subtropical colluvial environments. The depositional environment is therefore envisaged as an inland dune belt interrupted by a colluvial phase. The infrequent pebbly horizons are regarded as minor fluviatile incursions into the aeolian dune belt; the rare angular killas fragments could indicate minor wadi activity.

Beacon Cottage Farm Outlier

The British Geological Survey map of Newquay (Sheet 346) shows an extensive crescent-shaped outcrop of Tertiary sediments occupying all but the eastern side of St Agnes Beacon. This area encompasses the New Downs Sand Pits deposits (St Agnes Formation) and deposits around Beacon Cottage. Until 1987 most geologists assumed that the latter clays were not only contiguous with, but also coeval with those to the north of the Beacon, in the New Downs Sandpits. In 1987, however, Walsh *et al.* demonstrated that the two deposits were of different ages. Unfortunately, no exposures of the Beacon Cottage sediments have been visible for many years. The clays were worked for a long time as 'candle clay', a particularly unctuous clay sold to the mines of Cornwall for fixing candles to the walls of the underground workings and to the brims of the miner's hats.

It appears that this trade died out about the beginning of the Second World War.

Stratigraphy and Age of the Beacon Cottage Farm Sediments

As a result of the lack of exposures, the stratigraphy of the deposits in the Beacon Cottage Farm outlier is not well established. Unlike the St Agnes Formation, the Beacon Cottage sediments are dominated by ‘candle clay’, with thin basal sandy beds, which are often pebbly, and with a little sand overlying the clays. Augering by Jowsey *et al.* (1988) indicated a maximum thickness of 8.4 m, and they suggested that either the weathered Palaeozoic floor has palaeoslopes locally exceeding 45°, or that several post mid-Oligocene faults have affected the outlier.

Originally, it was thought that these sediments were probably Pliocene and rested on the ‘Pliocene (marine-cut) Platform’. However, a mid-Oligocene age has been established (Atkinson *et al.*, 1975) on the basis of the pollen. Although The Beacon Cottage floras show major differences from those of the Bovey and Petrockstow basins of Devon, they are the Cornish equivalent of the Bovey Formation.

Depositional Environment

As distinct from the New Downs Sandpits flora, the Oligocene Beacon Cottage floral assemblage contains palm pollen, indicating frost-free winters. It is suggested that, like the conditions at Bovey Tracey during this time, the climate was cooling with a lowland mean annual temperature about 12°C. With no marine fossils preserved, and with significant quantities of leaf cuticles and wood cells, a low-energy environment of deposition is indicated, possibly an inland lake.

Other Possible Oligocene and Miocene Outliers

The occurrence of gravels and clays in other outliers at about the same elevation as that at St Agnes, led Walsh *et al.* (1987) to suggest that they sit on a contemporaneous erosion surface (‘Reskajeage Surface’), even if their deposition may not have been synchronous. The most extensive outlier occurs on Crousa Common [SW 7730 2001], where the gravelly clay outcrop is roughly ovate, about 2 km long and 0.5 km wide, overlying the Gabbro-Troctolite of the Lizard Series. The clay appears to have been largely derived from post-depositional rotting of gabbro pebbles, whereas the quartz pebbles that are preserved may have had their source in the Gramscatho conglomerates found about 4 km to the north.

Another Tertiary outlier occurs near Polcrebo Downs [SW 6530 2201]. Although no permanent exposures of this deposit exist, augering and trial pits have revealed a greyish-white pebbly, clayey, coarse sand up to a maximum thickness of 3.3 m (Walsh *et al.*, 1987). As at Crousa, considerable post-depositional rotting of the sediments has taken place, which may have resulted from extreme Tertiary weathering.

During the 1980s, extensions to the cricket ground at Redruth [SW 696 409] revealed sands and clays at an elevation comparable to the St Agnes outlier, although the exact age of these sediments has yet to be determined.

PLIOCENE

St Erth Beds

Probably the best-known Tertiary deposits in Cornwall occur near the village of St Erth [SW 557 353]. Unlike other Tertiary deposits in Cornwall, the St Erth Beds have a prolific and diverse fauna. However, despite the variety of molluscs, ostracods, and benthonic and planktonic foraminifera, the exact age of the beds for a long time remained in dispute.

Stratigraphy and Age

Exposures in the pits at St Erth are no longer visible, but their sediments were described by Mitchell *et al.* (1973). Despite the amount of back-filling that had occurred since the pits were last worked, Mitchell managed, by reference to earlier workers, to decipher a limited stratigraphy. Even within the poor exposures that were available during the late 1960s there was considerable lateral and vertical variation. To generalize, for the northern part of the Vicarage Pit (*Figure 12.3*), below about 2 m of head, there is an intermittent coarse sand layer up to approximately 0.5 m thick. Below this there is up to 4 m of highly fossiliferous leached marine clay in which lenses of calcareous brown clay and blue clay occur. This clay overlies a fine sand which has an abnormally irregular upper contact with the clay, and in its upper 30 cm portion contains small rounded pebbles. Below this, the fine sand is about 4.5 m thick (Mitchell *et al.*, 1973). Earlier workers observed up to 6 m of this sand with coarser material in its lower layers. In the southern part of the Vicarage Pit, the clay is absent, and head rests directly on the fine sand. This lower sand is variegated laterally with yellow and orange-reds predominating.

A Pleistocene age was originally suggested for the St Erth Beds, but subsequently a Pliocene age, based largely upon benthonic foraminifera, was proposed (Mitchell *et al.*, 1973). More recently, Jenkins (1982), using planktonic foraminifera which are relatively rare in the St Erth sediment, has confirmed a late Pliocene age.

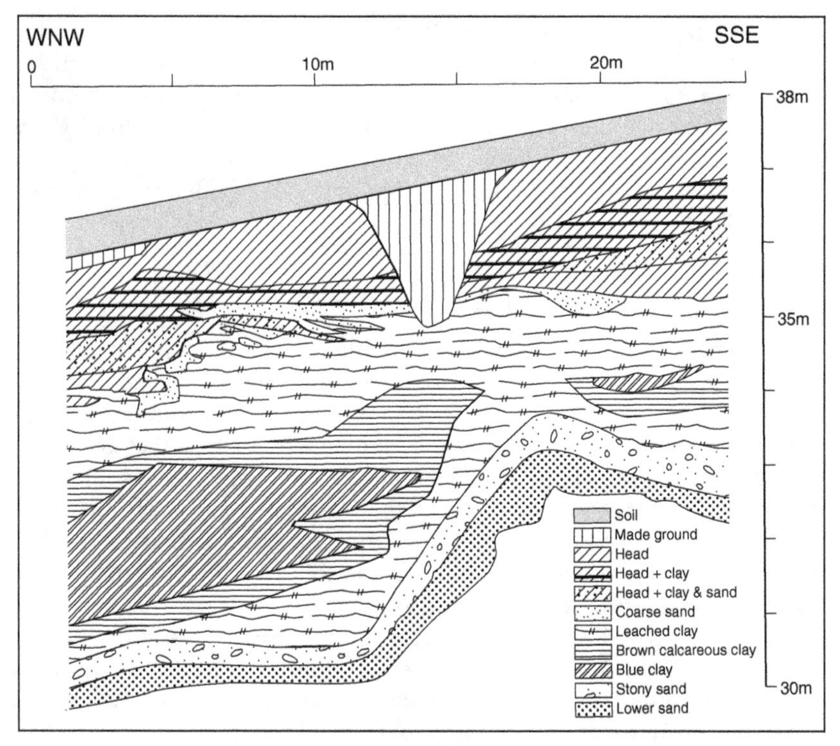

Figure 12.3. Cross section [SW 5570 3536–SW 5574 3535] of the St Erth Beds in the northern part of the Vicarage Pit (after Mitchell *et al.*,1973)

Depositional Environment

The St Erth Beds rest on what appears to have been a subaerially eroded rock surface. On this rock surface a thick layer of sand was deposited, the coarser sand being of beach origin, and the slightly finer sands being principally of dune origin. This suggests that there may have been a coastal set of dunes which were engulfed during rising sea level. Subsequently, deposition took place in the shallow waters of clayey lagoons, which eventually gave way to deeper water and the deposition of the highly fossiliferous clay. Mitchell *et al.* (1973), suggested a water depth of 10 m during the deposition of the clay, with an eustatic rise in sea-level before the cessation of clay deposition. After the deposition of a considerable thickness of clay, the direction of sea-level movement was reversed and a period of erosion ensued, removing an undefined amount of clay and any other overlying Tertiary sediments.

The palaeontological evidence suggests inner-shelf conditions and that the water depth at the time of deposition was less than 100 m, possibly somewhere between 60 m and 100 m. The relative paucity of planktonic foraminifera, and the low species diversity they exhibit, may indicate that the shallow water in which the St Erth Beds were laid down was on the fringe of the geographic range of the planktonic species. If conditions were similar to modern environments, water temperatures of 10–18°C are suggested, but the presence of one particular planktonic species, usually associated with sub-tropical conditions, provides evidence of temperatures slightly higher than 18°C.

TERTIARY WEATHERING PHENOMENA

The effects of pre-Oligocene deep weathering have been noted when considering the St Agnes Beds. To date, in Cornwall, it has not been possible to obtain sufficient evidence to produce a detailed description of the development of a Tertiary weathering mantle, such as that according to Devon by Isaac (1983). Undoubtedly, there would have been quite widespread tropical weathering over the Reskajeage surface prior to the deposition of the St Agnes Beds, and further east pre-Eocene tropical weathering produced the material that was subsequently deposited in the Dutson Basin. Other evidence available in Cornwall that demonstrates prolonged weathering during the Tertiary, is the occurrence of silcrete-like material analogous to the silcretes developed on Palaeozoic rocks in Devon. The latter, it has been suggested, were formed in the Palaeocene (Bristow, 1993), confirming Isaac's (1983) view that they were older than the Upper Eocene–Oligocene Bovey Formation. More evidence for a period of silcrete formation in the early Tertiary, comes from near Shebbear in Devon (Freshney *et al.*, 1979). Bristow (pers. comm.) has suggested that surface-silicified material such as this may be quite extensive in Cornwall, but has yet to be recognized outside a few localities. He cited St Breock Downs [SW 966 683], southwest of Wadebridge, as an area littered with large boulders, up to 2 m in size, of hard silicified sandstone riddled with irregular veins of fine-grained quartz. These blocks are unlike the softer underlying Lower Devonian Staddon Grits. Similar, large boulders with silcrete-like material also occur at Carnon Downs near Truro [SW 800 405].

Silcrete development suggests that during the Palaeocene, deep weathering occurred in Cornwall under a warm, but not unduly wet, climate. Silica, released by weathering of aluminium silicates, migrated in soil waters and was precipitated at suitable loci in the weathering mantle to form silcretes.

Chapter Thirteen

The Quaternary

The Quaternary Period is characterized by more extreme climatic fluctuations and lower mean temperatures than the Tertiary. Numerous cold and temperate stages have been recognized in Britain (*Table 13.1*), where the last three major cold stages produced ice sheets. Whether any of these ice sheets reached Cornwall is a matter of debate, but there is no doubt that the southern margin of at least the late Devensian ice sheet, about 18 ka BP (Before Present, which is conventionally taken as AD 1950), was close enough (*Figure 13.1*) to have produced a periglacial environment in Cornwall. Such conditions almost obliterated the evidence of earlier stages.

GLACIAL PHENOMENA

Glacial Deposits on the Mainland

Glacial deposits have an extremely limited distribution in Cornwall, and their recognition and interpretation have been controversial. One of the earliest references to glacial deposits was by Whitley (1882), who described a 'tough boulder clay, with marine shells' at St Erth [SW 556 352], but these beds have since been shown to be marine and of Pliocene age (Mitchell *et al.*, 1973).

Another deposit of possible glacial origin is the Trebetherick Boulder Gravel at Trebetherick Point [SW 925 780] and Tregunna [SW 967 739]. At Trebetherick it lies within a sequence resting on two well-developed rock platforms. At the base, a raised beach deposit partly enclosing some large (up to 2 m diameter) greenstone boulders is overlain by cross-bedded sand and sandrock. The overlying boulder gravel is up to 4 m thick, and is well bedded in its lower part. It consists of coarse gravel and well-rounded boulders in a matrix of comminuted slate and silty gravel that is reminiscent of the periglacial head that covers so much of Cornwall. No unambiguously far-travelled erratics have been described.

Table 13.1. STAGES OF THE BRITISH QUATERNARY

Series		Stage	Climate	Age (Ma BP)*	NW Europe
Holocene		Flandrian	t	0.01	Holocene
		Devensian	c	0.115	Weichselian
	Upper	Ipswichian	t		Eemian
		Wolstonian	c		Saalian
		Hoxnian	t	0.3	Holsteinian
	Middle	Anglian	c		Elsterian
		Cromerian	t	0.5	Cromerian
Pleistocene		Beestonian	c		
		Pastonian	t		
		Pre-Pastonian	c		
		Bramertonian	t		
	Lower	Baventian	c		
		Antian	t		
		Thurnian	c		
		Ludhamian	t		
		Pre-Ludhamian	t	2	

*Approximate age of commencement of Stage. Abbreviations: c, cold; t, temperate

The boulder gravel is overlain by head that is locally covered by blown sand.

It is perhaps not surprising that the curious assemblage of materials in the boulder gravel should have evoked such diverse interpretations. Reid *et al.* (1910) interpreted it as either ice-rafted or fluvial; Arkell (1943) as a beach deposit with an admixture of head; Mitchell (1960, 1972) and Clarke (1965, 1969, 1973) as a glacial till; Stephens variously as head (1961), glaciofluvial gravel (1970) and weathered till (1973); and Scourse (1991) as a river-ice deposit of probable late Devensian age from within the catchment of the Camel. Advocates of a glacial origin for the Trebetherick Boulder Gravel attributed it to an ice sheet whose southern margin lay close to, and in places against, the northern coast of Cornwall. However, the true nature of this deposit is still not resolved.

Glacial Deposits on the Isles of Scilly

Erratic-bearing deposits on the Isles of Scilly were interpreted as of glacial origin by Whitley (1882) and Barrow (1906). Mitchell & Orme (1967) recognized both till and outwash facies, and from their distribution identified an ice limit through the northern islands (*Figure 13.1*).

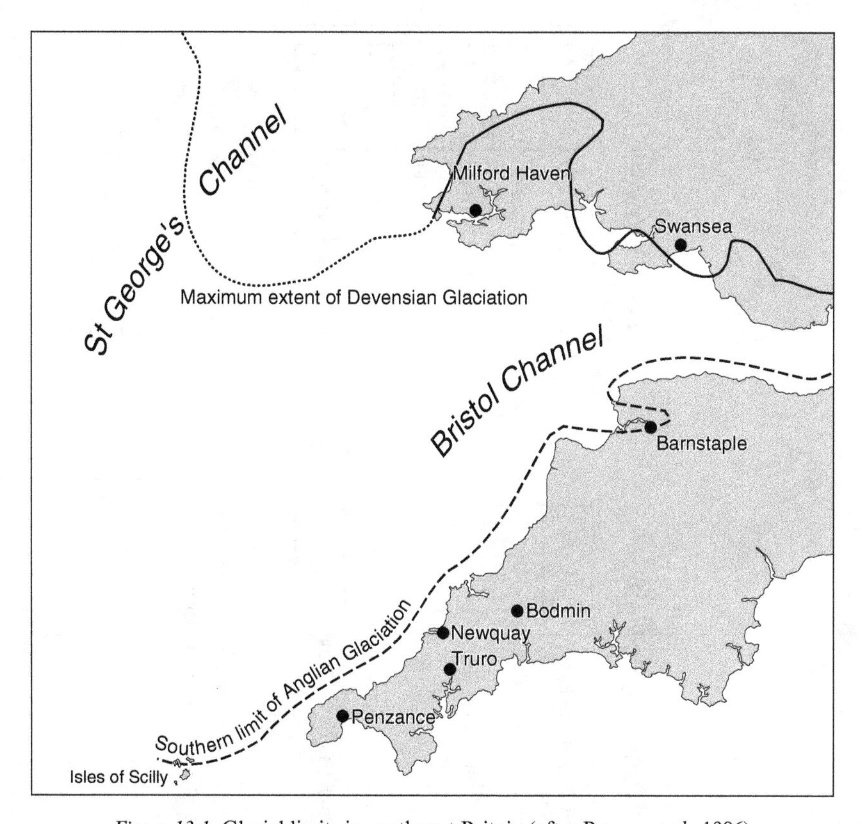

Figure 13.1. Glacial limits in southwest Britain (after Bowen *et al.*, 1986)

Scourse (1991) recognized eight members grouped into two interfingering formations (*Figure 13.2*). The glacial till (Scilly Till), within the Bread and Cheese Formation, is considered to be *in situ*. While the overlying Tregarthen Gravel and Hell Bay Gravel are interpreted as glaciofluvial outwash and solifluction deposits respectively, both carry erratic assemblages consistent with the till.

The best-studied exposure of the Scilly Till is at Bread and Cheese Cove, on St Martin's [SV 940 159]. It is a yellowish-brown, crudely stratified stony clay, with freshly striated and faceted clasts of varied lithologies, including flint, greywacke, quartzite, red sandstone, schist and local granite. Abundant siliceous sponge spicules are present, but no calcareous fossils. A derivation from the offshore area north of the Isles of Scilly has been suggested. This is consistent with two clast macrofabric analyses (Scourse, 1991) that are typical for lodgement till, and suggest glacier flow from the northwest. A third analysis resembles slumped till.

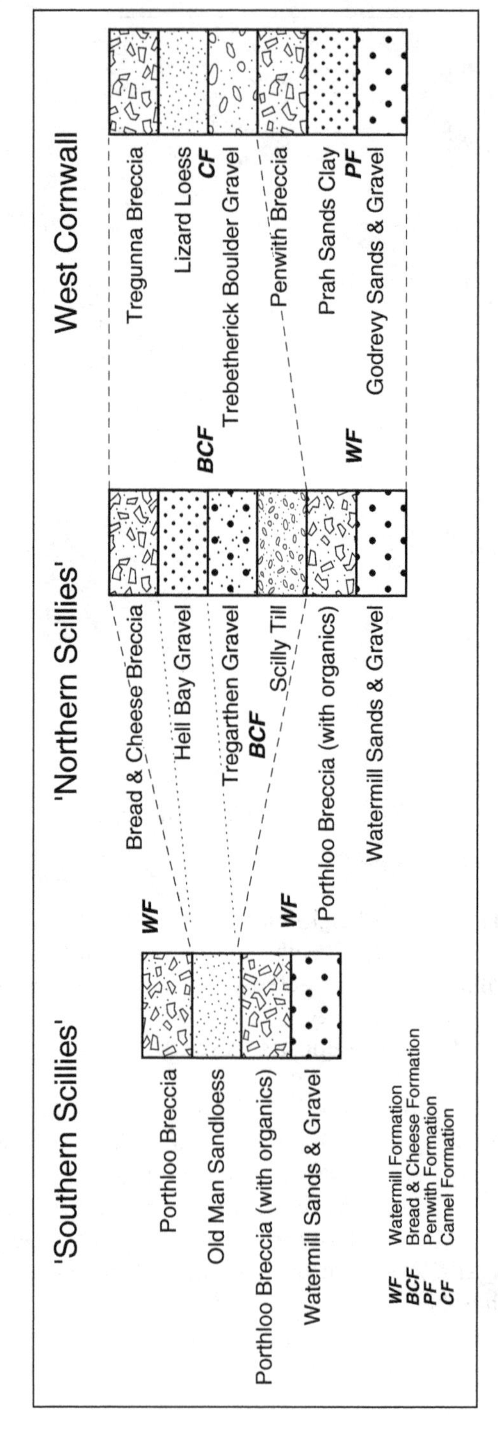

Figure 13.2. Successions for the Isles of Scilly and west Cornwall (after Scourse, 1987; 1991)

The Scilly Till also forms the core of both the Pernagie Bar, a marine tombolo that extends from Pernagie [SV 920 171] to Pernagie Isle [SV 918 175], and the White Island Bar which links St Martin's with White Island [SV 924 170–SV 925 171]. These bars may thus be partly morainic (Scourse, 1991). Other till occurrences reported by Mitchell & Orme (1967) were considered by Scourse (1991) to be Hell Bay Gravel. This horizon is probably a soliflual mixture of glacial and glaciofluvial material with Old Man Sandloess.

Erratics

Erratic blocks, inexplicable in terms of local or regional transport by agencies such as fluvial transport, mass movement or longshore drift (or some combination of these), are scattered around the coasts of South West England. The most celebrated Cornish example is the large erratic boulder of garnetiferous microcline gneiss (the 'Giants Rock') on the shore platform near Porthleven [SW 623 257] described by Flett & Hill (1912) and Stephens & Synge (1966). Although now wholly exposed, both platform and boulder were probably once covered by head, as in north Devon where exhumation is incomplete (Cullingford, 1982). The Porthleven erratic has been ascribed to both glacial (Mitchell, 1960) and iceberg transport (Stephens & Synge, 1966; Mitchell, 1972).

PERIGLACIAL PHENOMENA

Periglacial Landforms

The principal legacy of past periglacial conditions is the *slope form.* In the Tamar drainage basin, the commonest slope forms, at all altitudes and on all rock types, are smooth, convex upper slopes and concave or rectilinear lower slopes. Such slope forms, with a mostly periglacial debris mantle that normally thickens in a downslope direction, are well developed in the country between Bathpool [SX 285 747] and Trekenner [SX 343 784], and are characteristic of many present-day periglacial areas. Other relict periglacial landforms commonly associated with these slopes, especially in the upper parts of minor tributary valleys, are *debris aprons* and *terraces.* Examples of both can be seen in the upper Lynher valley between Trekernell Farm [SX 254 791] and Trenilk [SX 226 799].

Various *nivational forms* add to the variety of the periglacial legacy.The term *nivation* embraces a suite of erosional processes that act on susceptible lithologies beneath and near snowbanks, and include frost shattering, gelifluction and slopewash. Among the commonest resulting landforms are the enlarged rounded and/or bowl-shaped, sometimes dry valley heads whose dimensions are more in keeping with a nivational rather than a purely fluvial origin. Typical examples are seen near Trefursdon [SX 278 777] and Bray Shop [SX 331 749]. Even more common are the much smaller depressions or furrows, often bowl-shaped, known as *dells*, which are usually regarded as of niveofluvial origin. Dells are well seen east and southeast of Coad's Green [SX 295 767], including one that is 40–50 m across, 3–6 m deep, and has an undulating floor. Other nivational landforms present in Cornwall, rarer than those just described, include *cryoplanation terraces*, which are erosional bedrock benches found at the foot of frost-riven scarps or around tors. They often occur in staircases, forming *benched hillsides* such as those on metamorphic bedrock below Kit Hill and Hingston Down [SX 374 714–SX 410 714], where the benches are up to 600 m long and 100 m wide, and the risers up to 12–15 m high. The benches below Notter Tor [SX 271 737] are the only such features in the eastern Bodmin Moor area. They are much more common in west Cornwall where, according to Scourse (1987), almost all flat surfaces, apart from the Lizard and north Penwith platforms, can be interpreted as either solifluction or cryoplanation terraces.

Tors are a controversial legacy of Quaternary environmental changes, especially in the case of the granite tors of Dartmoor and Bodmin Moor. Linton (1955) advocated a two-stage origin involving warm-climate weathering of the granite and cold-climate stripping of the weathered material to expose the tors. Palmer & Neilson (1962) argued for a single-stage origin involving frost shattering and riving of exposed bedrock (*gelifraction*) combined with cold-climate downslope transfer (*gelifluction*) of the shattered debris. Gerrard (1978) demonstrated the importance of variations in joint-spacing in determining the distribution of the different types of tor, whilst Scourse (1987) emphasized the closeness of the association between the granite tors of the Isles of Scilly and the soliflucted debris whose removal exposed them. Scourse noted that rounded tors are confined to the northern islands, thought on other grounds to have been overridden by glacier ice. Away from the granites, two quartzite tors which protrude from the coastal slopes near Dodman Point [SW 987 407; SX 018 425] have been described by Bird (1963). It is probable that most Cornish tors are largely relict periglacial forms. However, tors in general are simply residual features that can be produced in a variety of morphoclimatic contexts.

Periglacial Deposits

The term *head,* first used in a geological context by De la Beche (1839), has been applied widely to periglacial slope deposits in southern England, although its original usage was non-genetic. Scourse preferred the term *breccia* as a non-genetic reference for all head-type deposits, but this has not yet gained wide acceptance. Except when referring to Scourse's lithostratigraphic units, *head* will be used. Head is very common throughout Cornwall, and its thickness varies greatly, up to a maximum of some 30 m. It was deposited under periglacial conditions wherever the angle of slope was sufficient to allow gelifluction and frost-creep to occur. In present-day periglacial environments the upper 0.5–1.5 m of water-saturated material (the *active layer*) moves downslope in spring and summer at typical rates of 1–5 cm per year. The limiting gradient may be as low as 1°.

The deposits resulting from such processes vary in nature with the bedrock, the depth and character of the previous regolith, and the pre-existing form of the land. On the Isles of Scilly and in west Cornwall Scourse (1987) emphasized the critical importance of bedrock type. The matrix of the stony Prah Sands Clay (*Figure 13.2*), for example, is the weathering product of felsite elvan dyke and associated metasediments. Heads derived from granitic and other massive bedrocks (such as gabbro and serpentine), including the Porthloo and Bread and Cheese Breccias, and some variants of the Penwith Breccia, are mostly very poorly sorted, with a wide range of clast sizes, a high matrix/clast ratio and a coarse granular matrix. In contrast, heads derived from slates, including the Tregunna Breccia and other variants of the Penwith Breccia, are usually clast-supported, with a very low matrix/clast ratio and a fairly well-sorted, silty matrix. These contrasting sedimentary features were produced by differing bedrock responses to gelifraction and frost-riving. Thus granites broke into large angular blocks and 'growan' (coarse sand and gravels), and the fissile slates into discoid clasts and silt. In later responses to the periglacial processes, the slaty heads proved frost susceptible and conducive to the development of ground ice bodies. In contrast, the granitic heads favoured the development of pore ice. This explains the comparative abundance of *periglacial structures* in slaty heads, where they are emphasized by the linear clast shape and dearth of matrix, and their rarity in granitic heads.

Frost-shattered bedrock to a depth of 1–2 m is ubiquitous throughout the Tamar drainage basin. The cover of head is near-continuous, except on some of the steepest slopes (where coarse scree deposits occur in a few places), some planed interfluves, and the flat floors of the main river valleys. The thickness of head is greatest on the lower slopes, and in a few localities bedded and laminated periglacial slopewash deposits form part of the head. Niveofluvial and niveosolifluxion deposits abound on the valley floors, and cold-climate fluvial sediments occur on some river terrace remnants, including those exposed near Latchley [SX 403 736]. Exposures in some fan remnants, for example near Lawhitton [SW 363 826], show a mixture of slopewash and niveofluvial debris.

Wind-blown silt, or *loess*, common in modern periglacial environments, has been identified in west Cornwall and the Isles of Scilly (Catt &

Staines, 1982; Scourse, 1987, 1991), The Old Man Sandloess and Lizard Loess (*Figure 13.2*), both mixed with non-loessic sands, have been dated by thermoluminescence to 18 600±3 700 and 15 900±3 200 BP, respectively (Wintle, 1981).

Periglacial Structures

Vertical stones are caused by differential frost heave rotating and tilting stones until they stand on end. They are common within the upper 0.2–0.4 m of head in the Tamar Basin, especially at higher altitudes. Good examples are exposed near Upton [SX 285 722] and at Landue Bridge [SX 347 797].

Involutions are regularly spaced deformations caused by cryostatic pressures in unfrozen material between a downward migrating freezing layer and perennially frozen ground beneath. They can also result from non-periglacial differential pore-water pressures when ice-rich sediments thaw. The term *cryoturbation structure* applies to relatively amorphous periglacial disturbances. Both types are common, especially in slaty heads, with good examples at Towan Beach [SW 873 332], described by James (1981), and Godrevy [SW 583 426], described by Stephens & Synge (1966).

Ice-wedge casts record the former presence of frost cracks or fissures caused by thermal contraction of the ground, and are valuable palaeoclimatic indicators because they form only in areas where the mean annual air temperature is −6°C or lower. Unfortunately, they have often been confused with tree-root casts, desiccation cracks and load structures. Two *thermal contraction cracks* over 1 m deep and 0.5 m wide at their tops occur in a coastal cliff section at Tregunna [SW 967 739]. They penetrate the Trebetherick Boulder Gravel, the underlying slaty Penwith Breccia (Clarke, 1973; Scourse, 1987), and extend to an unknown depth into the bedrock of the shore platform. The cracks contain silty material and vertical stones, and clasts marginal to them are upturned. Scourse (1987) argued that their small known size suggests formation as active-layer soil wedges, but that if their depth should prove to be greater than about 1.5 m, they are more likely to be ice-wedge casts. Small wedges up to 7 cm wide and 0.5 m deep occur at Towan Beach [SW 873 332], as described by James (1981).

Bedrock deformation by solifluction is best seen on slopes where steeply dipping weathered bedrock is overlain by head. The uppermost ends of the strata are then caused to be deflected downslope. A good example is exposed at Constantine Bay [SW 744 856], where geotechnical work has suggested that high pore fluid pressures resulting from confined aquifers in the active layer may have been the deformation mechanism (Cresswell, 1983).

SEA-LEVEL CHANGES

Ancient Marine Phenomena at and above Present Sea-level

Raised beach deposits are abundant throughout South West England, most commonly on the landward parts of shore platforms. In the Isles of Scilly the Watermill Sands and Gravel (*Figure 13.2*) at its type site on St Mary's [SV 925 123–SV 924 123] is a conglomerate of clast-supported rounded cobbles and boulders, overlain by well-sorted, medium sand. Its numerous occurrences range in altitude from 4.25–7.27 m OD. The mainland type site of the Godrevy Sands and Gravel at Godrevy [SW 583 426] exemplifies a common sequence: a 30 cm layer of rounded pebbles rests on a shore platform and is overlain by 1 m of beach and aeolian sand, over which lies 1.5 m of head (Stephens & Synge, 1966). The thicknesses vary laterally, and the sand is locally cemented into sandrock or aeolianite where it consists largely of comminuted shell fragments (James, 1975b). The temperate-climate molluscan fauna of the raised beach deposits implies an interglacial age. Amino-acid ratio measurements on shells of *Patella vulgata* from sandrock at Fistral [SW 799 624] and from conglomerate at Godrevy [SW 579 433], and on *Littorina saxatilis* from conglomerate at Trebetherick [SW 926 780], suggest that more than one temperate phase is represented in the raised beach deposits. The Trebetherick ratio is consistent with an Ipswichian age, and the other ratios with an earlier (but post-Hoxnian) temperate stage (Bowen *et al.*, 1985).

The *shore platforms* on which the raised beach deposits rest vary in width and degree of exhumation, and the altitude of the cliff-notch, where measured, varies from 4.4–10.8 m OD (James, 1975a & b, 1981). There are platforms at more than one level at several locations, and the 'present' intertidal platform is undoubtedly itself a fossil feature being exhumed and trimmed. They must be at least as old as the material resting on them.

Intertidal and Terrestrial Phenomena Below Present Sea-level

Submerged forests represented by fossilized trunks, roots and branches crop out on or beyond the foreshore at many places in Cornwall (French, 1985; Healy, 1995). The altitudes and ^{14}C ages of this material can be used to construct curves of relative sea-level change (Heyworth & Kidson, 1982), if the relationship of the dated material to its contemporary sea level is known. This latter requirement demands a combination of lithostratigraphic, biostratigraphic and chronometric techniques. In the barrier-protected embayment now occupied by Marazion Marsh [SW 506 326], Healy (1995) was able to show the environmental consequences of relative sea-level changes and changes in the effectiveness of the barrier over the last 7 ka using the pollen and diatom content of the sediments. The altitudes and ages of the dated samples from Marazion Marsh and some other sites are shown in *Figure 13.3* in relation to the regional sea-level curves of Heyworth and Kidson (1982). Some deviation occurs because of shortcomings in the curves and the local effect of the barrier.

PALAEOBOTANY

At five sites in the Isles of Scilly the Porthloo Breccia (*Figure 13.2*) includes interbedded organic sediments interpreted by Scourse (1991) as the infillings of small active-layer ponds. Pollen analysis indicates open grassland vegetation broadly consistent with a full-glacial climate, but with some thermophilous taxa reflecting the oceanic position of the Isles of Scilly. Radiocarbon determinations indicate deposition between $34\,500^{+885}_{-800}$ and $21\,500^{+890}_{-800}$ BP.

The work of Brown (1977) on Bodmin Moor, summarized in an archaeological context by Caseldine (1980), has provided a ^{14}C-dated

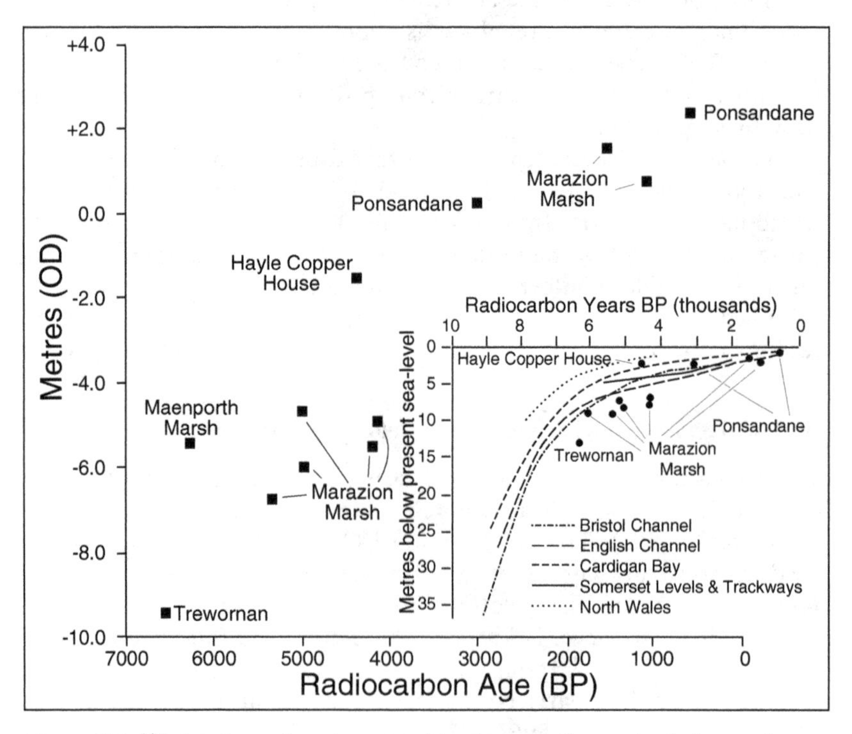

Figure 13.3. ^{14}C-dated samples relevant to Flandrian relative sea-level changes in west Cornwall (after Healy, 1995)

vegetational history for the period 13 088±300–6451±65 BP. Limnic sediments began to accumulate shortly before about 13 ka BP, when cold climatic conditions are indicated by the dominant vegetation of open grass heaths, snow beds and flushes. Climatic amelioration then allowed juniper scrub to invade, and tree birches spread about 12 ka BP, reaching their maximal extent towards the end of the Lateglacial Interstadial at 11 553±280–11 069±220 BP. A marked climatic deterioration about 11 ka BP then gave rise to the solifluction of upland soils and the development of grass sedge mires. At Hawks Tor [SX 147 744] solifluction deposits rest on organic Lateglacial Interstadial deposits, showing that solifluction was active during the Loch Lomond Stadial, the last cold phase of the Devensian, which ended about 10 ka BP with rapid warming. The early Flandrian saw the spread of tree birches and willow in the valleys, with *Empetrum* (crowberry) and juniper on the hillsides, followed by the replacement of the latter two taxa before 9 ka BP by hazel, with oak spreading in soon after. Oak, birch and hazel, the dominant woodland genera, probably colonized only the more sheltered upland sites.

THE QUATERNARY SEQUENCE

The fragmentary nature of much of the evidence summarized above explains why attempts to derive a sequence of Quaternary events have met with difficulty. Nowhere do all elements of the sequence exist together, and controversy over the number and ages of both raised beaches and heads present in an area is a recurrent theme in Quaternary studies in South West England. The most complete succession can be assembled in the Isles of Scilly, where Mitchell & Orme (1967) identified the following sequence:

Upper Head	Devensian
Raised Beach (Porth Seal)	Ipswichian
Glacial Deposit	Wolstonian
Main (Lower) Head	Wolstonian
Raised Beach (Chad Girt)	Hoxnian
Shore platform	

The assignment of the glacial deposit (Scilly Till) to the Wolstonian stage by Mitchell & Orme (1967) was based on the belief that the earlier of two raised beaches dates from the Hoxnian (by comparison with supposedly Hoxnian raised beaches at similar elevations in Ireland and elsewhere), and the presumption that the later one must therefore date from the next interglacial (the Ipswichian). However, Bowen (1973) recognized only one raised beach, of Ipswichian age, and argued that glacial material originally deposited during the Wolstonian was re-deposited over the raised beach by solifluction during the Devensian. Similarly, Scourse (1991) only recognized one raised beach member (Watermill Sands and Gravel) in his lithostratigraphy (*Figure 13.2*), having interpreted the higher of the raised beaches of Mitchell & Orme (1967) as a solifiual incorporation of material from the lower raised beach.

Although a Wolstonian date for the glacial limit on the Isles of Scilly was widely accepted, more recent work in the Midlands has suggested that the time between the Hoxnian and Ipswichian interglacials was occupied by more climatic episodes, with less extensive ice cover, than was formerly thought. It is thus more likely that an ice sheet would have reached the Isles of Scilly (and the northern coast of Cornwall) in Anglian rather than in Wolstonian times (Bowen *et al.*, 1986).

However, Scourse's (1991) determination of a ^{14}C age of 34–21 ka BP for organic deposits within the lower Porthloo Breccia (*Figure 13.2*), which predates the glacigenic deposits that he believed to be *in situ*, suggests that an ice sheet last reached the Isles of Scilly in the late Devensian. The aeolian Old Man Sandloess, thermoluminescence-dated to 18 600±3 700 BP and deposited in the southern part of the islands, is also consistent with a late Devensian glacial advance. Having denied the glacigenic origin of the Trebetherick Boulder Bed, Scourse (1991) regarded the ice limit on the Isles of Scilly as a terminal lobe of the late Devensian ice sheet, which failed to reach the mainland of Cornwall (*Figure 13.1*).

Finally, the controversy over whether and when glacier ice reached Cornwall and the Isles of Scilly remains alive. Eyles and McCabe (1989) have suggested that the supposed glacial deposits there are glaciomarine mud drapes laid down *beyond* the late Devensian ice sheet, whilst Scourse (1991) has pointed out that the widespread erratics found in the raised beach deposits of the Watermill Sands and Gravel imply that the late Devensian glacial event was not the only one to affect South West England.

Chapter Fourteen

History of Metalliferous Mining

Although archaelogical evidence exists for the ancient origin of metalliferous mining in Cornwall (Penhallarick, 1986), the documented history effectively starts with the regulations for the control of the tin trade and codes of mining law that were drawn up from the late 12th century. By the early 13th century, records of tin presented for coinage (that is the collection of royalties on metal output) also were being kept regularly. Thereafter the volume of surviving evidence gradually accumulates. It illustrates a pattern of development in the industry which is intimately bound up with the county's mineralization, its evolving demographic and economic structure, and the changing market demand for metals.

There is an extensive literature exploring all aspects of Cornish mining; this can be accessed through a number of helpful publications (Barton, 1969, 1978; Burt & Waite, 1988; Dines, 1956; Earl, 1994; Greeves, 1992; Morrison, 1983).

EARLY TIN MINING

During the medieval period, the Cornish population was sparse and scattered, and concerned primarily with providing for its own subsistence. Most people in the county spent much of their time in agriculture and supplemented their income from mining, fishing, shipping and some textile production (Hatcher, 1970). The amount of capital that was available for these activities was extremely limited and the technology simple. For tin production, this meant that most workings were small and shallow, confined to the washing of stream, or placer deposits, or open trenches on the surface outcrop of lodes. The term *tinner* was far more appropriate than miner. They commonly worked for themselves, and occasionally as wage labourers for others, and started and stopped operations with comparative ease and great frequency. Under such circumstances, total output depended as much on the availability of labour, itself influenced by the size of the population and the competing attractiveness of other occupations, as it did on the market price of tin, and fluctuated widely from year to year.

Some long-term trends of development, however, can be seen. The output of white tin (smelted metal), that had stood around 300–400 tons per annum (1 ton is 1.0016 tonnes) for much of the 13th century, began to increase rapidly in the second quarter of the 14th century to 700–800 tons by the early 1330s. Most of this was used in the manufacture of pewter and, to a lesser extent bronze and decorative ware, and a large proportion went for export. This expansion was cut short by visitations of the plague in the 1340s, 1360s, and possibly the 1370s. The population was greatly reduced and those that survived were drawn into more full-time agricultural pursuits. Tin production fell to just a few hundred tons a year and recovered only very slowly during the last decades of the century, notwithstanding a sharp increase in prices. By 1400 it was finally back at the levels of the 1330s and prospects for further growth seemed good. Nevertheless, fortunes were reversed from the 1420s, and the industry went into another half-century of decline. Again the problem appears to have been rising wages, caused this time, not by plague, but by the competing attractions of the local textile industry and agriculture. Large parts of the mining population migrated to the fertile east of the county, where farming was stimulated by the commercial demand coming from the expanding textile areas, and also possibly over the border into Devon, where Dartmoor tin mining saw a revival during the third quarter of the century.

The pattern of late-century recovery, early-century boom and then long decline, was repeated yet again in the late 15th and 16th centuries. By the 1490s, tin production was back at the levels seen in the 1330s and 1420s, and by the 1520s it had probably achieved a record level of over 1000 tons per annum (Hatcher, 1973). By this time the underground mining of lodes was becoming more common alongside stream and opencast workings, and although many still laboured in their own setts on a part-time basis to supplement agricultural incomes, it is clear that a full-time, wage-dependant class of tinner/miners, employed by wealthy merchants and entrepreneurs, was emerging rapidly. (Jenkin, 1948; Beare, 1586) The industry was beginning to acquire its own internal dynamic, but it remained very sensitive to problems elsewhere in the Cornish economy. This time, rising population and/or declining agricultural productivity, particularly in the western mining areas, meant that food and materials were in increasingly short supply, and the cost of living rose inexorably. Mining costs, however, were also rising as the more easily worked surface deposits were exhausted and lodes had to be pursued deeper, and wages, or returns to private effort, were held down. The problem was exacerbated by the control that the merchants and smelters had established over the tin trade. They made usurious advances to the tin producers and depressed prices to expand markets and profits. An attempt, at the end of the century, to break this stranglehold by a system of pre-emption purchases and low-cost loans to tinners from the Crown, failed to have any real effect. Tinners again found themselves getting poorer and looked for alternative methods of supplementing their income. As ever, these were predominantly found in agriculture, where prices were correspondingly high. The flight to agriculture, however, was not complete. There was a distinct fall-off in tin production after the boom years of the second quarter of the 16th century, but output settled on a new plateau of 500–700 tons per annum for most of the second half of the century. Even during the particularly depressed 1590s, when wage-dependant workers were seriously distressed, output levels were maintained around those seen 100 years earlier, in what was then a period of comparative prosperity.

The 17th century opened without the usual revival. Unsettled political conditions at home, finally resulting in the Civil War, upset domestic and foreign markets and appear to have depressed output to an all-time low in the 1640s. Clearly the Crown, with its taxes on production and monopolistic regulation of sales of tin, must also bear some considerable responsibility for this parlous state of affairs. When its influence was removed during the period of the Commonwealth, and Cromwell abolished the coinage and pre-emption purchases, the stannaries immediately boomed (Lewis, 1908). The prosperity was no doubt fuelled by the revival of trade and prices during the period, but was cut short by the Restoration in 1660 and the return of the coinage and price-rigging activities of the tin merchants. The only positive thing that can be said about this revival of the influence of the tin merchants is that they now appear to have been drawn mainly from South West England and have gradually excluded the traditional London interest. Profits were, therefore more likely to remain within the regional economy, although this was of little consequence to the miners.

The increasing costs of ever-deepening workings, and a renewed decline of tin prices, resulted in wage cuts and widespread deprivation during the last decades of the 17th century. Given this background, it is remarkable that tin output actually seems to have increased during the period at an unparalleled rate. Levels of production, which in the 1650s were still little different from those of the 1390s, had nearly tripled by 1700, and thereafter placed the industry on a higher plateau through to the second quarter of the 18th century. It may be that the industry was now strengthening its capitalistic structure and using improving technology to increase raised and smelted tonnages, but it could also be that a tightening of administration and reduced smuggling was resulting in a larger percentage of output being presented for coinage. The increase could thus be more perceived than real.

From the 1720s, the expansion of tin production continued apace. From less than 1500 tons annually at the beginning of that decade, it had risen to more than 2500 tons annually by the 1750s and 1760s, and to over 3000 tons by the late 1780s. As early as the first decades of the century, long before large-scale industrialization had begun elsewhere, some mines, like Poldice [SW 740 427], near Gwennap, were employing up to 1000 men and boys in underground operations, and were achieving an output of around £20 000 annually. It was the first period of unrestrained growth in the history of the industry. Unusually, however, the focus of attention within the overall structure of the industry began now to shift away from tin to copper. In the late 17th and early 18th centuries, the downward pursuit of tin, to depths of 600 ft (1 ft is 0.3048 m) or more in some mines, had begun to disclose underlying copper mineralization, and the economic structure of the industry was irrevocably changed.

THE BEGINNINGS OF COPPER MINING

The presence of copper ores in the county had long been known, but technical difficulties with their reduction initially prevented their successful exploitation. These difficulties finally were overcome in the 1690s, and a strong demand from a rapidly expanding brass industry, based in the Bristol area (Day, 1973), produced a rapid expansion of copper mining from the early 18th century. Borlase (1758) referred to 19 major copper producers in the county, of which Huel Virgin [SW 747 420] was the most productive and probably the most profitable. In just five weeks in the summer of 1757 it sold ore with a value of £15 300 for an expenditure of just over £200 in working costs. Overall, Borlase concluded that, 'the annual income to the county from copper equals very nearly at present that of tin, it being computed that for fourteen years last past, the copper of this county has produced cash one year with the other to the amount of one hundred and sixty thousand pounds'. The continued expansion of Cornish copper production suffered a severe setback during the last quarter of the century, however as a result of the over-stocking of the market and sharp price falls consequent on the exploitation of newly discovered rich deposits in Anglesey. Nevertheless, growth again resumed in the last years of the 18th century, and by the mid-19th century the county may have accounted for nearly half of the world's output of that metal. It had briefly become far more important than tin, although its impact on the economy of the county was diminished by the need to 'export' the ore for smelting to the neighbouring coal areas of South Wales.

To some degree, Cornish copper came from newly discovered deposits, worked by new mines, but elsewhere it was in joint production with tin from well-established workings. This production of several different minerals from relatively confined areas of heavily mineralized ground became an important feature of the industry that has survived to the present day. It created technical challenges for the ore dressing/benefication process, and provided a diversified financial base for mining companies who found themselves operating in several different, and sometimes unrelated, metal markets. Often this provided a form of insurance, with adverse price movements in one market being offset by advantageous developments elsewhere. On other occasions, however, the failure of one metal might depress output of others, as the overall economics of mining were unbalanced. Thus the collapse of the copper market in the late 18th century depressed output of tin, as the joint producers of those ores were forced to cut back their operations. Similarly, in the late 20th century, the drop in the tin market contributed to delays in the development of joint tin/tungsten projects, such as that at Hermerdon Ball [SX 572 588], near Plymouth.

With large-scale, capital-intensive, deep mining firmly established by the middle of the 18th century, Cornwall found itself in the vanguard of English and world industrialization. The financial structure of mining companies, their managerial systems, and organization of the labour process, provided a model for later comers in manufacturing industry to follow, while the local engineering industry, which supported mining, emerged as a world leader in steam engineering. If the story of the industrial revolution is that of transformation from an organic to a mineral-based economy, Cornwall helped to provide the raw materials, the strategic technology and the organizational experience for that new world to emerge.

LEAD AND SILVER PRODUCTION

With the gathering industrialization of the British and European economies, the demand for a widening range of minerals from Cornwall gradually increased. Lead and silver had been worked in South West England since the early medieval period, often on a large-scale, state-sponsored basis, which was more akin to the advanced mining districts of central Europe than the small-scale private tin ventures of the region. Most of these enterprises, however, were on the very eastern edge of the county, in the Tamar valley, and cannot strictly be counted as part of the Cornish tradition. When Borlase (1754, 1758) wrote in the mid-18th century, he noted that although lead had been found in many parts of the county, and worked on and off in a small way for several centuries, only one mine of note, at St Issey [SW 955 717] near Padstow, was then active. The output of lead started to rise, however, as lodes which previously had been worked only when they had a high argentiferous content, began to be exploited on their own account. By the early 19th century, Cornwall was probably one of the country's largest lead producers, with an annual output rivalling many of the better known lead districts of northern England and Wales (Douch, 1964; Burt, 1984). Silver output also undoubtedly increased along with lead, facilitated by improvements in refining techniques from the 1830s, and in terms of total value of output, became one of the county's leading economic minerals.

THE DIVERSIFICATION OF OUTPUT

Zinc

Zinc production, which was associated with lead and copper, but unnoticed by Borlase (1758), and by Pryce (1778) was gradually introduced from the early 19th century, and by the 1850s accounted for up to a third of national output. Production fell off in the later 1860s and early 1870s, but recovered in the late 1870s and early 1880s with a great surge of output from the West Chiverton [SW 793 508] and Duchy Peru [SW 796 556] mines near Perranzabuloe (Burt *et al.*, 1987). During those latter years, Cornwall was the largest single national producer, supplying a rapidly expanding market for galvanized iron, which was becoming one of the principal building materials for the tropical parts of the British Empire.

Iron And Manganese

Only iron and manganese failed to realize their hoped-for potential. The presence of iron ore in the county had been noticed by various commentators from the 17th century onwards, but effective commercial exploitation was prohibited by the greater availability of the metal in other parts of the country and the local scarcity and high cost of charcoal (Cantrill *et al.*, 1919). It was not until improved transport facilities became available in the early 19th century that production started to look viable. Even then, output remained small as speculators failed to find substantial deposits, other than the Great Perran Lode [SW 765 575], just south of Newquay in the Perranzabuloe area, and a number of more dispersed deposits around the St Austell granite. The most productive mines were Restormel Royal [SX 098 614] in the St Austell area and Trebisken [SW 779 563] at Perranzabuloe. They never featured as major national producers, however, and collapsed in the face of increasing competition from cheap Spanish ores during the last quarter of the 19th century.

Like iron, manganese ores were not found in quantity in Cornwall and the county never emulated the success of its Devonshire neighbour in this industry (Dewey & Dines, 1923; Burt & Wilkie, 1984). During a brief four-year period between 1878 and 1881, the Ruthers [SW 923 601] and Lidcott [SX 241 850] mines, near Newquay and Launceston respectively, produced a few hundred tons of ore which registered them as major national producers, but this was in an industry that had virtually ceased to exist domestically, and a heavy import dependency already had been firmly established.

Other Industrial Ores

In the same way that the early stages of industrialization had increased the demand for the principal metallic ores, the increasing sophistication of industrial technology in the second half of the 19th century widened the range of uses for hitherto rare and discarded minerals. In both volume and value terms, the most important of the these products was *arsenic* (Earl, 1983; Burt, 1988). Used in a range of industries, including the manufacture of pigments, dyes and glass, as well as the preservation of foods and destruction of pests, it became an important joint product of tin and copper mines (Dewey, 1920). Cornish mines, together with those just across the Devon border in the Tamar valley, became the principal source of world supply for most of the period through to the First World War. Most output came from mines in the central Camborne–Redruth area and, later, from the east of the county, near Callington. Surviving remains of the industry such as calciner flues can still be seen at many mine sites (Plate 8a).

While well over 100 mines derived a small income from the volume production of arsenic, a few companies found a high-value by-product in the form of the minerals wolfram and scheelite. *Tungsten* is used mainly in the manufacture of special steels, and its ores were commonly found with tin; their production therefore fluctuated with the production of that metal. They required complex chemical or magnetic dressing techniques to separate them from associated ores, and it was not until the last decades of the century that they generally began to be produced. Output occasionally topped 100 tons per annum during the 1880s and 1890s, but during the early years of the new century it rose to 200–300 tons. Throughout the whole period, Cornwall remained the largest domestic source of these ores, supported in a small way by mines in Devon and Cumberland. The largest producers were East Pool [SW 674 415], South Crofty [SW 669 413], Carn Brea [SW 676 410] and Tincroft [SW 670 409] in the central Camborne–Redruth area, and Clitters United [SX 421 720] and Kit Hill [SX 375 713] in the east of the county. Castle-an-Dinas [SW 950 616] at St Columb also became important later in the 20th century, and Cligga Point was operated during the Second World War. The successful exploitation of tungsten ores was not repeated on the same scale for other less common minerals. There was some very small-scale production of *antimony, bismuth, cobalt and uranium* in the late 19th and early 20th centuries, but it rarely amounted to more than a few tons, with a value of far less than £1000 before the First World War. No more than a few mines were ever involved in their production, although East Pool did see some limited production of all these ores (Burt *et al.*,1987).

COMPETITION AND DECLINE

While the late 19th century had seen a widening of the range of minerals in commercial production in Cornwall, it had also seen a marked decline in the main minerals produced in the county. First copper, then lead, and finally tin saw a major increase in foreign competition in home and overseas markets and a consequent sharp and sustained downward movement in prices. Unable to compete, levels of domestic output declined dramatically. Copper prices fell by two-thirds between 1855 and 1886, from £124 to £43 per ton, and ore output collapsed over the same period from 161 576 tons to just 7541 tons (Schmitz, 1979). To some extent this dramatic fall in copper output was a function of the structure of the more productive mineral lodes, which were either failing in depth or, in some cases, reverting back to tin as they were developed at even greater depths. Thus, just as there was a fortuitous discovery of copper mineralization under surface tin at the beginning of the 18th century, when the market for copper was beginning to mushroom, now there was a coincidental conversion back to tin at the very point when the copper market was collapsing.

There was no such luck for lead and zinc production. In some of the other British non-ferrous mining districts, zinc mineralization often increased as lead decreased at depth, but in Cornwall most zinc was derived from copper lodes. Cornish lead production, which already had been in slow decline during the high-price years of the third quarter of the century, completely collapsed between the late 1870s and late 1880s as prices were nearly halved from over £20 per ton to less than £12 per ton. Zinc production, which had reached an all-time peak in the county in 1881, had virtually ceased to exist by 1890, as available deposits failed to prove viable on their own account and price declines in other markets took the major producers down. Finally, tin ore production, which had been sustained at a high point of over 14 000 tons per annum in the 1880s and early 1890s, despite a fall in prices from a peak in the third quarter of the century, was halved during the mid-1890s as prices dropped sharply from well over £90 per ton in 1891/2 to just over £60 in 1896/7. Just as Cornish mines had pioneered the way into industrialization in the early 18th century, now they were beginning to pioneer the process of de-industrialization, which has become a general feature of the British economy in the late 20th century.

After the difficulties and turmoil of the late 19th century, the 20th century opened with stabilization and mild recovery among tin producers. Copper, lead and zinc mining had virtually disappeared from Cornwall, but arsenic and other ores continued to make a useful contribution to the overall economy of the industry. More importantly, the price of tin had recovered to over three times its mid-1890s low by the years preceding the First World War, and in 1913 the county's output was nearly 30% up on the 1899 level. Given the level of the price increase—in 1913 the price per ton of tin was around twice the average of the prosperous 1870s and 1880s—it is surprising that the revival was not even stronger. To some extent this may have resulted from the demise of copper. Prices of that metal remained below those achieved more than half a century earlier, and mines that previously had relied on an income from the joint production of both metals found it impossible to survive on tin alone. Clearly, however, the principal problem arose from the great damage that had been done to the industry by under-investment and the depletion of reserves during the depression years. This had resulted in the destruction of productive capacity, and the industry as a whole was no longer able to respond smoothly to new opportunities. Revived prices now simply brought stabilization on a new, lower plateau. This was to be the first example of a pattern that was to repeat itself several times as the 20th century progressed.

CONTRACTION AND COLLAPSE

The outbreak of hostilities in 1914 should, in theory, have proved a great boon for the mining industry. Competition from overseas producers was disrupted, demand increased, inflation took off, and market prices rose steeply. The average market price of tin in the UK, which had stood at £148 a tonne in 1914, had advanced to £324 by 1918, the highest point it was to reach before the late 1940s. Cornwall's tin mines, however, continued to struggle. Their labour force was not listed as a reserved occupation until the spring of 1916, and large numbers of the most skilled workers were lost to the Armed Forces early in the conflict, particularly after siege warfare became a characteristic of strategy on the western front. Two of the most productive mines—Dolcoath [SW 662 405] and East Pool—had lost 25 percent and 50 per cent of their pre-war labour force by the end of 1915. The prices received for the ore concentrates were pegged at an artificially low level, while the cost of essential supplies, such as coal, timber, and explosives, increased rapidly (Rowe, 1975). Essential investment in new equipment and maintenance was ignored, and exploratory and development work, which was vital for the proper long-term exploitation of the lodes, was all but abandoned. As in the difficult years of the 1890s, the industry was exploited rather than developed. Under the circumstances, it was fortunate just about to sustain pre-war production levels.

When the First World War ended, the mines were even less well equipped to respond to the revival of normal competitive conditions than they had been before. Decontrol of prices meant that they could begin to take advantage of higher market prices, but these immediately began to tumble while working costs stayed high. In particular, coal prices remained at a high level, and the exceptionally wet conditions of the winter of 1918/19 imposed a heavy financial burden on the ailing industry. The government, whose previous policies had done much to undermine the industry, now conspicuously failed to come to its assistance. There were to be no homes fit for heroes in Cornwall. The breakage of the Levant man engine in 1919, hurling 30 men to their deaths, was seen as the result of worn out and poorly maintained machinery, and epitomized the parlous state of the industry. In that year, talk was of the complete cessation of mining for the first time in a millennium, and by 1921 the gloomy prognosis was virtually fulfilled as all but one of the county's mines was abandoned or suspended (Trounson, 1984). The only resort for many miners during those years was to abandon the country for which they had so recently fought and, like their forefathers, seek better prospects through emigration.

Relief, however, came as the world industrial economies began to pick up and, after a short lag, tin prices with them. By 1925, 27 mines were again at work in the county and some of the richer and more efficient operations, such as East Pool, were satisfactorily in profit. Nevertheless, the recovery was short-lived, and by the end of the decade the industry had stabilized on a new production plateau of around 2500 tonnes annually. This was about half of the average output of the years of normality preceding the First World War, and a repeat of the gradual 'ratcheting down' of the industry after each period of major crisis. The numbers employed in the industry showed the same picture. In 1913 they stood at 8700, but by 1928 they were only 3269, or considerably less than half the earlier number. The way out of this impasse was clear to the industry. They must sink deeper mines to develop new and richer lodes that every indication suggested were present. East Pool had shown the way with successful investment in a new deep shaft after its old main shaft had collapsed in 1921. But the necessary capital, private or public, was not forthcoming. Existing lodes had to remain the productive core of the industry, with gradually diminishing returns.

Any hope of revived confidence and new investment for an expansionist future for the industry was again destroyed by the great international economic collapse of 1929–33. Mining activity was reduced to the almost non-existent levels of the worst years of the post-war depression. By the end of 1930 only East Pool was still producing tin, and total employment in the industry at the nadir of the slump fell to just 426. These years also saw the final demise of the centuries-old local tin smelting industry with the collapse of Consolidated Tin Smelters in 1931. Again recovery was slow and painful, and production shortly settled at a level below that seen after the previous recovery in the 1920s. Some mines, like Geevor [SW 375 345], which had suspended mining operations between late 1930 and 1932, recommenced, and by 1935 were reporting a complete recovery and sound profits (Noall, 1983). In that same year, a revived South Crofty showed the confidence to acquire the lease of the historic and recently closed Dolcoath sett, and to institute major development work. It began to show good returns by the end of the decade (Buckley, 1981). Generally, however, there was little interest from investors to get back into the industry and activity remained at a low level.

When the Second World War broke out there were only four tin mines and one tungsten producer working within the county. As in 1914, the potential benefits of hostilities for the industry seemed encouraging, particularly after the loss of the highly productive South East Asian tin fields to the Japanese in 1942. The actual outcome, however, was closely to parallel the First World War experience. Labour shortages, difficulties in obtaining raw materials, low investment, poor maintenance, the mining of proven reserves without provision for exploration and development, all reduced current output and mortgaged what was left of any future prospects for the industry. The government did make grants to the already ailing East Pool mine to sustain production even at a loss, but the withdrawal of those grants after the war meant its immediate demise. Other government funds promoted a feverish search for workable deposits of tungsten and other strategic minerals, but these had produced few tangible results before the end of hostilities (Trounson, 1989). When peace returned, only Geeevor and South Crofty tin mines survived, together with Castle-an-Dinas tungsten mine [SW 950 616], near St Columb. Total tin output had shrunk to less than half its already much diminished pre-war level and the 'downward ratchet' seemed finally to have all but destroyed the industry.

INTERNATIONAL ACTION

Through the 1950s and 1960s very little mining activity took place in the county outside of the two small remaining mines, Geevor and South Crofty. There was some considerable exploration, using diamond drilling technology developed in the inter-war years, but the companies involved were interested in potential future prospects, dependent on a significant rise in the tin price, rather than current development. However, one enormously important marketing innovation was initiated during this period which was to have great implications for the future. World tin producers had always been at the mercy of unrestricted market forces on their profitability. Now, like many other primary producers, they began to try to influence and control that market.

During the immediate post-war period there had been a rapid expansion of world tin production, led by a re-establishment of Malaysian output after the ravages of the war years. This had been facilitated by a major expansion of demand, largely generated by heavy United States buying for an increasingly mountainous strategic stockpile. In 1954 and 1955 purchasing for that stockpile was interrupted and there was an immediate and critical surplus in the market. A crisis threatened the entire international tin industry, and it resulted in the formation of the First International Tin Agreement in 1956. Funds were made available for the purchase of a buffer stock to soak up the surplus, and export controls were introduced to reduce supply. Restriction of supply to increase prices had been tried before with varying success for several metals, but now it looked as though it might work.

The First, and successive Second, Third, Fourth, Fifth and Sixth Tin Agreements, had a profound influence on tin producers and consumers through the 1960s, 1970s and early 1980s. Prices continued to fluctuate, and the industry had its short-term ups and downs, but broadly speaking, over the period as a whole, real prices were pushed up (Tin Crisis, 1987). This in turn strategically altered the context in which Cornish mining operated. Investors gradually gathered confidence in the future of the industry and began to look closely at the results of the exploratory activity undertaken in the post-war decades. Most significantly, major international mining companies were persuaded to take an interest in the county, and to use their immense financial and technical resources both to invest in the long-mooted development of known lodes at depth and to improve mining and ore concentration processes to increase efficiency and reduce costs. In that context, Consolidated Goldfields brought the revived Wheal Jane [SW 783 434] into full production in 1972 and, together with an expansion of activity at Mount Wellington [SW 561 295], Pendarves [SW 660 392], South Crofty, and Geevor, succeeded in doubling output during the high-price years of the early 1970s. Cornish tin production, which had stood between 1000 and 1500 tonnes per annum through the 1950s and 1960s, jumped to an average of well over 3000 tonnes during the 1970s and climbed to more than 5000 tonnes by 1985. Considerable success was also achieved in the production of other associated ores, most notably zinc, of which concentrates with more than 10 000 tonnes of metal were separated in 1981 and 1982 (Cornish Chamber of Mines, Annual Reports).

It was the greatest revival of mining in the county seen this century. Tin production was back to over 50% of the level achieved during the golden years of the 19th century and was supplying around 40% of national consumption, saving the country about £40 million a year in imports. By the 1980s zinc output was twice the level ever previously achieved, and even copper and tungsten were again being produced in noticeable quantities. The count of mines only stood at five, but they held much larger setts than previously worked. Trounson (1984) calculated that in terms of their 19th-century persona, the equivalent of 68 mines were at work in 1984. Only the total numbers employed remained seriously reduced, estimated in October 1985 at 1555. This was the consequence of greatly changed and improved technology, but the higher earnings that this brought helped to make a regional economic impact far greater than would have been achieved by that number earlier in the century. Higher prices and the return of confidence had finally created the conditions for the resumption of the vertical development of Cornwall's mineral wealth, and a new era of prosperity for the county seemed to be at hand.

The International Tin Council proved unable, however, to restrict supply sufficiently to sustain the artificially high price levels, and on the 24 October 1985 the Buffer Stock Manager informed the London Metal Exchange that he could not pay his debts. The London Metal Exchange immediately suspended trading and prices outside of the market collapsed. Although the government provided loans to help tide the largest producers over the short term, the effect on Cornish mining was disastrous. The large, multinational mining operations lost interest in the county and the Rio Tinto Zinc Corporation sold its subsidiary, Carnon Consolidated, in a management buy out. The great prospect of the 1970s, Wheal Jane, was shut, and even Geevor, a stalwart of the industry through the difficult years earlier in the century, was forced to abandon operations. By 1993, tin prices had fallen to a low of less than £3000 a tonne, compared with a high of over £10 000 before the collapse in 1985. This was mainly because of a flood of supply from the liquidation of the huge stocks accumulated by the International Tin Council, unrestrained exports from Brazil and China, now the world's leading producers, and a decline in demand as aluminium replaced tin cans for a wide range of uses. The industrial economies also went into recession. Only South Crofty was left in work, and that was operating at a loss. Its ore production had been reduced by around 50% since 1987 and its labour force to just 244. Nevertheless, costs were reduced, the efficiency of milling operations were greatly increased to maintain the output of concentrates, and confidence in the future was sustained by the prospect of developing still deeper lodes and a revival of demand and prices. The mine's optimism was supported by the market in August 1994, with the successful launch of a new share issue to ensure that accumulated debts to the government and Rio Tinto Zinc were written off and that development and further operations could be guaranteed until 1997. In that year, however, continued depressed prices initiated moves to close the mine and in spite of strong local opposition, it ceased production in 1998.

ESTIMATES OF TOTAL OUTPUT

Finally, some estimate may be made of the total quantities of the principal minerals that have been extracted from Cornwall. If the coinage returns for *tin* are calculated for the period from the beginning of the 14th century to the 1830s, with estimates for gap years, and are then totalled with official government returns from then down to the present day, it produces an indicated output of over 1.33 million tons. Since the coinage returns related to metal presented for taxation purposes, and therefore probably under-represented real output, and since they take no account of production during the preceding thousand years or more, it seems reasonable to conclude that the real level of output from the county may have approached 1.75 million tons of metallic tin. At a current price (January 1995) of £3900 per ton, this would be valued at around £6.8 billion.

By comparison with tin, the output of *copper* can be traced with some accuracy during its relatively brief 200-year history, from about 1700 to about 1900. Returns of the sale of ore at ticketing in Cornwall and Swansea, with assayed metal content, suggest that a little less than 1 million tons of metallic copper was derived from southwestern ores, probably two-thirds came from Cornwall and one third from Devon, with a value at today's price of around £1900 per ton of £1.6 billion. Estimating *lead* production is more problematical. The are no figures for production before 1845, by which time the industry may well have been past its peak. However, indicated output of metallic lead between that date and the 1890s, by which time regular production had ceased, was just over 170 000 tons. If a similar or slightly greater tonnage had been produced during earlier years, this would suggest a total tonnage of around 400,000 tons, valued at today's price of £430 per ton at £0.17 billion. On a similar principle, if the recorded output of *silver* in Cornwall between 1851 and 1913 was 5.45 million ounces, derived almost entirely from lead mines, it is likely that total production from the county may have been 12 million ounces or more, valued at today's prices of £3.00 an ounce at £0.36 billion.

Zinc production also started in Cornwall before regular statistics were collected, although most output probably came after the mid-19th century. Nevertheless, it is difficult to calculate the amount of metal produced, because no assay figures were given before 1881 when the industry was already in sharp decline. If the average metal content for the 1880s is used as a guide, however, the 80 749 tons of dressed ore sold between 1856 and 1913 would have yielded around 32 300 tons of metallic zinc. If this is added to the 93 000 tons of metal produced since the revival of the industry in the 1970s, the total amounts to 124 600 tons of metallic zinc, which at today's price of £740 per ton, would be valued at about £0.9 billion. Together with the production of *iron, manganese, arsenic,* and the numerous other metalliferous minerals raised in Cornwall, this gives an overall value of output from the county approaching £10 billion in current prices. Unfortunately, most of this endowment has long since been consumed in, or expropriated from, the county, and very little remains of the wider infrastructural development it once supported to compensate future generations for the dereliction that the industry has left behind. William Pryce (1778) warned more than 200 years ago that Cornwall was suffering, 'the exportation of their raw staple, in order to give other countries the benefit of its manufacture'. Sadly he seems to have gone unheeded.

Chapter Fifteen

The Contemporary Extractive Industry

The value of the minerals obtained from Cornwall must be one of the highest of any county in Great Britain. Extractive activity can be divided into: (1) metals, predominantly Sn and Cu, (2) china clay and chinastone, and (3) constructional raw materials. These resources are mainly associated with the granites of the Cornubian orefield with 80–90% of the mineral wealth coming from Cornwall. Today, the main source of wealth is china clay (*Table 15.1*).

The employment figures (*Table 15.2*) reveal the importance of the extractive industry in the economy of Cornwall.

METALLIFEROUS MINERAL PRODUCTION

Following comparatively low levels of production in the period 1930–1970 (*Figure 15.1*), there was a considerable increase in Sn production after 1970, to 2000–5000 tonnes per annum (metal basis). This resulted from the opening of new mines. Particularly notable, was the development of Wheal Jane [SW 783 434] in 1971, which provided a substantial addition to Sn output, with significant Zn and Cu production. Sn output was further augmented as other new mines (*Figure 15.2*) were opened at Wheal Pendarves [SW 646 379] (1971) and Mount Wellington [SW 760 423] (1976). The last is on an extension of the Wheal Jane orebody.

In addition, alluvial deposits and old dumps were exploited by Hydraulic Tin at Bissoe [SW 778 413], Tolgus Tin near Redruth, and Brea Tin near Carn Brea. Pilot-scale exploitation of placer Sn deposits on the sea-bed in St Ives Bay by the Marine Mining Corporation used a specially adapted ship. Some reprocessing of old mine dumps was also undertaken by these companies, as well as by the Cornish Tin Smelting Company at Roscroggan [SW 647 418]. The latter plant also processed ore from Wheal Pendarves.

A great deal of exploration for Sn was carried out by large international mining houses and by smaller groups, when projects such as the

Table 15.1. PRODUCTION AND VALUE OF MINERAL EXTRACTED IN CORNWALL, 1994

Commodity	Production (tonnes)	Value (£)
Tin (metal)	1900	7 million
China clay	2 640 000	210 million
Constructional raw materials	2 400 000	10 million
TOTAL	5 041 900	227 million

Based on figures from the *United Kingdom Minerals Yearbook 1994*

Table 15.2. EMPLOYMENT IN THE EXTRACTIVE INDUSTRY OF CORNWALL, 1993

Metalliferous	200
China clay & chinastone	3600
Constructional raw materials	800
TOTAL	4600

Based on Griffin (1994) and sources from within the industries

Redmoor mine [SX 356 711] near Callington, received serious consideration. Some major feats of underground mining engineering were also accomplished, including the sealing of the sea-floor breach into the old Levant mine [SW 368 346]. This opened up an extensive new area of mineralized ground for Geevor mine [SW 374 345] to exploit. As a result of all these developments, Sn production rose to just over 5000 tonnes in 1984-1985.

However, the Sn price had been kept at artificially high levels during the 1970s and early 1980s by the activities of the International Tin Council. This collapsed in 1985, and the price of Sn metal plummeted from around £10 000 per tonne to between £3000 and £4000 per tonne.

By 1997 the only mine left in production was South Crofty [SW 665 412], whose underground workings extended beneath Redruth and Camborne to a depth of 1 km. Following a successful placing of a new share issue by South Crofty in 1994, the future of this mine in the short term seemed reasonably assured. But in August 1997 its closure was announced.

Secondary engineering facilities producing everything the mining industry needed, from beam engines to mineral separators, have grown up and become important industries in their own right. They still form an important sector of the Cornish economy.

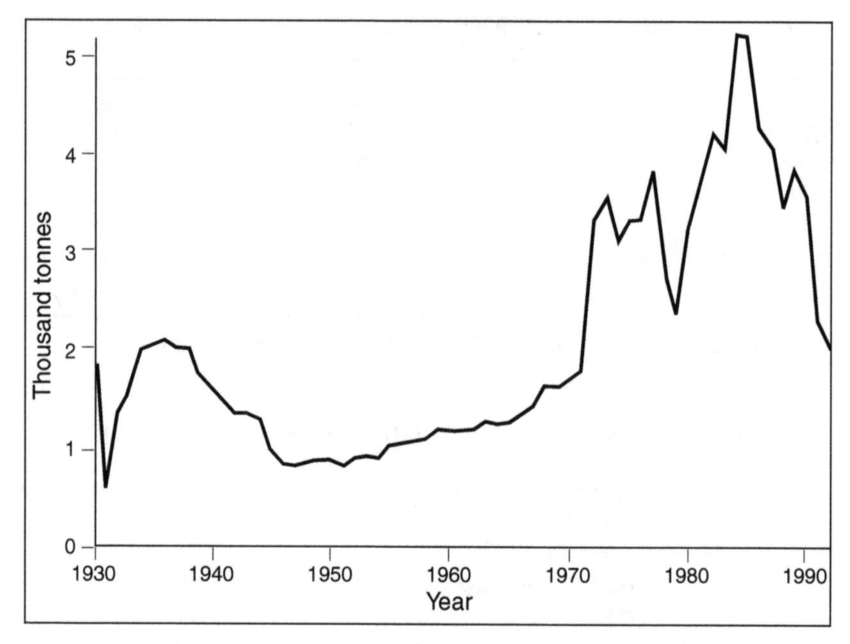

Figure 15.1. Cornish Sn production (metal basis) 1930–1992

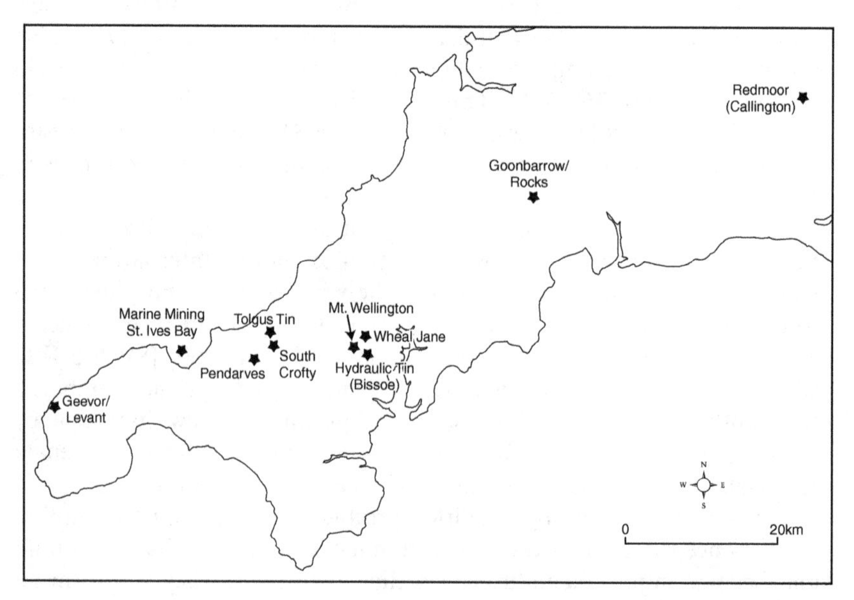

Figure 15.2. Locations at which Sn concentrate production has been attempted since 1945

DEVELOPMENT OF THE CHINA CLAY AND CHINASTONE INDUSTRIES

The historical development of these industries is dealt with in Chapter 10, but rationalization within the industry has left two producers in Cornwall.

Table 15.3. CHINA CLAY PRODUCERS

Company	Approximate capacity (tpa)
English China Clays	2 800 000
Goonvean & Rostowrack	220 000

Figures quoted on dry basis

Production in 1994 is estimated to have been 2 640 000 tonnes, of which 2 240 000 tonnes (85%) was exported. The paper industries of Finland, Germany and the Scandinavian countries were the most important export destinations.

Competition for china clay (kaolin) markets has intensified with the appearance of entirely new producers of paper-coating kaolins in the Amazon Basin (Cadam, on the Rio Jari; 500 000 tpa) and in Queensland, Australia (Comalco, at Weipa; 150 000 tpa). There are vast resources of high-quality kaolin in the Amazon Basin and further new plants on the Rio Capim have recently been commissioned. Ultrafine ground and precipitated $CaCO_3$ made from marble, chalk and limestone is also competing strongly for paper and other filler markets (Bristow, 1995).

In Cornwall, the environmental impact of the china clay industry has become a matter of increasing importance in recent years. Although adding to the costs of production, significant advances have been made in softening the impact of the workings by landscaping, and by establishing vegetation on tips and residue dams. Backfilling of china clay pits may seem to be an obvious way of minimizing environmental impact, but it presents problems. The shape of china clay deposits (*Figure 15.3*) means that the future resources of china clay tend to lie under the pit, so premature backfilling would sterilize reserves. Abandoned pits also perform an invaluable role as water reservoirs, storing the large quantities of water used in the production process.

The cause of conservation of mineral resources is also served if more saleable mineral can be extracted from the clay matrix. Major advances have been made in the recovery of the large curled stacks of kaolinite (*Figure 10.2*), which form up to 70% of some refining residues. Once recovered, this coarse kaolinite can be ground to produce more commercial kaolin. Since the particle size distribution of the fillers produced by this route can be closely controlled, these fillers are especially

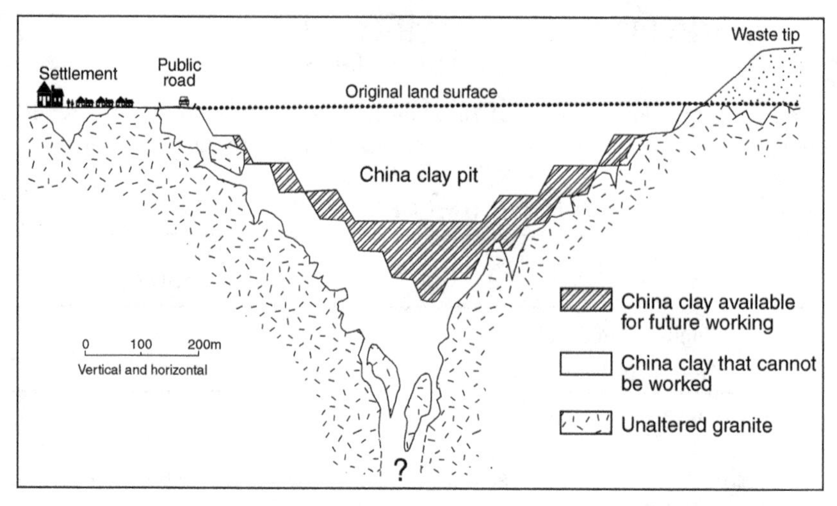

Figure 15.3. Diagrammatic cross-section of a china clay deposit

attractive to the paper industry. Because of the commercial success of these kaolins, work has now begun on reprocessing residues in some of the older waste lagoons.

While some progress can be reported on the treatment and reduction in the quantity of mica residues produced, the same cannot be said of the 24 million tonnes of waste sandy material that accumulate each year. Much of this material is part gravel by particle size definition. The total industry waste stockpile is estimated at over 600 million tonnes, with about 450 million tonnes in the St Austell area alone.

Only a relatively small proportion of this waste is suitable for use as concreting sand, because the sand is generally too coarse and deficient in fines. Concrete made from this material can require up to 20% more cement than conventional sands. However, in appropriately designed structures, clay works sand can be used as a bulk fill material for the construction of embankments and road foundations. The dam wall of Colliford dam [SX 180 710] on Bodmin Moor is entirely composed of china clay sand from nearby Park china clay pit [SX 195 708]. Local beaches constructed from clay works sand washed down the rivers have also proved remarkably stable, suggesting another possible application for china clay waste in beach make-up and coastal protection schemes.

Crushed waste rock (stent) has been used for many major road schemes in the county. Unfortunately, numerous investigations by the industry, local authority and central government have shown that it is uneconomic to transport china clay waste out of Cornwall.

However, china clay waste also includes other interesting materials. It is possible to recover a mica containing 2% Li_2O from residues, and the St

Austell resource is probably the largest reserve of Li in the European Community. Nevertheless, at the present time this is not competitive. Mica is used as an industrial filler, which when ground to the correct fineness has the ability to convert plastics into rigid engineering materials. It is used extensively in the automotive industry. Topaz is another mineral with interesting possibilities in the future as a source of mullite for refractories and F compounds. Finally, there are small quantities of heavy minerals containing U, Ti and Rare Earth metals and, at some locations, Sn and W minerals, but concentrations are too low to allow economic recovery at present.

Chinastone is still produced in small amounts for the ceramics industry. About 8000 tonnes was produced from a single quarry [SW 954 564] near Nanpean in 1994.

RAW MATERIALS FOR THE CONSTRUCTION INDUSTRIES

The location of the principal quarries in Cornwall and the types of stone produced are shown in *Figure 15.4*.

Greenstone is a convenient local term for a variety of basic igneous rocks. Principal greenstone quarries, active at the moment, include Greystone Quarry [SX 363 805] near Launceston (Camas Ltd) and Dean Quarry [SW 800 202] in the Lizard (Redland Aggregates Ltd). There are also several other smaller quarries in the Lizard and Launceston areas. Greenstones generally yield good materials for roadmaking, including the all-important wearing surface which must use a stone with a sufficiently good Polished Stone Value to minimize the risk of skidding. Slightly weathered or rotted greenstone can be used for the base course of roads, but the freshest rock is reserved for the wearing course.

One of the earliest building stones used in Cornwall was Cataclews stone, which was originally obtained from a dolerite intruded into Upper Devonian rocks at Cataclews Point [SW 873 761], west of Padstow (Reid *et al.*, 1910). An altered ultrabasic rock at Polyphant [SX 258 827] has also been worked since the 11th century for an easily carved stone that can be used for delicate interior work. Many churches in east Cornwall and farther afield (including Canterbury and Exeter cathedrals) have interior features made of Polyphant stone. Launceston Priory and Launceston parish church also contain much Polyphant stone, but it has not weathered well, presumably because the stone contains a substantial proportion of talc. Small quantities of serpentine are extracted from a quarry near Kynance Cove for the manufacture of ornaments in Lizard Town.

Large granite quarries for aggregate are worked at Hingston Down (ARC Southern) [SX 409 720], Luxulyan (Camas Ltd) [SX 053 592],

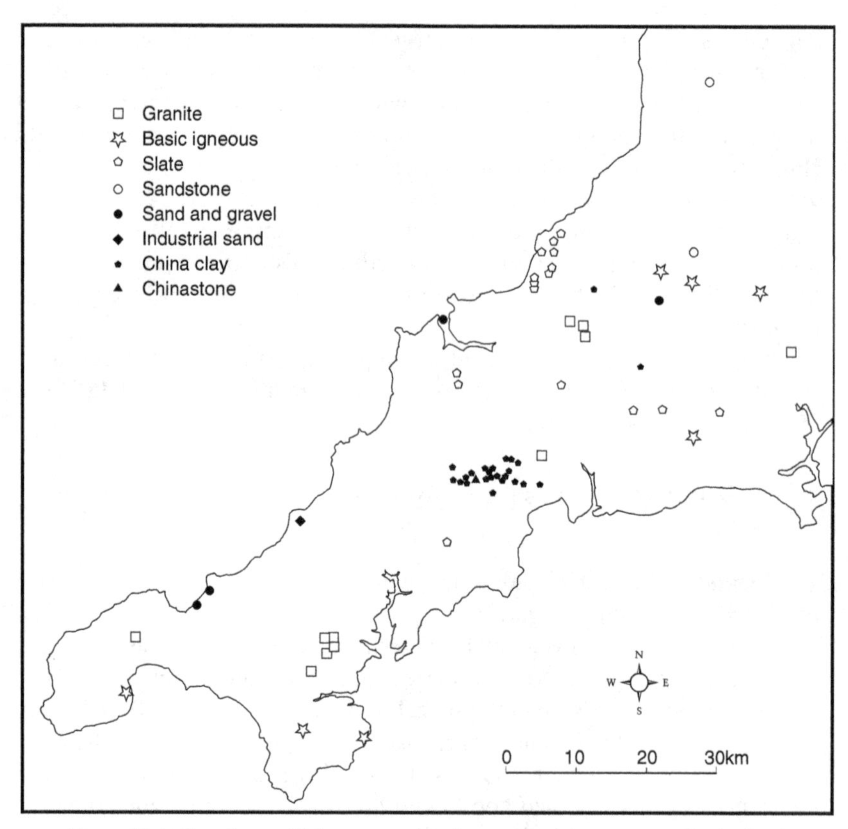

Figure 15.4. Sites from which constructional raw materials are currently obtained

Carnsew Quarry [SW 764 355], near Falmouth and at Castle-an-Dinas [SW 489 341] in the Land's End granite (Penryn Granite Ltd). There are also many small granite quarries in the Carnmenellis granite and a few on Bodmin Moor. Crushed granite makes an excellent aggregate for general usage.

Granite is also worked for dimension stone on the west side of Bodmin Moor at De Lank [SX 101 753] and Hantergantick [SX 103 757] quarries. De Lank granite was used in the construction of the Eddystone, Bishop Rock and Beachy Head lighthouses, and Tower and Blackfriars bridges in London. High-quality dimension stone, renowned for its strength, was formerly extracted from a number of quarries in the the Luxulyan valley. This stone was used in the construction of the British Museum, the old London Bridge and Plymouth breakwater.

Another rock of granitic composition that has been extensively used as a dimension stone for building is Pentewan stone (a felsite dyke). This was formerly obtained from small quarries [SX 025 475] on the cliff just northeast of Pentewan and also inland in a quarry [SX 022 478] north of the village. Pentewan stone is easily worked and of excellent durability, and was used in the construction of St Austell parish church and Antony House. However, much of what is nowadays called 'Pentewan stone' may well have come from other quarries extracting similar material at Sticker [SW 985 504; SX 974 505] and Polgooth [SW 997 503]. Other felsitic dyke material has also been worked in the west of the county and in the vicinity of Bodmin Moor to be used for building purposes.

Outside of the granite areas, slate is extensively used for building, both for walls and roofing. Delabole quarry [SX 073 840] near Tintagel, is the largest and best known source of roofing slate, which has a pleasant pale grey colour. The quarry is said to have been continuously worked since Tudor times and a considerable export trade was already in existence by 1602 (Reid *et al.*, 1910). A group of slate quarries in the Wadebridge area has yielded large quantities of 'St Issey stone' used for constructing 'Cornish hedges' and for a variety of purposes where a natural stone finish is desired. There are also many small slate quarries throughout Cornwall that are used locally for walling and construction. Most are producing Devonian slate, but in the Launceston area Carboniferous roofing and building slates were formerly worked.

Sandstone is extracted on a small scale from a few localities in north Cornwall, and was formerly worked from quarries in the Gramscatho Group in mid-Cornwall, as well as the Staddon Grits in southeast Cornwall. Its main use is as a walling stone.

Apart from a few pockets in the Launceston area, limestone is almost absent from Cornwall, and none is presently worked. In the past, agricultural lime was frequently obtained from beach sands containing seashells fragmented by wave action. Sands of this type are found in the Hayle and Padstow estuaries and in the Bude area. A canal was built in the early 19th century (opened 1823) from Bude to Holsworthy to take such sand to the farms of mid-Devon situated on the sour 'Culm' soils. Where these Quaternary sands are locally cemented by $CaCO_3$ (Godrevy Point, Padstow and Fistral Bay), they have had limited use as a building stone.

Sand and gravel have not been worked extensively in Cornwall, although sands in the Hayle and Padstow estuaries and adjoining areas are still used as a source of fine sand for building purposes. During the 1960s and 1970s, when alluvial Sn was being worked, sand and gravel were an important by-product, as for example in Hydraulic Tin's workings at Bissoe.

Bricks were formerly produced from small plants scattered all over Cornwall that mostly worked superficial clays. None are now produced in Cornwall, but a brickworks at Millbrook [SX 435 528], near Torpoint, was active up to the late 1960s. Specialist brickworks, supplying tiles for china clay pan kilns, were active in the St Austell area (Wheal Remfry brickworks [SW 928 578]) until around 1972.

Waste materials from the china clay industry now represent some of the most widely used raw materials by the construction industry in South West England. In the 1980s about 1.5 million tonnes of this waste was used in the construction industry, but the present figure is appreciably lower, due to recession in the building industry.

It is worth recording some cautionary events concerning the use of unconventional raw materials in the local construction industry. One concerns the employment of old metalliferous mine waste to make concrete blocks known as 'mundic blocks'. Mundic means iron pyrites. When exposed to a damp atmosphere pyrite slowly oxidizes, releasing sulphuric acid which dissolves the cement, so that blocks eventually crumble. This process takes a long time to become apparent and has only become a serious problem in recent years. A number of properties have had to be demolished and rebuilt in consequence. Cases of concrete cancer have also been reported in Cornwall. The mechanism involves a normal or high alkali cement being used in conjunction with a reactive chert in the sand or aggregate. This produces a gel that expands when wet and cracks the concrete.

The extractive mineral industry remains one of Cornwall's most important industries, and looks set to maintain this role in the future.

Chapter Sixteen

Environmental Geology

The geology of Cornwall influences the quality of life in many ways. Most obviously, the landscape provides continuing enjoyment to both residents and visitors; a feature enhanced by the availability of attractive building stones and roofing slate. However, not all interactions between the local geology and the inhabitants are so apparent, and not all are beneficial. In this chapter, some of the most important aspects of the environmental geology of Cornwall are reviewed.

Perhaps most critical, the availability of water and water quality are considered first. Rainfall is high, but most of the rocks do not provide conditions suitable for the occurrence of major sources of stored, underground water, and there is rapid run-off. The demands of the tourist industry in dry summers therefore present a particular problem. Additionally, the chemical reaction between rainwater, which even away from sources of pollution is naturally acidic and the rocks and minerals, can give rise to problems, especially in areas where mine waste is present.

Mine waste not only causes contamination of water, but also creates local chemical environments that can inhibit the growth of vegetation. Together with the uneven landscape created by the tips, this gives a derelict appearance to the land (Johnson *et al.*, 1996). Fortunately, such sites are usually localized. Toxic elements present in the soil are also of particular concern, for they may be ingested directly or through the food chain.

Radon is also now recognized as a potential health hazard. This is a radioactive gas that emanates from rocks bearing U. It can be trapped by buildings where its decay products may be inhaled and remain in the lungs. Because these daughter products are radioactive, they damage the lung tissue and can cause lung cancer. Unfortunately, Cornwall has some of the highest concentrations of Rn in buildings found in Britain.

Finally, and on a happier note, the varieties of rocks in Cornwall demand distinctive engineering solutions when constructions, excavations, or foundations are required. Often this means that engineering works blend well with their environment. Rock variability is particularly important in coastal areas, where cliff stability and the recession of the land by erosion are of direct concern. Land stabilization in these circumstances requires detailed knowledge of the local geology.

HYDROLOGY AND WATER QUALITY

Cornwall has a maritime climate with an annual rainfall of 1000–1850 mm distributed fairly evenly throughout the year, and a mean evaporative loss equivalent to about half the rainfall. Surface drainage is abundant, with 1983 km of river characterized by well-developed dendritic river systems draining upland areas, and long estuaries. The rocks of Cornwall, dominated by granite and slate, are generally regarded as 'impermeable'(IGS, 1977) and do not represent major aquifers. These rocks do, however, have a fracture permeability that can sustain limited yields of groundwater for individual farms or dwellings. Rotten granite (growan) and some recent sediments, such as river gravels, also yield local supplies, particularly in granite areas.

Water chemistry is variable, reflecting the underlying lithology and mineralization, and is affected locally by mining, and agriculture. Surface-water quality is generally good, with around 85% of rivers being categorized as Class 1. However, a number of rivers, including the Carnon River and its tributaries and those draining the china clay district around St Austell (DOE, 1985), have been identified as of poor quality or Class 3. The Red River was ranked as Class 4, or bad, but has subsequently been improved. In the metalliferous mining districts of west Cornwall some waters carry enhanced levels of Fe, Mn, Zn, Cu and other metals, including Cd. The release of acidic and heavy-metal-contaminated mine waters to rivers represents a particular environmental hazard, as illustrated by a major pollution incident at Wheal Jane in January 1992 (NRA, 1994). An engineered wetland, in the form of a reed bed, was subsequently constructed to create the geochemical conditions necessary for removal of metals from the mine overflow.

Physical pollution of rivers by sedimentary material was a problem in the past, but the rivers of the St Austell area are now protected from pollution by china clay waste by settling ponds. The release of tailings from South Crofty mine to the Red River had also ceased (prior to its closure), the material being transferred to a tailings pond at Wheal Jane. Water pollution incidents related to agriculture generally comprise the release of livestock slurry and silage effluent into rivers. With their high biological oxygen demand, these are particularly harmful to fish. Groundwater nitrate levels, due to the use of fertilisers, are generally low, but may exceed 20 mgl^{-1} locally.

Granite groundwaters generally have lower values of total dissolved solids (TDS) than waters draining the country rocks (55–370 mgl^{-1} and

75–525 mgl^{-1} respectively). The pH is also lower in granite waters (typically 4–7) than in those of the country rocks (typically 5–8). In mineralized areas, goundwater acidity is due to the presence of sulphides which are oxidized to sulphate and form sulphuric acid. Elsewhere, low pH is due to dissolved CO_2 and organic acids, the latter particularly affecting moorland waters. Groundwaters encountered at a depth of 690 m at South Crofty may reach 43°C, and give salinity values up to 19 100 mgl^{-1} TDS. Stable isotope data show that the salinity is due to water–rock interaction and not to the presence of seawater (Burgess *et al.*, 1982). Shallow granite groundwaters are cool at 10–12°C and give low TDS of around 150 mgl^{-1}. Most groundwaters are characterized by high Rn activities of up to 1 $MBqm^{-3}$. Hydrogeological studies in the Carnmenellis granite to depths of up to 2.6 km at the Rosemanowes geothermal site, have shown that even at depth the granite has a significant fracture permeability. Hydraulic conductivity values for the Carnmenellis granite are typically in the range 10^{-6}–10^{-9} ms^{-1} (Heath & Durrance, 1985), most of this being attributable to a small number of major fractures. The mean yield of boreholes in the Carnmenellis granite is 37 m^3 per day, compared with 34 m^3 per day for the surrounding country rocks (Leveridge *et al.*, 1990).

ENVIRONMENTAL IMPACT OF METALLIFEROUS MINING

Some 2000 years of mining in Cornwall have left an extensive legacy of pollution. The most obvious visual evidence is mine waste or spoil. Recent estimates indicate that this occupies an area of 4888 ha, confirming that the county has the greatest area of contaminated land in the UK.

Extent and Nature of Contamination

Mine waste is highly variable both in extent and visual appearance. The spoil may vegetate naturally, but more often either part or all of the site presents a scene of bare rock and soil, with a sparsely developed cover of vegetation. The variable character of the waste can be explained by the nature of mining operations. Where the waste is dominated by cobbles and boulders, it is the result of exploration and development activities such as shaft sinking, and does not present a chemical hazard. However, very fine-grained tailings, which are the unwanted products of mineral processing, may contain high levels of contaminants. In some instances the tailings were discharged into local rivers, and so extended the pollution beyond the mining district. For example, South Crofty mine used to discharge its tailings into the Red River, and a bright red slurry of haematite and other minerals was carried some 9.5 km into St Ives Bay

[SW 583 423]. Because a mine will often have been worked at different periods, the character of these materials may vary at a single site. Recent activity, such as exploiting the waste for aggregate or fly tipping, may also add to the environmental impact.

The key environmental problems associated with mine waste in Cornwall are high concentrations of heavy metals, low pH, and an inadequate soil structure with a low level of plant nutrients (Johnson *et al.*, 1996). The main contaminants are As and Cu, although Pb and Zn may be of localized importance (Atkinson *et al.*, 1990). Most of our knowledge of these metals comes from reconnaissance sampling, which suggests that the concentrations are heterogenous. However, detailed sampling of the Tresavean mine waste [SW 725 395] indicated that there are distinctive zones of high metal concentrations, in part related to the remnants of the mineral processing plant.

Mine waste in Cornwall is normally very acid, with pH values typically about 5. The acidity is caused by the oxidation of pyrite, which tends to be a ubiquitous gangue mineral associated with Sn and Cu ores. There are wide ranges of values, with pH as low as 2 in water draining from some mine adits, but it is quite common to find mine spoil with pH in excess of 8. This is difficult to explain.

The physical structure of mine waste is often unfavourable for plant growth. The coarse material tends to drain too rapidly, and the fine-grained tailings are often compacted. Furthermore, plant nutrients such as N and P are present at very low levels.

The pollution caused by heavy metals extends much further than the visible mounds of spoil. In a discussion on the heavy metal content of soils and streams in Cornwall, Abrahams & Thornton (1987) estimated that more than 1000 km^2 is either highly or moderately contaminated. This wider pattern of pollution is partly due to the dispersion of metals from mine waste by agencies such as gravity, water and wind. Recent studies suggest that dispersion by gravity, surface water and groundwater is of limited extent, and that wind dispersion is restricted to periods of prolonged dry weather, which may only occur once every three or four years. A study of dispersion from the extensive and poorly vegetated mine waste at Poldice [SW 743 434] indicated that the pattern of dispersion is restricted to a plume extending for up to 150 m downwind from the source. Soil contamination may even be caused by mine waste that has been removed, leaving behind a fingerprint of metal pollution.

High metal concentrations in the soil can also be due to other factors. These include naturally elevated abundances due to the weathering of rocks and mineralized zones. In addition, the widespread use of calciners to remove As from Sn concentrate has resulted in extensive contamination downwind from the plants. The ruins of the calciners form the most intense zones of As contamination.

Pollution is also caused by acid mine water draining from the adits of abandoned mines. This was dramatically illustrated by the release of ochreous water from an adit at the recently closed Wheal Jane mine in January 1992, which contaminated the Fal estuary. This led to Cd concentrations of 600 ugl^{-1} in the River Carnon. This was 20 times the normal background value. Subsequent remedial action has reduced metal concentrations to their previous levels. Although mine adits represent multiple sources of acid drainage in west Cornwall, the elevated metal concentrations are normally rapidly diluted downstream.

Impact of Mine Waste on the Community

An aspect of importance to people living in Cornwall is the extent to which the presence of extensive areas of contamination affect the health of the community. The impact of metal pollution on farmland is one area of concern. Most of the mine waste in Cornwall is surrounded by pasture-land and there is, therefore, a risk of cattle and sheep ingesting contaminated grass and soil. This topic has been reviewed by Thornton *et al.*, (1983). They concluded that crops and livestock can be farmed successfully in areas contaminated by heavy metals because natural mechanisms inhibit the uptake of metal by plants. However, cattle and sheep also ingest substantial quantities of soil, and the take-up of metals from this source requires further research.

The population distribution is also important. For example, the Camborne–Redruth district is one of the largest urban areas in Cornwall, and coincides with the most intense zone of Sn and As mineralization. The average soil concentration of As in this area is 450 ppm, about 50 times the average for normal soils. However, whether or not this represents a hazard to human health is not clear. The health implications of public exposure to As have been reviewed by Mitchell & Barr (1995). They concluded that there is insufficient evidence to substantiate a link between low-level As intake with disease in the human population. A computer-based model is currently being developed by the Department of the Environment, which will form an improved scientific basis for assessing the extent of any health risks associated with metal toxicity (Ferguson & Denner, 1993).

Reclamation of Mine Waste

There is a school of thought that believes mine waste forms part of a unique industrial landscape and should be left to vegetate naturally as a distinctive ecological habitat (Johnson *et al.*, 1996). However, there is a case for reclamation where sites have not recovered and present an eyesore within a county eager to develop tourism. Where sites are prone to erosion, vegetation would limit the extent of secondary contamination.

Initial reclamation was carried out by Cornwall County Council, but more recently various District Councils and the Kerrier Groundwork Trust have been the principal agencies using funds provided by central government. Various approaches to reclamation have been adopted. The Kerrier Groundwork Trust has used metal-tolerant grasses, such as *Agrostis tenuis* (Parys Mountain) and *Festuca rubra litoralis,* in its reclamation of a number of sites including Marriot's shaft [SW 675 397] and at East Basset stamps [SW 693 400]. Here it has deliberately left unvegetated those zones known to be of mineralogical interest. One of the main sites reclaimed by Kerrier District Council is at Tresavean [SW 725 395]. Here topsoil was used as a barrier layer over the areas of high metal concentration, and the former mineral processing site is now used as a football pitch. To date, tree planting has only been variably successful. It is now clear that site preparation, choice of species, planting technique and site management are critical factors for a successful tree planting programme.

In some cases metal values are so high that a more complex approach has to be adopted. For example, the development of Geevor mine [SW 375 348] as a heritage site by Cornwall County Council included the problem posed by a ruined calciner. Soil associated with the collapsed labyrinth of the calciner contained As concentrations up to 30%. In this case a protective barrier layer had to be used to seal off the contamination.

Future Strategy

The scale of contamination in Cornwall is very large, but the intensity of metal contamination has not been clearly linked to a serious public health problem. Hence it is probable that funding from central government will only be available to treat the zones of very high metal concentration and carry out remedial work in areas that are particular eyesores. It is therefore important to rank the areas where work should be carried out, and to ensure that funding is effectively used.

More research is needed immediately to establish the effect of metal contamination on public health. This should involve the development of improved systems of co-operation between research institutes and local authorities carrying out programmes of remedial work. Two aspects demand urgent attention. Firstly, the identification and treatment of all calcination plants because of their high As concentrations, and secondly, the public living in contaminated areas must be advised on factors such as the problems associated with eating home-grown vegetables.

RADON

Among the natural geological hazards of Cornwall is the potential health risk represented by the natural radioactive gas Rn in homes and workplaces. The distribution and behaviour of Rn in the Cornish environment are related to the underlying geology, which controls both the generation and transport of the gas into buildings. Cornwall and Devon are considered by the National Radiological Protection Board (NRPB), which is responsible for radiological protection in the UK, to be 'Affected Areas' as far as Rn in homes is concerned.

Properties and Origin of Radon

A noble gas, Rn is produced by the decay of U and Th in rocks and soils. Of the three major isotopes (*Table 16.1*), ^{222}Rn with its longer half-life is the most important, both in terms of potential risk to human health and as a natural geological tracer. ^{222}Rn is produced by the decay of its immediate parent ^{226}Ra and, in turn, decays by alpha-particle emission to ^{218}Po; the final stable daughter is ^{206}Pb. Exposure of living tissue to alpha-particles is particularly hazardous. Although Rn itself is a gas, its daughters are metals that can become attached to particles in the air. These particles can then be inhaled and retained within the lungs, exposing the tissues directly to alpha-particles and thus giving an increased risk of lung cancer.

The surface distribution of ^{222}Rn reflects both the occurrence of its initial parent, ^{238}U, in the underlying rocks and soils, and the presence of migration pathways via faults and fractures. Rn surveys can, therefore, be used as an aid to U exploration, to fault and fracture mapping, and to the understanding of groundwater flow in fracture systems. The presence of Rn anomalies can indicate either U enrichment in the rocks or soils of the area or the presence of faults or fractures. High Rn concentrations can build up in soil gas and in groundwaters. The migration of Rn from soils, or the presence of high Rn concentrations in groundwaters used as domestic water supplies, can give rise to elevated concentrations indoors. In Cornwall, most high Rn levels are associated with the granites.

Table 16.1. PRINCIPAL RADON ISOTOPES

Rn isotope	Initial parent	Immediate parent	Immediate daughter	Common name	Half-life
^{222}Rn	^{238}U	^{226}Ra	^{218}Po	Radon	3.825 days
^{220}Rn	^{232}Th	^{224}Ra	^{216}Po	Thoron	55.6 sec
^{219}Rn	^{235}U	^{223}Ra	^{215}Po	Actinon	3.92 sec

Radon Surveys

A number of studies have been carried out in South West England, both in relation to Rn as a health hazard, and to its use as a tracer of fracture-controlled geological processes. These studies have shown that a major factor controlling Rn distribution is groundwater flow, ascending groundwater giving rise to high concentrations in the surface environment. Springs appear to be the main cause of significant Rn anomalies in streams, and groundwater-controlled Rn anomalies may contribute to high Rn levels in houses both within and beyond the margins of the granite (Heath, 1991). Another important factor is ground permeability, since Rn is transported more rapidly in permeable than in impermeable ground. Higher levels of Rn thus occur in areas underlain by fractured granite.

Durrance & Heath (1993) proposed a reconnaissance technique for the identification of areas of high Rn risk using stream surveys. A detailed study of Rn abundance in streams was carried out in west Cornwall and the results compared with published data on Rn in houses in the area. The results of the survey are shown in *Figure 16.1*. The good correlation observed between mean stream Rn concentrations and the published indoor Rn concentrations (NRPB 1990a) demonstrated the potential usefulness of regional stream Rn surveys in identifying areas at particular risk. Priorities for detailed surveys can then be established.

Where data are available, no correlation has been observed in any of the studies carried out in Cornwall between Rn and either U or Ra in solution. The Rn is thus transported as a gas unsupported by its parent nuclides.

National Radiological Protection Board Surveys

Cornwall has been the focus of Rn surveys by the NRPB since the 1970s. On the basis of the measurements made, it is estimated that there are about 60 000 dwellings in Cornwall and Devon (12% of the total number of homes in the two counties) in which the inhabitants receive an annual radiation dose (that is, 'effective dose equivalent') of more than the recommended maximum of 20 mSv due to Rn and Rn decay products (NRPB 1990a). This compares with a maximum recommended dose of 5 mSv per year for new dwellings. To ensure compliance with international radiological protection standards, the NRPB has set an 'Action Level' of 200 Bqm^{-3} of Rn in indoor air above which remedial action should be taken to reduce concentrations (NRPB, 1990a). The distribution of homes in Cornwall and Devon exceeding the Action Level is shown in *Figure 16.2*. In view of the widespread distribution of elevated Rn concentrations in South West England, the NRPB recognized the whole of Cornwall and Devon as an 'Affected Area' (NRPB 1990b) and provides free Rn surveys

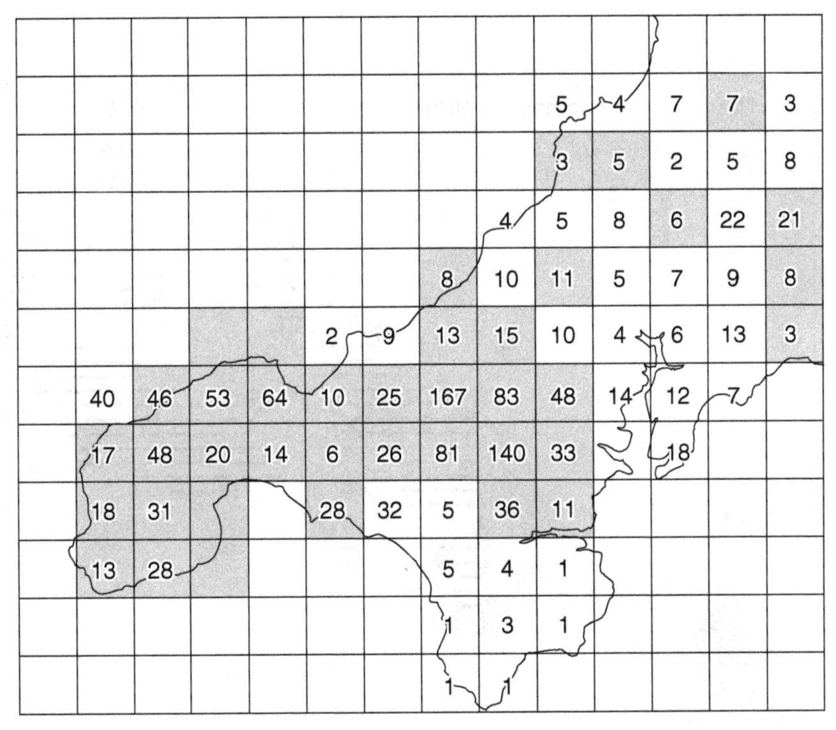

Figure 16.1. Mean stream Rn activities (Bql^{-1}) by 5 km grid squares. The stippled areas are those in which <30% of homes have Rn levels exceeding the NRPB Action Level of 200 Bqm^{-3} (after Durrance & Heath, 1993)

for all householders in the two counties. Although there is a clear association of high Rn levels and the outcrop of the granites (*Figure 16.2*), the presence of suitable migration pathways is also important in determining local risks. Thus not all houses on the granite will have high Rn levels. Conversely, high Rn levels can also occur on other rock types if migration pathways are present. Specific surveys are, therefore, required to determine whether or not a particular dwelling is at risk.

ENGINEERING GEOLOGY

Geology in Cornwall, as elsewhere, has a strong influence on certain aspects of civil engineering and construction activities. Driving through the county on any of the newer sections of road, such as the A30, this influence is apparent in the shape of cuttings and in the rocks sometimes exposed in them. During road construction, geology is also important in the choice and design of foundations for bridges and in the materials for embankments.

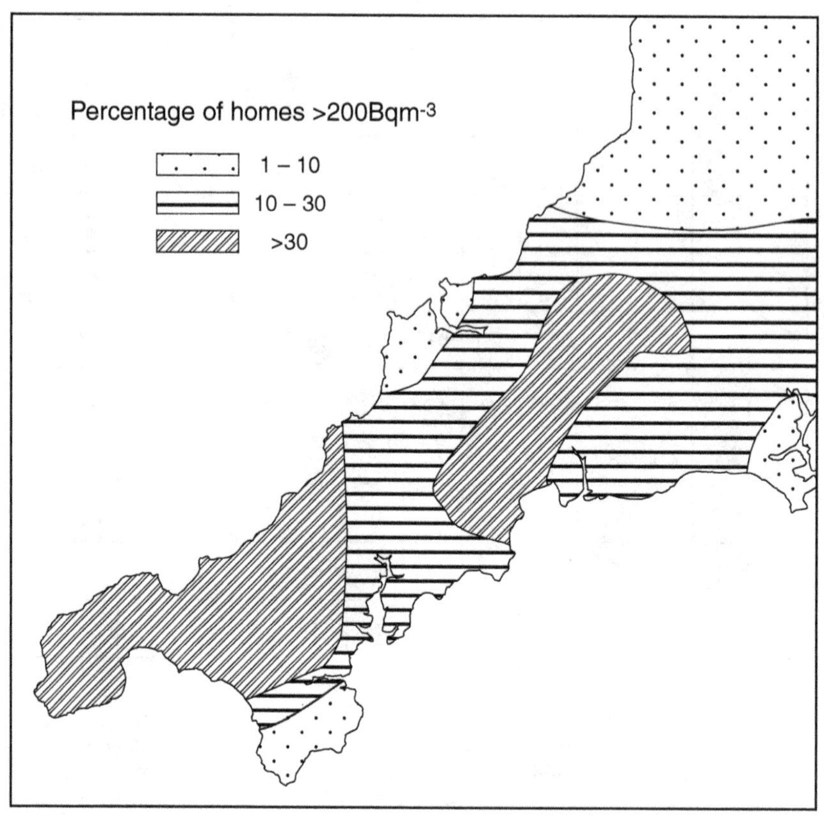

Figure 16.2. Estimated proportion of homes exceeding the Action Level in areas of Devon and Cornwall (after NRPB, 1990b)

Properties of Engineering Rocks in their Unweathered State

Engineering rocks occur beneath a cover of *engineering soils* composed of easily excavated materials such as clay or gravel. Cornwall is regarded as a *hard rock* area, and almost all the Palaeozoic lithologies in their unweathered state are categorized as *strong to extremely strong.* That is they are engineering rock materials with compressive strengths greater than 100 MPa. The weakest materials are the shales and mudstones of the Upper Carboniferous Crackington and Bude formations in north Cornwall.

Much greater distinction between the various rock types, in terms of engineering behaviour *in situ,* has been produced by tectonic activity and the way in which this has influenced the geological structure of different formations. The same intense tectonic deformation and regional metamorphism of all the pre-granite Palaeozoic strata, which have tended to strengthen them as materials, have consequently weakened the rock mass by the formation of fractures or discontinuities. The orientation, spacing, persistence and shear strength along these discontinuities dictate the overall strength and stability of the rock mass. This is especially important on coastal slopes and in engineering excavations. Also, the slates of the region have a low shear strength parallel to their fabric, particularly where this has developed as a planar fracture cleavage. The reorientation of clay particles, as they were metamorphosed to sericite and chlorite, into an axial-planar parallelism resulted in the development of the cleavage and the lower frictional resistance along its fracture surfaces. This structurally-controlled shear strength is at its lowest in the low-grade metamorphic mudrocks of the Upper Carboniferous in north Cornwall.

As almost all of the Palaeozoic strata have been asymmetrically or recumbently folded, bedding-parallel discontinuities can be found dipping at any angle, from horizontal to vertical, thus making slope stability of *in situ* rock highly variable and difficult to predict. The strike direction is less variable and thus more predictable. In west and south Cornwall the strike is generally ENE–WSW, resulting in dip directions to the north-northwest and south-southeast. Where overturning of folds is to the north-northwest, then south-southeasterly dips dominate, and vice versa. However, north of the Rusey Fault Zone (*Figure 6.1*), the strike is generally E–W with overturning to the south, giving northerly dips, but as the folds become upright towards Bude, dips are both to the north and south. Tensile and shear joints in these rocks have also produced closely to widely spaced sets of discontinuities across the bedding and cleavage planes, resulting in blocky rock masses rather than large slabs. Cliff recession is often controlled by the attitude of these blocks.

In contrast to the Palaeozoic slates, the main granite outcrops show no intense deformation, and hence lack dipping discontinuities. Moreover, even the sub-horizontal and subvertical joints, usually widely spaced in fresh granite, are not conducive to sliding in excavated or natural slopes. The coarse granular texture also provides a roughness to these tensile fractures that creates a large angle of friction.

Finally, finer-grained igneous bodies, sandstone-dominated strata, and the high-grade metamorphic rocks seen in the Lizard Complex, are intermediate in structural integrity. They have mass strengths between those of granites and slates.

Properties of Engineering Soils

All civil engineering activity begins at the ground surface, and in many cases only extends to a depth of a few metres (for example most foundations and roads constructed at the original ground level). Apart from topsoil, which is usually set aside for future landscaping, the ground profile encountered normally begins with engineering soil and will at some depth become fresh, unweathered rock. Where there has been no Tertiary or Quaternary sedimentation over the rock, then the soil–rock profile represents the gradational weathering by physical and chemical processes of the underlying rock material. In a consistent climate over several millions of years, equilibrium profiles would develop, maybe tens of metres deep. Such profiles formed during the tropical–subtropical environments present in Cornwall during the Tertiary, but they have been disturbed by the variability of the Quaternary climate, with its preponderance of periglacial conditions. Freeze–thaw cycles, in particular, have greatly affected the weathered zone. That these effects vary with position and aspect, and are found in soils that have different mineralogies and textures depending on their parent rock, means that the engineer is faced with a very complex situation. One of the main difficulties is to specify the level in a weathered rock profile at which the engineering properties reach the required value for accepting a certain vertical load. Another difficulty, frequently a cause of contractual dispute, is defining the boundary between engineering soil (easier and hence cheaper to excavate) and engineering rock. A five-fold graded scale of weathering, from *unweathered* to *completely weathered* helps overcome this problem, the boundary between soil and rock lying within the *moderately weathered* grade.

In many profiles in Cornwall, particularly on slopes greater than 3–4° (that is off the flattest of the plateaux), periglacial activity has actually assisted the engineer in one respect. Often the products of the weathered profile, from completely weathered soil to slightly weathered fractured rock, have been mixed in the top few metres by freeze–thaw disturbance (congeliturbation) and downslope creeping (congelifluction or solifluction) to form head. This can have consistent properties over several metres depth, and a sharp boundary with the underlying rock. The composition of most head deposits in Cornwall, reflecting their hard-rock origins, is of angular rock fragments loosely supported by a matrix of sandy to clayey silt. However, a more clay-rich head, and completely weathered parts of *in situ* rock profiles, are found on the mudrock-dominated strata of north Cornwall. These give rise to poorly drained, low-permeability conditions and lower shear strength soils. In such areas the benefits of periglacial action to the engineer are negated, or even reversed, because the former development of widespread solifluction leaves old, easily reactivated shallow landslides (Grainger & Harris, 1986). Elsewhere, soils are reasonably well drained where found on slopes.

Interestingly, in many head deposits the total silt component of the matrix appears greater than would be expected from weathering alone. An additional windblown component of silt known as loess, has been invoked to explain this (Catt, 1977). The loess is of periglacial origin, probably from an easterly source.

Alluvium of Pleistocene and Recent age occurs in and around the main coastal inlets, estuaries and rivers. Fluctuating sea-level, as well as the variable Quaternary climate, enabled rivers to develop channels to base-levels tens of metres below present sea-level. As glacial melting subsequently raised the base-level, sedimentation kept pace and infilled these channels with normally consolidated alluvium that has never been loaded by more than the present overburden. All types of engineering soil may be encountered in profiles through these deposits, from gravel to sand, silt, clay and peat, with unpredictable lateral and vertical variations. Their generally low shear strength, high compressibility, high variability, high water table and irregular depth to bedrock, result in very difficult foundation conditions. Site investigations for construction over such deposits are consequently more detailed and expensive.

High-level moorland peats may also be thick and form very compressible soils. However, they rarely present more than a minor problem to construction, as a very distinct base with solid rock or sandy soil can be reached by piled foundations. Alternatively, the whole peat deposit may be removed and replaced during road construction; for example, on the A30 northeast of Jamaica Inn [SX 192 775].

Engineering Considerations

Inland, the design of appropriate foundations for buildings is relatively straightforward. Here, the principal engineering problems arise from the thickness and properties of the soil. Larger temporary or permanent excavations, such as for pipelines or road cuttings, are designed according to soil and rock properties. With road cuttings, account must be taken of geological structure relative to the orientation of the excavation. Problems associated with mining activity, such as cavities and contamination of soils and water, are also important to the civil engineer.

At the coast there are special engineering considerations, which relate to the shape of the land and the occurrence of active marine erosion or deposition. Away from the relatively small and localized areas of alluvial, estuarine, barrier beach or dune deposition, a cliffed coastline is generally present. Indeed, this is the feature of the Cornish landscape that many people find most memorable. However, every cliff is the backscarp of an active or dormant landslide and reflects the slow erosion of much of the coastline. The frequency of cliff falls or landslides is very low in many hard-rock headlands and beach-protected cliffs, but is significant on the engineering time-scale of tens of years in other areas. The most rapid general rate of erosion occurs in north Cornwall, where the weaker rock material and highly fractured structure of the Upper Carboniferous formations result in some spectacular landslides. Yet even here, the long-term rate of erosion is small compared to the 1 m per year recorded for some coasts in eastern England. Judging by the width of the rock shore-platform in north Cornwall, and the few thousand years of the current sea-level, a maximum rate of 1 m recession in ten years is considered probable. There are other more localized occurrences of coastal erosion problems. These often appear in relatively low cliffs where there is a considerable thickness of head, perhaps developed over raised-beach deposits resting on a rock shore-platform. Where habitation is close to the cliff top, the landslide risk is increased even though the rate of erosion may be only moderate.

APPENDIX

Geological and Geomorphological Conservation Sites

Three categories of conservation site are recognized in Cornwall: Sites of Special Scientific Interest, Geological Conservation Review sites, and Regionally Important Geological/Geomorphological Sites. Together these highlight areas of greatest geological interest.

Sites of Special Scientific Interest (SSSI). SSSIs are of national and often international significance. They have legal status under the 1981 Wildlife and Countryside Act. English Nature (the governmental advisor on nature conservation in England) is responsible for their designation and monitoring. Many Cornish SSSIs cover large areas and have been designated for both their biological and geological interest. Probably, Cornwall has more hard-rock SSSIs than any other English county.

Geological Conservation Review (GCR) sites. The Nature Conservancy Council (the predecessor to English Nature with a country-wide remit) instituted a national review of all potential geological conservation sites in 1977. The selection of sites into geologically defined subject areas has largely been completed, and a series of reference volumes is being published dealing with the various networks. Currently, only one volume applicable to Cornwall has been published (Floyd, Exley & Styles, 1993), although others will follow. Production of these volumes is a function of the GCR section of the Joint Nature Conservation Committee, a sister organization to English Nature with a national overview remit. Site lists for the unpublished networks are available.

Regionally Important Geological/Geomorphological Sites (RIGS). RIGS are typically of regional significance and are approved by a local voluntary committee. The Cornwall RIGS Group is affiliated to the Cornwall Wildlife Trust. Although RIGS sites do not have the same legal status as SSSIs, they are registered with the relevant District Planning Authorities, and are taken into account when planning applications are considered. RIGS can be proposed by anyone.

The following list is arranged in categories corresponding as closely as possible to the network classification used in the Geological Conservation Review. It should be noted that inclusion of a site does not mean that the public has a right of access to it, or that it is a safe place to visit. Where the site is on private land, the permission of the owner must be sought beforehand. Fortunately, many of the coastal, and a few inland sites, lie on land belonging to conservation bodies such as the National Trust. Sampling is often restricted, and it is advisable to seek further advice or authorization from English Nature (for SSSIs and GCRs), or the Cornwall Wildlife Trust (for RIGS) before collecting specimens.

DEVONIAN

SSSI/GCR Sites

Bedruthan Steps and Park Head SSSI [SW 850 700]. Bedruthen and Trevose slate formations (Lower–Middle Devonian). Fossiliferous slates with primitive fish.

Bull Cove SSSI, near Rame Head [SX 422 485]. In Ramehead and Whitsand Bay Fossiliferous beds near the top of the Dartmouth Group (Lower Devonian). The earliest marine Devonian fauna in Britain.

Daymer Bay [SW 929 776]. Included in the Rock Dunes SSSI and the Trebetherick Point SSSI. Fossiliferous Polzeath Slate Formation (Upper Devonian) is found both on the north and south sides of the Bay.

Gerrans Bay–Camels Cove SSSI [SW 886 370–SW 935 388]. Includes GCR sites at Gerrans Bay, Pendower to Shannick Point, and Nare Head to The Straythe. This 5 km stretch of coast shows the Middle and Upper Devonian sediments of the Gramscatho Basin, including the olistostromic Roseland Breccia Formation.

Gunwalloe–Church Cove [SW 654 653–SW 663 204]. In the Baulk Head to Mullion SSSI. Gramscatho Beds (turbidites) and Roseland Breccia Formation (olistostrome).

Harbour Cove SSSI, near Padstow [SW 915 768]. Type locality for the Harbour Cove Slate Formation (Upper Devonian).

Little Dennis–Porthallow SSSI [SW 787 256–SW 797 232]. Coastal section through the Meneage Mélange (Roseland Breccia Formation). Pillow lava and radiolarian chert at Nelly's Cove [SW 797 235].Normal fault dipping south, juxtaposes the Lizard Complex at Porthallow.

Meneage Coastal Section SSSI [SW 800 240]. Meneage Formation (Middle/ Upper Devonian) of the Gramscatho Group. Showing a transition from plane-laminated sediments to chaotic mélanges.

Mullion Island [SW 660 175]. Included in the Mullion Cliff to Predannack Cliff SSSI. Conodonts from siliceous limestones associated with pillow lavas indicate an Upper Devonian (Frasnian) age.

Pentire Peninsula SSSI, at Pentire Point [SW 923 804] and The Rumps [SW 933 812]. Middle and Upper Devonian slates associated with volcanic rocks. At Gravel Caverns [SW 931 798] a late Devonian conglomerate contains a derived ammonoid fauna.

Perhaver SSSI, near Gorran Haven [SX 014 418]. Coastal section exposes Roseland Breccia Formation, including pillow lavas at Great Perhaver Point.

Poldhu–Polurrian Cove [SW 664 200–SW 668 188]. GCR site in Baulk Head–Mullion SSSI. Coastal section through the Meneage Mélange (Roseland Breccia Formation). At the south end a southward-dipping normal fault juxtaposes the Lizard Complex.

Polyne quarry SSSI, near Polperro [SX 225 531]. Fossiliferous slates of the Meadfoot Group (Lower Devonian).

Rosenun Lane SSSI, near Liskeard [SX 249 617]. Middle Devonian slates with trilobites.

Stourscombe quarry SSSI, near Launceston [SX 344 839]. Stourscombe Formation (Upper Devonian). Slates with a rich clymenid fauna.

Tintagel. See Variscan structures.

Trevone Bay SSSI. Incorporating GCR sites at Pentonwarra Point [SW 890 760] and Marble Cliff [SW 890 763]. Trevose Slate Formation. At the Point a rich fauna of late Middle Devonian ammonoids is found, and at Marble Cliff an inverted alternating sequence of limestones and slates yields Upper Devonian (Frasnian) conodonts.

Viverdon quarry SSSI, near Callington [SX 374 675]. Upper Devonian slates.

Whitsand Bay [SX 380 530]. In Rame Head and Whitsand Bay SSSI. Dartmouth Group.

RIGS

Black Cliff, near Hayle [SW 553 387]. Sandstone-dominated turbidites in the Porthtowan Formation (?Upper Devonian).

Black Head–Ropehaven, near St Austell [SX 039 484]. Meadfoot Group (Lower Devonian). Slates with orthoceratids and corals.

Cant Hill, Camel Estuary [SW 950 742]. Fossiliferous Trevose Slate Formation (Middle Devonian).

Caragloose Point, Roseland [SW 947 399]. Rudite with granite clasts in Roseland Breccia Formation (Upper Devonian).

Carlyon Bay (W), near St Austell [SX 054 520]. Meadfoot Group (Lower Devonian). Slates with orthoceratids, spiriferids and corals.

Carne quarries, Roseland [SW 913 381]. Olistoliths of Ordovician quartzite in Roseland Breccia Formation (Upper Devonian).

Fishing Cove, near Godrevy [SW 599 428]. 60 m thick olistostrome in Porthtowan Formation (Upper Devonian).

Gamas Point, near Pentewan [SX 023 472]. Conformable relationship between the Meadfoot Group (Lower Devonian) and the sediments of the Gramscatho Basin.

Great and Little Hogus [SW 512 306]. Volcaniclastic debris flows within Mylor Slate Formation (Upper Devonian).

Jacka Point, Roseland [SW 939 393]. Polymict conglomerates in Roseland Breccia Formation (Upper Devonian).

Landlake quarries, near Launceston [SX 328 823]. Petherwin Formation (Upper Devonian). Old limestone quarries yielding ammonoids, trilobites and brachiopods.

Lantic Bay [SX 146 509]. Meadfoot Group (Lower Devonian). Sedimentary structures in slates with thin limestones.

Lower Merope Island, near Trevone [SW 895 770]. Harbour Cove Slate Formation (Upper Devonian; Frasnian). Ammonoid bands on the south side of the island.

Nansough quarry, near Ladock [SW 875 510]. Turbidites in the Treworgans Sandstone Member (?Middle Devonian).

Oldwit Lane, near Launceston [SX 318 819]. Lezant Slate Formation (Upper Devonian; Famennian). Ammonoid fauna in slate.

Port Arthur, Padstow [SW 922 747]. Fossiliferous Trevose Slate Formation (Middle Devonian).

Portquin [SW 970 806]. Trevose Slate (Upper Devonian) with ammonoids.

Southwest Constantine Bay [SW 857 745]. Trevose Slate Formation (Middle Devonian). Slates with ammonoids.

Stepper Point, near Padstow [SW 910 780]. Polzeath Slates and Harbour Cove slate formations (Upper Devonian). Pyritized ammonoids on the south side of Butter Hole [SW 908 779].

Underwood quarry, near Launceston [SX 302 872]. Liddadon Formation (Upper Devonian). Slates with ammonoids and trilobites.

Watergate Bay, near Newquay [SW 841 651]. Mudstones belonging to the Dartmouth Group (Lower Devonian). Slates with pteraspidomorph fish.

Whipsiderry, near Newquay [SW 832 635]. Conformable transition between the Dartmouth and Meadfoot groups (Lower Devonian).

CARBONIFEROUS

SSSI/GCR Sites

Boscastle to Widemouth SSSI [SX 140 970]. Widemouth–Crackington section. Type area for the Crackington Formation (Upper Carboniferous); shales and turbiditic sandstones with fossiliferous horizons containing ammonoids.

Bude Coast SSSI [SS 200 070]. Bude Formation (Upper Carboniferous); sandstone and shales with occasional fossiliferous bands containing ammonoids.

Tintagel. See Variscan structures.

Viverdon quarry SSSI, near Callington [SX 374 675]. Lower Carboniferous overlying Devonian, due to thrusting.

Yeolmbridge quarry, [SX 322 875]. Type locality for the Yeolmbridge Formation (Upper Devonian; Famennian–Lower Carboniferous; *Gattendorfia* Stufe). Slates with thin limestones.

RIGS

Penfoot quarry, near Tregadillett [SX 301 832]. Yeolmbridge Formation (Lower Carboniferous). Slates with ammonoids (*Gattendorfia* Stufe).

Upton coast, Widemouth–Bude [SS 197 020–SS 200 061]. Bude Formation (Upper Carboniferous; Westphalian). Folded and faulted sandstones, siltstones and shales.

Vinegar Hill, near St Mellion [SX 397 642]. Crocadon Sandstone Formation (Lower Carboniferous). Deltaic sandstones thrust over Upper Devonian slates.

VARISCAN STRUCTURES

SSSI/GCR Sites

Boscastle to Widemouth SSSI [SX 092 916–SS 194 018]. Incorporates GCR sites at Widemouth to Saltstone Strand, Boscastle, Millook to Foxhole Point, and Rusey Cliff to Buckator (Rusey Fault Zone). Crackington and Bude formations (Upper Carboniferous; Namurian–Westphalian). Strongly folded and faulted slates and sandstones.

Bude Coast SSSI [SS 200 060–SS 200 079]. Folded and faulted Bude Formation (Upper Carboniferous; Westphalian).

Cotty's Point, near Perranporth [SW 757 551]. In Penhale Dunes SSSI. Demonstrates the structural chronology on the northern margin of the Gramscatho Basin.

Godrevy Head to St Agnes SSSI [SW 886 370–SW 935 388]. Includes a GCR site at Godrevy Point [SW 580 430] and Strap Rocks [SW 579 417]. Folding and faulting in Mylor Slate Formation.

Greystone Quarry SSSI [SX 366 805]. Active quarry showing Greystone Formation overthrust by Lezant Slate Formation. Normal faulting with Pb/Ag/Cu mineralization.

Pentire Peninsula SSSI [SW 934 798–SW 938 815]. Includes a GCR site between Polzeath [SW 930 790] and Pentire Point [SW 923 805]. Structures associated with the Padstow Confrontation.

Porthleven Cliffs East SSSI [SW 634 250]. Folding in Mylor Slate Formation.

Rosemullion SSSI, south of the Helford River [SW 795 281–SW 782 267]. Complex folding in Gramscatho Group.

Stepper Point SSSI [SW 915 783]. Complex thrust structures lying south of the Padstow Confrontation (also see RIGS site of the same name in Devonian section which covers a different area).

Tintagel Cliffs SSSI [SX 043 857–SX 095 918]. Incorporates GCR sites at Smiths Cliff to Tintagel Island [SX 050 890] and Trebarwith Strand [SX 949 864]. Low-angle faults repeat Upper Devonian and Lower Carboniferous sequences.

Trevone Bay SSSI. Incorporates a GCR site at Bridge [SW 892 764]. Parasitic folding in Upper Devonian slates and limestones.

Trevose Head to Constantine Bay SSSI [SW 858 753–SW 860 766]. Includes a GCR site from Booby's Bay to Trevose Head [SW 857 751–SW 850 766]. Complex folding and faulting in Middle and Upper Devonian slates, limestones and volcanic rocks.

RIGS

Carlyon Bay (E), near St Austell [SX 068 522]. Folding and volcanic rocks.

Carne quarries (SW 913 381). See Devonian.

Fishing Cove (SW 599 428). See Devonian.

Lantic Bay (SX 146 507). See Devonian.

Porthcadjack Cove, near Portreath [SW 641 447]. Contractional fault systems and 'The Great Cross Course' (dextral strike-slip fault affecting the mineral lodes).

Tregardock Beach, near Delabole [SW 041 840]. Folding and late-stage extensional faulting. Major rotational slip affects cliffs.

Upton Coast, south of Widemouth (SS 197 020–SS 200 061). See Carboniferous.

IGNEOUS ROCKS–LIZARD COMPLEX

SSSI/GCR Sites

Coverack Cove–Dolor Point SSSI [SW 784 187–SW 785 181]. Contact between peridotite and gabbro.

Dean Quarry [SW 803 207]. In Coverack to Porthoustock SSSI. Active quarry in gabbro of the Lizard Complex.

Kennack Sands [SW 734 165]. Included in the Kennack to Coverack SSSI. Kennack gneiss and 'primary' peridotite.

Kynance Cove SSSI [SW 684 133]. Two types of serpentinized peridotite.

Lankidden/Carrick Lûz [SW 756 164]. In the Kennack to Coverack SSSI. Sheared gabbro associated with a transform fault.

Lizard Point [SW 695 116–SW 706 115]. In the Caerthillian to Kennack SSSI. Old Lizard Head metasediments and metabasalts, and Landewednack Hornblende Schists.

Mullion Island [SW 660 175]. In the Mullion Cove to Predannack SSSI. Frasnian tholeiitic pillow lavas.

Polbarrow–The Balk [SW 717 135–SW 715 128]. In the Caerthillian to Kennack SSSI. Kennack Gneiss Group and the thrust contact between the Lizard peridotite and the Landewednack Hornblende Schists.

Porthallow–Porthkerris Cove [SW 797 232–SW 806 230]. In Meneage Coastal Section SSSI. Boundary fault of Lizard Complex at Porthallow, with highly deformed mafic and ultramafic rocks of the Traboe cumulate complex to the south.

Porthoustock Point [SW 810 217]. In Coverack–Porthoustock SSSI. Sheeted dolerite dyke swarm, Lizard Complex.

IGNEOUS ROCKS–PRE-GRANITE VOLCANIC ROCKS

SSSI/GCR Sites

Botallack Head–Porth Ledden, near St Just-in-Penwith [SW 362 339–SW 355 322]. In the Aire Point to Carrick Du SSSI. Thermally metamorphosed basic sills and pillow lavas within the aureole of the Land's End granite.

Carrick Du–Clodgy Point, near St Ives [SW 507 414–SW 512 410]. Pillow lavas in the Mylor Slate Formation affected by low-grade contact metamorphism.

Clicker Tor quarry SSSI, near Menheniot [SX 285 614]. Ultramafic intrusion into Middle Devonian slate.

Cudden Point–Prussia Cove SSSI, Mount's Bay [SW 548 275–SW 555 278]. Metadolerite and metagabbro.

Dinas Head–Trevose Head [SW 847 761–SW 850 766]. In the Aire Point to Carrick Du SSSI. Adinoles developed at the contact between a dolerite and the enclosing sediments.

Greystone Quarry SSSI [SX 366 805). Active quarry in dolerites of Greystone Formation.

Gurnard's Head, near Zennor [SW 432 387]. In the Aire Point to Carrick Du SSSI. Dolerite with a pillowed top.

Nare Head–Blouth Point, near Veryan [SW 920 371]. In the Gerrans Bay to Camels Cove SSSI. Basic pillow lavas and other igneous rocks contained within the Roseland Breccia Formation.

Penlee Point SSSI, near Newlyn [SW 474 269]. Contact metamorphism of a dolerite sill by the Land's End granite.

Pentire Point–Rumps Point [SW 923 805–SW 935 812]. In the Pentire Peninsula SSSI. Upper Devonian pillow lavas.

Polyphant quarry SSSI, near Launceston [SX 262 822]. Picrite in a thrust slice which has been carbonatized to a talc–carbonate–chlorite rock.

Porthleven Cliffs East SSSI [SW 628 254–SW 634 250]. Dolerite intrusions into Mylor Slate Formation.

Tater-du SSSI, near Lamorna [SW 440 230]. Metamorphism of basic lavas, within the aureole of the Land's End granite.

Tintagel Head–Bossiney Haven [SX 047 892–SX 066 895]. In the Tintagel Cliffs SSSI. Basaltic lavas and volcaniclastic rocks of the Tintagel Volcanic Formation.

Trevone Bay SSSI [SW 890 762]. Potassic metadolerite showing pumpellyte facies metamorphism.

RIGS

Black Head, near St Austell [SX 040 480]. Quartz-dolerite intrusion.

Carlyon Bay (E), near St Austell [SX 067 523]. Volcaniclastic rocks and dolerites at Fishing Point in Meadfoot Group (Lower Devonian).

Duporth, near St Austell [SX 036 512]. Dolerite sill intruded into Lower Devonian slates, which has been extensively carbonatized to a talc–chlorite–carbonate rock.

Gannel quarry, near Newquay [SW 795 612]. Weathered lamprophyre dyke.

Lemail, near Wadebridge [SW 022 731]. Lamprophyre dyke, 3 m wide, dipping 45°NNW in railway cutting.

Parson's Beach, Helford River [SW 788 272]. Lamprophyre dyke.

Stepper Point, near Padstow [SW 915 785]. Dolerite intrusion into Upper Devonian slates.

Tubbs Mill Quarry, Roseland [SW 962 432]. Devonian volcanics with a MORB geochemistry in the Veryan nappe.

CORNUBIAN GRANITE BATHOLITH AND ASSOCIATED IGNEOUS ROCKS

SSSI/GCR Sites

Cape Cornwall, near St Just-in-Penwith [SW 352 318]. In the Aire Point to Carrick Du SSSI. Contact between the Land's End granite and metasediments, showing pods of pegmatite, aplite and borosilicate, and evidence of K metasomatism.

Carn Grey Rock and quarry SSSI, near St Austell [SX 033 551]. Granite of an intermediate character between the eastern and western parts of the St Austell granite.

Cligga Head SSSI, near Perranporth [SW 738 537]. Greisen-bordered sheeted-vein system in a small cusp of granite, with W/Sn/Cu/Fe mineralization.

De Lank Quarries SSSI [SX 101 753]. Coarse biotite granite of small megacryst type, with a strong, approximately N–S-trending, subvertical tectonic foliation. This is emphasized by orthoclase megacrysts and irregular bands of strained quartz.

Kingsand–Sandway Point SSSI [SX 440 509]. Rhyolitic lava contemporary with the intrusion of the granites.

Megiliggar Rocks, near Porthleven [SW 609 266]. In the Tremearne-Par SSSI. Pegmatite–aplite–granite sheets cutting Mylor Slate Formation in the roof of the Tregonning granite.

Porthmeor Cove, near Zennor [SW 425 376]. In the Aire Point to Carrick Du SSSI. Two small outlying cusps of the Land's End biotite granite showing pegmatite and granite veins in the roof zone.

Praa Sands [SW 573 280]. In the Folly Rocks SSSI. Felsitic elvan dyke showing evidence of multiple intrusion, possibly in a fluidized state.

Rinsey Cove, near Helston [SW 593 269]. In the Porthcew SSSI. Contact zone of the Tregonning granite, with a pelitic roof pendant and evidence of metasomatism.

Roche Rock SSSI [SW 991 596]. Tor-like mass of schorl composed of tourmaline and quartz, probably representing an accumulation of borosilicate fluid in an outlying cusp of the St Austell granite.

St Mewan Beacon SSSI, near St Austell [SW 985 535]. Quartz–topaz–tourmaline rock forming a small tor-like mass at the southern contact of the St Austell granite with the adjoining Lower Devonian metasediments.

St Michael's Mount SSSI [SW 515 298]. Granite cusp occupies the southern part of the island, shows greisen-bordered vein swarm on the southern wave-cut platform.

Tregarden quarry, Luxulyan [SX 054 591]. (Called Luxulyan Quarry SSSI by English Nature, but do not confuse with abandoned dimension stone quarries farther down the Luxulyan valley.) Active quarry displaying the older, eastern biotite granite of the St Austell pluton; the tourmaline–orthoclase rock called 'luxullianite' is found here.

Tregargus quarries SSSI [SW 949 541]. Quarries in non-megacrystic lithium–mica–topaz granite and fluorite granite, formerly worked for chinastone.

Wheal Martyn SSSI, near St Austell [SX 003 556]. A small exposure in the grounds of the Wheal Martyn Museum shows relatively fresh lithium mica granite, representative of the parent rock from which china clay formed.

RIGS

Beacon Hill, Penryn (A39) by-pass [SW 783 337]. Cutting shows foliated Carnmenellis granite and some Fe mineralization.

Carn Brea, near Redruth [SW 683 407]. Tor exposures of megacrystic biotite granite.

Chywoon Quarry, near Long Downs [SW 748 348]. A N–S-trending pegmatite passes through the quarry.

Goonbarrow, near Bugle [SX 007 583]. Active china clay pit on the site of former Beam Sn mine; intrusive relationships, stockscheider pegmatite, Sn/W mineralization in sheeted vein system, felsitic elvan dykes. Late magmatic/hydrothermal features and kaolinization.

Helman Tor, near Luxulyan [SX 063 615]. Tor of coarse megacrystic biotite granite. Xenoliths and veins of luxullianite.

Tremore Valley West [SX 010 649] and Tremore Valley East [SX 011 648], near Lanivet. Quarries on both sides of valley show the Tremore elvan dyke (a porphyry used as an ornamental stone) intruded into calc-flintas (Lower Devonian impure limestones affected by contact metamorphism).

Venton Cove [SW 527 303]. Hydrothermal tourmaline breccia containing clasts of killas, greenstone and granite, intruded into killas; associated with an elvan dyke.

Wheal Remfry, near Indian Queens [SW 924 574]. Active china clay pit containing a hydrothermal breccia composed of clasts of granite and killas set in a tourmaline/quartz matrix; also a felsitic elvan and intrusive contacts with the adjoining Devonian metasediments.

Wicca Pool, near Zennor [SW 463 398]. Granite veining in slates of the Land's End granite aureole.

Withnoe, Whitsand Bay [SX 404 517]. Plug-like intrusion of rhyolite, with inclined dykes of felsite.

MINERALOGY

SSSI/GCR Sites

Aire Point to Carrick Du SSSI. Includes GCR mineralogy sites at Botallack Mine [SW 364 331]–Wheal Owles [SW 366 325] which yielded important suites of secondary Cu minerals, together with some Pb and U secondaries (botallackite), and skarn mineralization. In Priest's Cove [SW 352 317], a pegmatite with coarse tourmaline is found.

Belowda Beacon SSSI [SW 972 627]. Former mine workings expose topaz–tourmaline–quartz rock with large euhedral topaz crystals.

Cligga Head. SSSI. Includes a GCR mineralogy site at Cligga Head mine [SW 739 538]. Veins contain cassiterite, wolframite, topaz, chalcopyrite and arsenopyrite.

Godrevy–St Agnes SSSI. Includes a GCR mineralogy site at Wheal Coates [SW 699 500]. Towanrath lode in the cliffs shows brecciated hornfels with a matrix of quartz and haematite, occasionally impregnated by sulphides; perfect pseudomorphs of cassiterite after feldspar were formerly obtained.

Great Wheal Fortune SSSI [SW 627 288], near Helston. Mineralization in the Mylor Slate Formation, with cassiterite crystals.

Hingston Down quarry and Consols SSSI [SX 410 718]. Type locality for arthurite (Cu–Fe arsenate). Native Bi, molybdenite, scheelite, wolframite, arsenopyrite, opal and bertrandite (beryllium silicate) have also been found.

Lidcott Mine SSSI, 9 km west of Launceston [SX 241 851]. Small Mn mine which yielded rhodonite, pyrolusite, wad and rhodochrosite.

Mulberry Downs quarry SSSI, near Lanivet [SX 019 658]. Openwork exploited stockwork of veins containing cassiterite, together with small amounts of arsenopyrite, Cu ores and W minerals.

Penberthy Croft Mine SSSI, St Hilary [SW 552 324–SW 556 325]. Prolific source of rare and unusual Cu and Pb arsenates and phosphates.

Penhale Dunes SSSI [SW 765 575]. Includes the Fe/Zn Gravel Hills mine.

Perran Beach–Holywell Bay [SW 760 591–SW 762 578]. In Penhale Dunes SSSI and Kelly Head SSSI. Fe, Zn and Pb mineralization.

Porthgwarra to Pordenack SSSI [SW 357 238]. Sn mineralization in granite at Nanjizal Cove.

St Michael's Mount SSSI [SW 515 298]. Greisen-bordered vein swarm in granite on the south side of the island shows cassiterite, wolframite and green oxidation minerals of Cu.

South Terras Mine SSSI [SW 933 522; SW 934 523; SW 934 524]. Former U mine with secondary U minerals, and Ni and Co arsenides in old dumps.

Trelavour Downs SSSI, near Nanpean [SW 960 575]. Coarsely crystalline pegmatite containing large sheaves of Li-rich biotite.

Tremearne Par SSSI–Megiliggar Rocks [SW 609 266]. Pegmatites contain apatite and unusual phosphate minerals (zwieselite, ixiolite, amblygonite and triplite).

Trevaunance Cove SSSI, near St Agnes [SW 723 517]. Sn (including wood tin on the beach) and Cu mineralization with fluorite.

Wheal Alfred SSSI, near Hayle [SW 580 370]. Complex Sn/Cu/Zn/Pb mineralization with important suite of secondary minerals, including pyromorphite, mimetite and agardite.

Wheal Gorland SSSI, St Day [SW 732 429]. Includes Wheal Unity. An old Cu/Sn/W mine with many rare minerals, including Cu/Pb phosphates and arsenates and other secondaries (cornwallite, olivenite, cuprite, liroconite, clinoclase and pharmacosiderite).

Wheal Penrose SSSI, near Porthleven [SW 634 252]. Old Pb/Zn mine with interesting secondary minerals. Blende with anglesite, phosgenite and pyromorphite.

RIGS

County Bridge Quarry, north of Goonhilly satellite tracking station [SW 722 220]. Disused quarry in serpentine which has produced native Cu, cuprite and other minerals.

Gilson's Cove mine, near Portquin [SW 967 804]. Pb, Zn and Cu minerals.

Gryll's Bunny, near St Just-in-Penwith [SW 364 335]. Within Cape Cornwall to Clodgy Point SSSI; the only 'tin-floor' exposed in South West England.

Harrowbarrow mine near Gunnislake [SX 401 701]. Old Pb/Ag mine.

Kithill [SX 376 713]. Mine waste with cassiterite, wolframite and some molybdenite.

Redmoor mine, near Callington [SX 356 711]. Workings which formerly produced Sn, Pb, Cu and Zn.

Wheal Basset, near Redruth [SW 690 398]. Type locality for bassetite and vochtenite.

Wheal Carpenter, near Hayle [SW 584 353]. Sn and Ag mine (argentiferous galena); type locality for bayldonite.

Wheal Cock [SW 364 340]. Within Cape Cornwall to Clodgy Point SSSI; early mine near Botallack which produced a rich suite of minerals, including chalcomenite (Cu/Se).

Wheal Drea, near St Just-in-Penwith [SW 365 322]. Spoil tip with native Cu, siderite, vivianite and achroite.

Wheal Hazard, near St Just-in-Penwith [SW 363 334]. Within Cape Cornwall to Clodgy Point SSSI; mine workings exhibiting granite contact and mineralization, including heulandite.

Wheal Johnny, near Camborne [SW 627 412]. Cu/Ag/As mineral suite in dumps.

Wheal Phoenix (Stowe's Section), near Minions [SX 262 722]. Surface tips and underground workings with rare minerals, including chalcosiderite.

Wheal Uny, near Redruth [SW 695 410]. Only known access to the Great Flat Lode (Cu/Sn).

POST-OROGENIC (PRE-QUATERNARY) SEDIMENTS

SSSI/GCR Sites

Kingsand–Sandway Point SSSI [SX 441 511]. Early Permian/Late Carboniferous red breccias and sandstones with the trace fossil *Beaconites.*

St Agnes Beacon pits SSSI [SW 705 510]. Sands and clays exposed in working sand pits, with Miocene palynomorphs.

St Erth sandpits SSSI [SW 557 351]. Disused sand pits with gastropods and microfossils of late-Pliocene age.

RIGS

Crousa Common, near St Keverne [SW 771 198]. Quartz gravel with a clayey matrix, of possible Palaeogene age.

PLEISTOCENE–QUATERNARY

SSSI/GCR sites

Dozmary Pool, SSSI, Bodmin Moor [SX 192 743]. Palynological record of the Flandrian vegetational history of Bodmin Moor.

Godrevy, near Hayle [SW 582 423 to SW 581 430]. Within the Godrevy Head to St Agnes SSSI. Raised beach composed of cemented sandrock overlain by head; amino acid dating suggests penultimate interglacial.

Giant's Rock, near Porthleven [SW 623 257]. Within the Porthleven Cliffs SSSI. A large glacial erratic block of garnetiferous gneiss resting on intertidal shore platform.

Hawkstor Pit SSSI, Bodmin Moor [SX 150 746]. Palynomorphs indicate late-glacial Devensian and Holocene deposits.

Pendower beach [SW 902 382]. Within the Gerrans Bay–Camel Cove SSSI. Raised beach deposit. Thermoluminesence dating suggests the penultimate interglacial (oxygen isotope stage 7).

North Naven [SW 356 309]. Within Aire Point–Carrick Du SSSI. Raised boulder beach overlain by head.

Trebetherick Point SSSI, Camel estuary [SW 926 770]. Raised beach overlain by boulder gravel of possible glacigenic origin.

RIGS

Carlyon Bay (E), near St Austell [SX 067 522–SX 076 524]. Raised beach surmounted by head.

Fistral Bay, near Newquay [SW 799 625]. Sandrock raised beach material overlain by head. Penultimate and last interglacial material may be represented.

Praa Sands, Germoe [SW 572 280–SW 587 275]. Most complete late Quaternary sections available in west Cornwall, includes raised beach, head and Holocene sediments.

The Hutches, near Mawnan Smith [SW 791 286]. Wave cut platform and raised beach deposits.

GEOMORPHOLOGY

SSSI/GCR Sites

Gwithian to Mexico Towans SSSI near Hayle [SW 570 395]. Wind-formed dunes composed of sand with a high shell content.

Loe Bar SSSI, including Porthleven Beach East SSSI [SW 629 254–SW 646 235]. Beach and bar formed of shingle and sand with a high flint content.

Tintagel Cliffs SSSI [SX 042 857–SX 095 918]. Coast features include cliffs, caves, geos, arches and stacks.

Whitsand Bay [SX 380 530]. Within Rame Head and Whitsand Bay SSSI. Cliff line little affected by coastal retreat. Beach volume small and contemporary input of sediment neglible.

RIGS

Bog Inn, near St Just [SW 393 319]. Large depression possibly formed by periglacial action.

Carlyon Bay [SX 055 520]. Beach formed of sandy china clay waste washed down from the Carclaze area by a stream discharging onto the beach.

Helman Tor, near Luxulyan [SX 062 616]. The only large classic tor in the eastern part of the St Austell granite, rock bowls and rainwater grooves are visible.

Luxulyan Valley [SX 056 572]. Corestones of granite formed by a past episode of deep weathering, from which the intervening saprolite has been removed.

References

ABRAHAMS, P.W. & THORNTON I. 1987. Distribution and extent of land contaminated by arsenic and associated metals in mining regions of south England. *Transactions of the Institution of Mining and Metallurgy (Section B: Applied Earth Science)*, **96**, 1–8.

AINSWORTH, N.R. 1987. Biostratigraphy of the Lower Cretaceous, Jurassic, and uppermost Triassic of the North Celtic Sea and Fastnet Basin. *In*: BROOKS, J. & GLENNIE, K.W. (eds), *Petroleum geology of North West Europe,* Graham & Trotman, London, 611–622.

ALDERTON, D.H.M. 1993. Mineralization associated with the Cornubian granite batholith. *In:* PATTRICK, R.A.D. & POLYA, D.A. (eds). *Mineralization in the British Isles*, Chapman and Hall, London, 270–354.

ALEXANDER, A.C. & SHAIL, R.K. 1995. Late Variscan structures on the coast between Perranporth and St Ives, south Cornwall. *Proceedings of the Ussher Society,* **8,** 398–404.

ALEXANDER, A.C. & SHAIL, R.K. 1996. Late- to Post-Variscan structures on the coast between Penzance and Pentewan, south Cornwall. *Proceedings of the Ussher Society,* **9,** 72–78.

ALLMAN-WARD, P., HALLS, C., RANKIN, A.H. & BRISTOW, C.M. 1982. An intrusive hydrothermal breccia body at Wheal Remfry in the western part of the St Austell granite pluton, Cornwall, England. *In*: EVANS, A.M. (ed) *Metallization Associated With Acid Magmatism*. Wiley, Chichester, 1–28.

ANDREWS, J.R. 1993. Evidence for Variscan dextral transpression in the Pilton Shales, Croyde Bay, north Devon. *Proceedings of the Ussher Society*, **8**, 198–199.

ANDREWS, J.R., BAKER, A.J. & PAMPLIN, C.F. 1988. A reappraisal of the facing confrontation in north Cornwall: fold- or thrust-dominated tectonics? *Journal of the Geological Society of London*, **145**, 777–788.

ANDREWS, J.R., BAKER, A.J. & PAMPLIN, C.F. 1990. Discussion on a reappraisal of the facing confrontation in North Cornwall: fold- or thrust-dominated tectonics? *Journal of the Geological Society of London*, **147**, 408–410.

ANDREWS, J.R., BULL, J.M., JOLLY, R., ROBERTS, S. & SANDERSON, D.J. 1995. The Carrick-Luz shear zone: Moho or leaky transform? (Abstract). *Proceedings of the Ussher Society*, **8**, 457.

ARKELL, W.J. 1943. The Pleistocene rocks at Trebetherick Point, north Cornwall; their interpretation and correlation. *Proceedings of the Geologists' Association,* **54**, 141–170.

ATKINSON, K., BOULTER, M.C., FRESHNEY, E.C., WALSH, P.T. & WILSON, A.C. 1975. A revision of the geology of the St. Agnes Outlier, Cornwall. *Proceedings of the Ussher Society,* **3**, 286–287.

ATKINSON, K., EDWARDS, R.P., MITCHELL, P.B. & WALLER, C.P. 1990. Roles of Industrial minerals in reducing the impact of metalliferous mine waste in Cornwall. *Transactions of the Institution of Mining & Metallurgy (Section A: Mining Industry)*, **99**, 158–172.

AUSTIN, R.L., DREESEN, R., SELWOOD, E.B. & THOMAS, J.M. 1992. New conodont information relating to the Devonian stratigraphy of the Trevone Basin, north Cornwall, south-west England. *Proceedings of the Ussher Society*, **8**, 23–28.

AUSTIN, R.L. & MATTHEWS, S.C. 1967. In: *Annual Report of the Institute of Geological Sciences for 1966*, 88.

AUSTIN, R.L., ORCHARD, M.J. & STEWART, I.J. 1985. Conodonts of the Devonian System from Great Britain. *In*: HIGGINS, A.C. & AUSTIN R.L. (eds) *Stratigraphical Index of Conodonts*. British Micropalaeontological Society, Ellis Horwood, Chichester, 93–166.

BADHAM, J.P.N. & KIRBY, G.A. 1976. Ophiolites and the generation of ocean crust: data from the Lizard Complex, Cornwall. *Bulletin de la Société géologique de France*, **18**, 885–888.

BARNES, R.P. 1983. The stratigraphy of a sedimentary mélange and associated deposits in south Cornwall. *Proceedings of the Geologists' Association, London,* **94**, 217–229.

BARNES, R.P. 1984. Possible Lizard-derived material in the underlying Meneage Formation. *Journal of the Geological Society of London,* **141,** 79–85.

BARNES, R.P. & ANDREWS, J.R. 1981. Pumpellyite–actinolite grade regional metamorphism in south Cornwall. *Proceedings of the Ussher Society*, **5**, 139–146.

BARNES, R.P. & ANDREWS, J.R. 1984. Hot or cold emplacement of the Lizard Complex? *Journal of the Geological Society of London*, **141**, 37–39.

BARNES, R.P. & ANDREWS, J.R. 1986. Upper Palaeozoic ophiolite generation and obduction in south Cornwall. *Journal of the Geological Society of London,* **143**, 117–124.

BARROW, G. 1906. The geology of the Isles of Scilly. *Memoir of the Geological Survey of Great Britain.*

BARTON, C.M. 1994. Geology of the Liskeard district (Cornwall). 1:10,000 Sheet SX26SE. *Technical Report of the British Geological Survey, Onshore Geology Series, WA/94/30.*

BARTON, C.M., GOODE, A.J.J. & LEVERIDGE, B.E. 1993. Geology of the St Germans district (Cornwall). 1:10,000 sheets SX35NW, SX35SW, SX35NE and SX35SE. *Technical Report of the British Geological Survey, Onshore Geology Series, WA/93/93.*

BARTON, D.B. 1969. *A history of tin mining and smelting in Cornwall.* Bradford Barton, Truro.

BARTON, D.B. 1978. *Copper mining in Cornwall and Devon.* Bradford Barton, Truro.

BASSETT, M.G. 1981. The Ordovician brachiopods of Cornwall. *Geological Magazine,* **118,** 647–664.

BEARE, T. (transcribed and edited by BUCKLEY, J.A.) 1586. *The Bailiff of Blackmoor.* Penhellick Publications, Camborne.

BEER, K.E., BURLEY, A.J. & TOMBS, J.M. 1975. The concealed granite roof in south-west Cornwall. *Institute of Geological Sciences: Mineral Reconnaissance Programme Report.* No 1.

BEESE, A.P. 1984. Stratigraphy of the Upper Devonian argillite succession in north Cornwall. *Geological Magazine*, **121**, 61–69.

BIRD, E.C.F. 1963. Coastal landforms of the Dodman district. *Proceedings of the Ussher Society*, **1**, 56–57.

BIRPS & ECORS, 1986. Deep seismic reflection profiling between England, France and Ireland. *Journal of the Geological Society of London*, **143**, 45–52.

BLUCK, B.J., COPE, J.C.W., & SCRUTTON, C.T. 1992. Devonian. *In*: COPE, J.C.W., INGHAM, J.K. & RAWSON, P.F. (eds) *Atlas of Palaeogeography and Lithofacies.* Geological Society of London, Memoir **13**, 57–66.

BOIS, C., LEFORT, J.-P., LE GALL, B., SIBUET, J.-C., GARIEL, O., PINET, B. & CAZES, M. 1990. Superimposed Variscan, Caledonian and Proterozoic features inferred from deep seismic profiles recorded between southern Ireland, southwestern Britain and western France. *Tectonophysics*, **177**, 15–37.

BORLASE, T. 1754. *Antiquities, historical and monumental, of the County of Cornwall.* Oxford.

BORLASE, W. 1758. *The natural history of Cornwall.* Oxford.

BOWEN, D.Q. 1973. The Pleistocene succession of the Irish Sea. *Proceedings of the Geologists' Association*, **84**, 207–271.

BOWEN, D.Q., ROSE, J., McCABE, A.M. & SUTHERLAND, D.G. 1986. Correlation of Quaternary glaciations in England, Ireland, Scotland and Wales. *Quaternary Science Reviews*, **5**, 299–340.

BOWEN, D.Q., SYKES, G.A., REEVES, A., MILLER, G.H., ANDREWS, J.T., BREW, J.S. & HARE, P.E. 1985. Amino acid geochronology of raised beaches in south west Britain. *Quaternary Science Reviews*, **4**, 279–318.

BOYD, S.R., HALL, A. & PILLINGER, C.T. 1993. The measurement of $d^{15}N$ in crustal rocks by static vacuum mass spectrometry: Application to the origin of ammonium in the Cornubian batholith, southwest England. *Geochimica et Cosmochimica Acta*, **57**, 1339–1347.

BRISTOW, C.M. 1968. The derivation of the Tertiary sediments in the Petrockstow Basin, North Devon. *Proceedings of the Ussher Society*, **2**, 29–35.

BRISTOW, C.M. 1989. The Fal Valley lineaments. *Journal of the Camborne School of Mines*, **89,** 34–41.

BRISTOW, C.M. 1990. Ball clays, weathering and climate. *Proceedings of the 24th Forum on the Geology of Industrial Minerals, Greenville, South Carolina.* South Carolina Geological Survey, Columbia.

BRISTOW, C.M., 1993. Silcrete duricrusts west of the Bovey Basin. *Proceedings of the Ussher Society,* **8**, 177–180.

BRISTOW, C.M. 1994. Environmental aspects of mineral resource conservation in southwest England. *In*: O'HALLORAN, D., GREEN, C., HARLEY, M., STANLEY, M. and KNILL, J. (eds) *Geological and Landscape Conservation*, Geological Society of London, 79–86.

BRISTOW, C.M. 1995. Kaolin–how self-sufficient is Europe?. *Mineral Industry International, Bulletin of the Institution of Mining and Metallurgy*, **1023**, 8–13.

BRISTOW, C.M. 1996. *Cornwall's Geology and Scenery: An Introduction.* Cornish Hillside Publications, St Austell.

BRISTOW, C.M. & EXLEY, C.S. 1994. Historical and Geological Aspects of the China Clay Industry of South-west England. *Transactions of the Royal Geological Society of Cornwall*, **21**, 247–314.

BRISTOW, C.M. & ROBSON, J.L. 1994. Palaeogene basin development in Devon. *Transactions of the Institution of Mining and Metallurgy* (Section B: Applied Earth Sciences), **103**, 163–174.

BRITISH GEOLOGICAL SURVEY, 1984. Penzance. England and Wales Sheet 351 and 358, Solid and Drift Geology. 1:50 000. (British Geological Survey, Keyworth, Nottingham).

BRITISH GEOLOGICAL SURVEY, 1994a. Tavistock. England and Wales Sheet 337. Solid and Drift Geology. 1:50 000. (British Geological Survey, Keyworth, Nottingham).

BRITISH GEOLOGICAL SURVEY, 1994b. Trevose Head and Camelford. England and Wales Sheet 335 and 336. Solid and Drift Geology. 1:50 000. (British Geological Survey, Keyworth, Nottingham).

BROMLEY, A.V. 1989. *Field guide to the Cornubian Orefield*, Sixth International Symposium on Water-Rock Interaction, Malvern, August, 1989, International Association of Geochemistry and Cosmochemistry.

BROMLEY, A.V. & HOLL, J. 1986. Tin mineralisation in southwest England. *In:* WILLIS, B.A. & BARLEY R.W. (eds) *Mineral Processing at a Cross-roads.* NATO ASI Ser. No.117, Martinus Nijhoff, Dordrecht, 159–262.

BROOKS, M., DOODY, J.J. & AL-RAWIL, F.R.J. 1984. Major crustal reflectors beneath SW England. *Journal of the Geological Society of London*, **141,** 97–103.

BROOKS, M. & JAMES, D.G. 1975. The geological results of seismic refraction surveys in the Bristol Channel, 1970–1973. *Journal of the Geological Society of London*, **131**, 163–182.

BROOKS, M., HILLIER, B.V. & MILIORIZOS, M. 1993. New seismic evidence for a major geological boundary at shallow depth, N Devon. *Journal of the Geological Society of London*, **150**, 131–135.

BROWN, A.P. 1977. Late-Devensian and Flandrian vegetational history of Bodmin Moor, Cornwall. *Philosophical Transactions of the Royal Society of London*, B **276**, 251–320.

BUCKLEY, J.A. 1981. *A history of South Crofty.* Dyllansow Truran, Redruth.

BURGESS, W.G., EDMUNDS, W.M. ANDREWS, J.N., KAY, R.L.F. & LEE, D.J. 1982. *The origin and circulation of goundwater in the Carnmenellis granite: the hydrogeological evidence*. Institute of Geological Sciences, London.

BURNE, R.V. 1995. The return of 'The fan that never was': Westphalian turbidite systems in the Variscan Culm Basin: Bude Formation (South West England). *In*: PLINT, A.G. (ed.) *Sedimentary facies analysis.* International Association of Sedimentologists, Special Publication **22**, 101–135.

BURT, R. 1984. *The British lead mining industry*. Dyllansow Truran, Redruth.

BURT, R. 1988. Arsenic—its significance for the survival of South-western metal mining in the late 19th and 20th centuries. *Journal of the Trevithick Society*, **15**, 5–26.

BURT, R., WAITE, P. & BURNLEY, R. 1987. *Cornish mines.* University of Exeter Press, Exeter.

BURT, R. & WAITE P. 1988. *Bibliography of the history of British metal mining.* University of Exeter Press, Exeter.

BURT, R. & WILKIE I. 1984. Manganese mining in South-west England. *Journal of the Trevithick Society,* **11,** 18–40.

BURTON, C.J. & TANNER, P.W.G. 1986. The stratigraphy and structure of the Devonian rocks around Liskeard, east Cornwall, with regional implications. *Journal of the Geological Society of London*, **143**, 96–105.

CAMM, G.S. 1995. *Gold in the counties of Cornwall and Devon*. Cornish Hillside Publications, St Austell.

CANTRILL, T.C., SHERLOCK, R.L. & DEWEY, H. 1919. *Iron ores: sundry unbedded ores of Durham, east Cumberland, north Wales, Derbyshire, the Isle of Man, Bristol district and Somerset, Devon and Cornwall.* Special reports of the mineral resources of Great Britain. IX. *Memoir of the Geological Survey of Great Britain.*

CASELDINE, C.J. 1980. Environmental change in Cornwall during the last 13,000 years. *Cornish Archaeology*, **19**, 3–16.

CATHELINEAU, M. 1988. Cation site occupancy in chlorites and illite as a function of temperature. *Clay Minerals*, **23**, 471–485.

CATT, J.A. 1977. Loess and coversands. *In*: SHOTTON, F.W. (ed) *British Quaternary Studies—Recent Advances*, Oxford University Press, Oxford, 221–229.

CATT, J.A. & STAINES, S.J. 1982. Loess in Cornwall. *Proceedings of the Ussher Society*, **5**, 368–376.

CHANDLER, P. & ISAAC, K.P. 1982. The geological setting, geochemistry and significance of Lower Carboniferous basic volcanic rocks in central south-west England. *Proceedings of the Ussher Society,* **5**, 279–288.

CHANDLER, P., DAVEY, R.F., DURRANCE, E.M. & JADY, R.J. 1984. A gravity survey of the Polyphant Ultrabasic Complex, East Cornwall. *Proceedings of the Ussher Society,* **5**, 116–120.

CHAPMAN, T.J. 1989. The Permian to Cretaceous structural evolution of the Western Approaches Basin (Melville sub-basin), UK. *In:* COOPER, M.A. & WILLIAMS, G.D. (eds). *Inversion Tectonics,* Geological Society of London, Special Publication No. 44, London, 177–200.

CHAROY, B. 1986. The genesis of the Cornubian batholith (South-west England): the example of the Carnmenellis pluton. *Journal of Petrology*, **27**, 571–604.

CHEN, Y., CLARK, A.H., FARRAR, E., WASTENEYS, H.A.H.P., HODGSON, M.J. & BROMLEY, A.V. 1993. Diachronous and independent histories of plutonism and mineralization in the Cornubian Batholith, southwest England. *Journal of the Geological Society*, **150**, 1183–1191.

CHESLEY, J.T., HALLIDAY, A.N., & SCRIVENER, R.C. 1991. Samarium-Neodymium direct dating of fluorite mineralisation. *Science,* **252**, 949–951.

CHESLEY, J.T., HALLIDAY, A.N., SNEE, L.W., MEZGER, K., SHEPHERD, T.J. & SCRIVENER, R.C. 1993. Thermochronology of the Cornubian batholith in southwest England: Implications for pluton emplacement and protracted hydrothermal mineralization. *Geochimica et Cosmochimica Acta*, **57**, 1817–1835.

CLARK, A.H., CHEN, Y., FARRAR, E., WASTENEYS, H.A.H.P., STIMAC, J.A., HODGSON, M.J., WILLIS-RICHARDS, J. & BROMLEY, A.V. 1993. The Cornubian Sn–Cu (–As, W) Metallogenetic Province: product of a 30 m.y. history of discrete and concomitant anatectic, intrusive and hydrothermal events. *Proceedings of the Ussher Society*, **8**, 112–116.

CLARKE, B.B. 1965. The superficial deposits of the Camel estuary and suggested stages in its Pleistocene history. *Transactions of the Royal Geological Society of Cornwall*, **19**, 257–279.

CLARKE, B.B. 1969. The problem of the nature, origin and stratigraphical position of the Trebetherick boulder gravel. *Proceedings of the Ussher Society*, **2**, 87–91.

CLARKE, B.B. 1973. The Camel estuary Pleistocene section west of Tregunna House. *Proceedings of the Ussher Society*, **2**, 551–553.

CLAYTON, R.E., SCRIVENER, R.C., & STANLEY, C.J. 1990. Mineralization and preliminary fluid inclusion studies of lead-antimony mineralisation in north Cornwall. *Proceedings of the Ussher Society,* **7**, 258–262.

COCKS, L.R.M. 1993. Triassic pebbles, derived fossils and the Ordovician to Devonian palaeogeography of Europe. *Journal of the Geological Society of London,* **150**, 219–226.

COCKS, L.R.M., McKERROW, W.S. & VAN STALL, C.R. 1997. The margins of Avalonia. *Geological Magazine*, **134**, 627–636.

COLLINS, J.H. 1912. Observations on the west of England mining region. *Transactions of the Royal Geological Society of Cornwall*, **14**, 1–683.

CONYBEARE, W.D. & PHILLIPS, J. 1822. *Outline of the geology of England and Wales*. London.

COOPER, J.A.G. 1987. A chert microfauna from the Gramscatho Group of the Lizard Peninsula, Cornwall. *Proceedings of the Geologists' Association,* **98,** 75–76.

COPE, J.C.W. 1987. The Pre-Devonian geology of south-west England. *Proceedings of the Ussher Society*, **7**, 468–473.

COPE, J.C.W. & BASSETT, M.G. 1987. Sediment sources and Palaeozoic history of the Bristol Channel area. *Proceedings of the Geologists' Association*, **98,** 315–330.

COPE, J.C.W., INGHAM, J.K., & RAWSON, P.F. (eds) 1992. Atlas of Palaeogeography and Lithofacies. *Geological Society of London, Memoir* **13**.

CORNISH CHAMBER OF MINES. *Annual Reports.* Truro.
COSGROVE, M.E. & ELLIOTT, M.H. 1976. Supra-batholithic volcanism of the south west England granites. *Proceedings of the Ussher Society*, **3**, 391–401.
COWARD, M.P. & McCLAY, K.R. 1983. Thrust tectonics of South Devon. *Journal of the Geological Society of London.* **140**, 215–228.
COWARD, M.P. & SMALLWOOD, S. 1984. An interpretation of the Variscan tectonics of S W Britain. *In*: HUTTON, D.H.W. & SANDERSON, D.J. (eds), *Variscan Tectonics of the North Atlantic Region.* Geological Society Special Publication, **14**, 89–102.
COWARD, M.P. & TRUDGILL, B. 1987. Basin development and basement structure of the Celtic Sea Basins (S.W. Britain). *Bulletin Société Géologique Français,* **3**, 423–436.
CRESSWELL, D. 1983. Deformation of weathered profiles, below head, at Constantine Bay, north Cornwall. *Proceedings of the Ussher Society,* **5**, 487.
CULLINGFORD, R.A. 1982. The Quaternary. *In*: DURRANCE, E.M. & LAMING, D.J.C. (eds) *The geology of Devon.* University of Exeter Press, Exeter, 249–290.
DARBYSHIRE, D.P.F. & SHEPHERD, T.J. 1985. Chronology of granite magmatism and associated mineralization, SW England. *Journal of the Geological Society of London*, **142**, 1159–1177.
DARBYSHIRE, D.P.F. & SHEPHERD, T.J. 1987. Chronology of granite magmatism in south-west England: the minor intrusions. *Proceedings of the Ussher Society*, **6**, 431–438.
DARBYSHIRE, D.P.F. & SHEPHERD, T.J. 1994. Nd and Sr isotope constraints on the origin of the Cornubian batholith, SW England. *Journal of the Geological Society of London*, **151**, 795–802.
DAVIES, G.R. 1984. Isotopic evolution of the Lizard Complex. *Journal of the Geological Society of London*, **141**, 3–14.
DAVIS, P.G. 1990. A late Gedinnian–early Siegenian palynomorph assemblage from the Dartmouth Beds of north Cornwall (Abstract). *Proceedings of the Ussher Society,* **7**, 307.
DAY, G.A. 1986. The Hercynian evolution of the South West British Continental Margin. *In:* Reflection seismology: the continental crust. *American Geophysical Union, Geodynamics Series,* **14**, 233–241.
DAY, G.A. & EDWARDS, J.W.F. 1983. Variscan thrusting in the basement of the English Channel and SW Approaches. *Proceedings of the Ussher Society,* **5**, 432–436.
DAY, J. 1973. *Bristol brass.* David & Charles, Newton Abbot.
DEARMAN, W.R. 1963. Wrench-faulting in Cornwall and South Devon. *Proceedings of the Geologists' Association,* **74**, 265–287.
DEARMAN, W.R. 1964a. Refolded folds in the Dartmouth Slates at Portwrinkle, south Cornwall. *Proceedings of the Ussher Society*, **1**, 79–81.
DEARMAN, W.R. 1964b. Some observations on boudinage structures in Cornwall. *Proceedings of the Ussher Society*, **1**, 81–83.
DEARMAN, W.R. 1969. Tergiversate folds from South-west England. *Proceedings of the Ussher Society,* **2**, 112–115.
DEARMAN, W.R., FRESHNEY, E.C., SELWOOD, E.B., SIMPSON, S., STONE, M. & TAYLOR, R.T. 1971. Symposium on the structure of SW England. *Proceedings of the Ussher Society,* **2**, 220–263.
DEARMAN, W.R., LEVERIDGE, B.E., RATTEY, R.P. & SANDERSON, D.J. 1980. Superposed folding at Rosemullion Head, South Cornwall. *Proceedings of the Ussher Society,* **5**, 33–38.
DE LA BECHE, H.T. 1839. *Report on the geology of Cornwall, Devon and West Somerset*. Longman, London.

DEPARTMENT OF THE ENVIRONMENT, 1985. *River quality in England and Wales 1985. A report of the 1985 survey*. HMSO, London.

DEWEY, H. 1914. The geology of north Cornwall. *Proceedings of the Geologists' Association,* **25**,154–179.

DEWEY, H. 1920. Arsenic and Antimony Ores. Special reports on the mineral resources of Great Britain, **XV**. *Memoir of the Geological Survey of Great Britain.*

DEWEY, H. 1925. The mineral zones of Cornwall. *Proceedings of the Geologists' Association*, **36**, 107–135.

DEWEY, H. & DINES, H.G. 1923. Tungsten and Manganese Ores. Special reports on the mineral resources of Great Britain, **1**. *Memoir of the Geological Survey of Great Britain.*

DINES, H.G. 1934. The lateral extent of the ore-shoots in the primary depth zones of Cornwall. *Transactions of the Royal Geological Society of Cornwall*, **16**, 279–296.

DINES, H.G. 1956. *The metalliferous mining region of South-West England.* 2 vols, HMSO, London.

DOODY, J.J. & BROOKS, M. 1986. Seismic refraction investigation of the structural setting of the Lizard and Start complexes, SW England. *Journal of the Geological Society of London*, **143**, 135–140.

DOUCH, H.L. 1964. *East Wheal Rose.* Bradford Barton, Truro.

DUNHAM, K.C., BEER, K.E., ELLIS, R.A., GALLAGHER, M.J. & NUTT, M.J.C. 1978. *United Kingdom*, Vol. 1 *Northwest Europe*. *In:* BOWIE, S.H.U., KVALHEIM, A. & HASLAM, H (eds) *Mineral Deposits of Europe.* Institute of Mineralogy, Metallurgy and Mineralization Society, London, 263–317.

DURNING, B. 1989. A new model for the development of the Variscan facing confrontation at Padstow, north Cornwall. *Proceedings of the Ussher Society*, **7**, 141–145.

DURRANCE, E.M. 1985. Lower Devonian acid igneous rocks of south Devon: Implications for Variscan plate tectonics. *Proceedings of the Ussher Society,* **6,** 205–210.

DURRANCE, E.M., BROMLEY, A.V., BRISTOW, C.M., HEATH, M.J. & PENMAN, J.M. 1982. Hydrothermal circulation and post-magmatic changes in the granites of south-west-England. *Proceedings of the Ussher Society*, **5**, 304–320.

DURRANCE, E.M. & HEATH, M.J. 1993. Radon in West Cornwall, England. *Proc. 2me Colloque International sur la Géochimie des Gaz (C.I.G.G.)*, Université de Franche-Comté, Besançon/Montbéliard, France, 5–9 July 1993.

EARL, B. 1983. Arsenic winning and refining methods in the west of England. *Journal of the Trevithick Society,* **10,** 9–29.

EARL, B. 1994. *Cornish mining: the techniques of metal mining in the west of England, past and present.* Trevithick Society.

EDMONDS, E.A., WRIGHT, J.E., BEER, K.E., HAWKES, J.R., WILLIAMS, M., FRESHNEY, E.C. & FENNING, P.J. 1968. Geology of the Country around Okehampton. *Memoir of the Geological Survey of Great Britain.*

EDMUNDS, W.M., ANDREWS, J.N., BURGESS, W.G., KAY, R.L.F. & LEE, D.J. (1984). The evolution of saline and thermal groundwaters in the Carnmenellis Granite. *Mineralogical Magazine*, **48**, 407–424.

EDWARDS, J.W.F., DAY, G.A. & LEVERIDGE, B.E. 1989. Thrusts under Mount's Bay and Plymouth Bay. *Proceedings of the Ussher Society,* **7,** 131–135.

EDWARDS, R.P. 1976. Aspects of trace metal and ore distribution in Cornwall. *Transactions of the Institute of Mining and Metallurgy* (Section B: Applied Earth Sciences), **85**, 83–90.

ENFIELD, M.A., GILLCRIST, J.R., PALMER, S.N. & WHALLEY, J.S. 1985. Structural and sedimentary evidence for the early tectonic history of the Bude and Crackington Formations, north Cornwall and Devon. *Proceedings of the Ussher Society*, **6**, 165–172.

EVANS, C.D.R. 1990. *United Kingdom offshore regional report: the geology of the western English Channel and its western approaches*. British Geological Survey, HMSO, London.

EVANS, K.M. 1981. A marine fauna from the Dartmouth Beds (Lower Devonian) of Cornwall. *Geological Magazine*, **118**, 517–523.

EVANS, K.M. 1985. The brachiopod fauna of the Meadfoot Group (Lower Devonian) of the Torbay area, south Devon. *Geological Journal*, **20**, 81–90.

EXLEY, C.S. 1976. Observations on the formation of kaolinite in the St Austell granite. *Clay Minerals*, **11**, 51–63.

EXLEY, C.S., STONE.M. & FLOYD, P.A. 1983. Composition and petrogenesis of the Cornubian granite batholith and post-orogenic volcanic rocks in southwest England. *In*: HANCOCK, P.L. (ed.) *The Variscan fold belt in the British Isles*. Adam Hilger, Bristol, 153–185.

EYLES, N. & McCABE, A.M. 1989. The Late Devensian (>22,000 BP) Irish Sea Basin: the sedimentary record of a collapsed ice sheet margin. *Quaternary Science Reviews*, **8**, 307–351.

FEHN, U. 1985. Post-magmatic convection related to high heat production in granites of Southwest England: a theoretical study. *In*: *High Heat Production (HHP) granites, hydrothermal circulation and ore genesis*. Institution of Mining and Metallurgy, 99–112.

FERGUSON, C. & DENNER, J. 1993. Soil guideline values in the UK: a new risk based approach. *In*: ARENDT, F., ANNOKKEE, G.J., BOSMAN, R. & VAN DEN BRINK, W.J. (eds) Contaminated Soil '93. Kluwer Academic Publishers, Netherlands.

FLETT J.S. 1946. Geology of the Lizard and Meneage (2nd ed.). *Memoir of the Geological Survey of Great Britain*.

FLOYD, P.A. 1984. Geochemical characteristics and comparison of the basic rocks of the Lizard Complex and the basaltic lavas within the Hercynian troughs of SW England. *Journal of the Geological Society of London*, **141**, 61–70.

FLOYD, P.A., EXLEY, C.S. & STYLES, M.T. 1993. *Igneous rocks of south-west England*. Chapman & Hall, London.

FOX, H. 1905. Devonian fossils from the parish of St. Minver, North Cornwall. *Geological Magazine*, **52**, 145–150.

FRENCH, C. 1985. *Cornish Biological Records*. Institute of Cornish Studies Series, Issue No. 8.

FRESHNEY, E.C. 1965. Low-angle faulting in the Boscastle area. *Proceedings of the Ussher Society*, **1**, 175–180.

FRESHNEY, E.C., EDMONDS, E.A., TAYLOR, R.T. & WILLIAMS, B.J. 1979. Geology of the country around Bude and Bradworthy. *Memoir of the Geological Survey of Great Britain*.

FRESHNEY, E.C, EDWARDS, R.A., ISAAC, K.P., WITTE, G., WILKINSON, G.C., BOULTER, M.C., & BAIN, J.A. 1982. A Tertiary basin at Dutson, near Launceston, Cornwall, England. *Proceedings of the Geologists' Association*. **93**, 395–402.

FRESHNEY, E.C., McKEOWN, M.C. & WILLIAMS, M. 1972. Geology of the coast betweem Tintagel and Bude. *Memoir of the Geological Survey of Great Britain*.

FRESHNEY, E.C. & TAYLOR, R.T. 1972. The Upper Carboniferous stratigraphy of North Cornwall and West Devon. *Proceedings of the Ussher Society*, **2**, 464–471.

GAUSS, G.A. 1973. The structure of the Padstow area, North Cornwall. *Proceedings of the Geologists' Association*, **84**, 283–313.

GAUSS, G.A. & HOUSE, M.R. 1972. The Devonian successions in the Padstow area, North Cornwall. *Journal of the Geological Society of London*, **128**, 151–172.

GAYER, R.A. & JONES, J.A. 1989. The Variscan foreland in South Wales. *Proceedings of the Ussher Society*, **7**, 177–179.

GERRARD, A.J.W. 1978. Tors and granite landforms of Dartmoor and eastern Bodmin Moor. *Proceedings of the Ussher Society*, **4**, 204–210.

GIBBONS, W. & THOMPSON, L. 1991. Ophiolitic mylonites in the Lizard complex: ductile extension in the lower crust. *Geology*, **19**, 1009–1012.

GOLDRING, R. 1962. The Bathyal Lull: Upper Devonian and Lower Carboniferous sedimentation in the Variscan geosyncline. *In*: COE, K. (ed) *Some Aspects of the the Variscan fold belt.* Manchester University Press, Manchester, 75–91.

GOLDRING, R & SEILACHER, A. 1971. Limulid undertracks and their sedimentological implications. *Neues Jahrbuch für Geologie und Paläontologie, Abhandlungen,* **137**, 422–442.

GOODAY, A.J. 1974. Ostracod ages from the Upper Devonian purple and green slates around Plymouth. *Proceedings of the Ussher Society*, **3**, 55–62.

GOODE, A.J.J. 1973. The mode of intrusion of Cornish elvans. *Institute of Geological Sciences, Report 73/7.*

GOODE, A.J.J. & LEVERIDGE, B.E. 1991. Geological notes and details for 1:10 000 sheets SW 86 NW (part), NE (part); SW 87 NE, SW (part) and SE; SW 96 NW (part) and NE (part); SW 97 NW (part) and SW (part). (Trevose Head area, north Cornwall district). *Technical Report*, WA/92/11. British Geological Survey.

GOODE, A.J.J. & MERRIMAN, R.J. 1977. Notes on marble and calc-silicate rocks from Duchy Peru borehole, near Perranporth, Cornwall. *Proceedings of the Ussher Society,* **4,** 57–60.

GOODE, A.J.J. & MERRIMAN, R.J. 1987. Evidence of crystalline basement west of the Land's End granite, Cornwall. *Proceedings of the Geologists' Association*, **98**, 39–43.

GOODE, A.J.J. & TAYLOR, R.T. 1988. Geology of the country around Penzance. *Memoir of the British Geological Survey.*

GRACIANSKY, P.C. de & POAG, C.W. 1985. *Initial Reports of the Deep Sea Drilling Project,* **80,** United States Government Printing Office, Washington.

GRAINGER, P. & HARRIS, J. 1986. Weathering and slope stability on Upper Carboniferous mudrocks in south-west England. *Quarterly Journal of Engineering Geology*, **19**, 155–173.

GREEN, D.H. 1964a. The metamorphic aureole of the peridotite at Lizard, Cornwall. *Journal of Geology*, **72**, 543–563.

GREEN, D.H. 1964b. A restudy and reinterpretation of the geology of the Lizard Peninsula, Cornwall. *In*: HOSKING, K.F. & SHRIMPTON, G.J. (eds) *Present views of some aspects of the geology of Cornwall and Devon.* Royal Geological Society of Cornwall Commemorative Volume, 87–114.

GREEVES, T.P.C. 1992. Adventures with fiery dragons–the Cornish tinner in Devon from the 15th to the 20th century. *Journal of the Trevithick Society,* **19**, 2–17.

GRIFFIN, C.G. 1994. Cornwall County Council Minerals Local Plan, Consultation Draft. Cornwall County Council, Truro.

HALL, A. 1988. The distribution of ammonium in the granites from southwest England. *Journal of the Geological Society of London*, **145**, 37–41.

HANCOCK, J.M. 1989. Sea level changes in the British region during the late Cretaceous. *Proceedings of the Geologists' Association,* **100,** 565–594.

HARDING, R.R. & HAWKES, J.R. 1971 The Rb:Sr and K/Rb ratios of samples from the St Austell granite, Cornwall. *Institute of Geological Sciences Report 71/6.*

HARRISON, R.K., SNELLING, N.J., MERRIMAN, R.J., MORGAN, G.E. & GOODE, A.J.J. 1977. The Wolf Rock, Cornwall: new chemical, isotopic age and palaeomagnetic data. *Geological Magazine,* **114,** 249–264.

HART, M.B. 1985. Oceanic anoxic event 2 on-shore and off-shore S.W. England. *Proceedings of the Ussher Society,* **6,** 183–190.

HART, M.B. 1990. Cretaceous sea level changes and global eustatic curves: evidence from S.W. England. *Proceedings of the Ussher Society,* **7,** 268–272.

HART, M.B. & BALL, K.C. 1986. Late Cretaceous anoxic events, sea-level changes and the evolution of the planktonic foraminifera. *In*: SUMMERHAYES, C.P. & SHACKLETON, N.J,. (eds), *North Atlantic Palaeoceanography,* Geological Society of London, Special Publication, **21,** 67–78.

HART, M.B. & DUANE, A. 1989. Late Cretaceous development of the Atlantic Continental Margin off south-west England. *Proceedings of the Ussher Society,* **7,** 165–167.

HARTLEY, A. 1991. Debris flow and slump deposits from the Upper Carboniferous Bude Formation of SW England: implications for Bude Formation facies models. *Proceedings of the Ussher Society,* **7,** 424–426.

HARTLEY, A.J. & WARR, L.N. 1990. Upper Carboniferous foreland basin evolution in SW Britain. *Proceedings of the Ussher Society*, **7,** 212–216.

HARVEY, M., STEWART, S., WILKINSON, J., RUFFELL, A. & SHAIL, R. 1994. Tectonic evolution of the Plymouth Bay Basin. *Proceedings of the Ussher Society,* **8,** 271–278.

HARWOOD, G.M. 1976. The Staddon Grits—or Meadfoot Beds? *Proceedings of the Ussher Society,* **3,** 333–338.

HATCHER, J. 1970. *Rural economy and society in the Duchy of Cornwall 1300–1500.* Cambridge University Press, Cambridge.

HATCHER, J. 1973. *English tin production and trade before 1550.* Clarendon Press, Oxford.

HAWKES, J.R. 1981. A tectonic 'watershed' of fundamental consequence in the post-Westphalian evolution of Cornubia. *Proceedings of the Ussher Society,* **5,** 128–131.

HAWKES, J.R., HARDING, R.R. & DARBYSHIRE, D.P.F. 1975. Petrology and age of the Brannel, South Crofty and Wherry elvan dykes, Cornwall. *Bulletin of the Geological Survey of Great Britain*, **52,** 27–42.

HAWKINS, J. 1832. On a very singular deposit of alluvial matter on St Agnes Beacon, and a granitical rock which occurs in the same situation. *Transactions of the Royal Geological Society of Cornwall,* **4,** 135–144.

HAZLETON, R. & GAWLINSKI S. 1982. Post drilling report on the Egloskerry programme. Unpublished Report MRD/44/82, Open File, BGS Archive.

HEALY, M.G. 1995. The lithostratigraphy and biostratigraphy of a Holocene coastal sediment sequence in Marazion Marsh, west Cornwall, U.K. with reference to relative sea-level movements. *Marine Geology,* **24,** 237–252.

HEATH, M.J. 1991. Radon in the surface waters of south west England and its bearing on uranium distribution, fault and fracture systems and human health. *Quarterly Journal of Engineering Geology*, **24,** 183–189.

HEATH, M.J. & DURRANCE, E.M. 1985. *Radionuclide migration in fractured rock: hydrogeological investigations at an experimental site in the Carnmenellis granite, Cornwall.* Report EUR 9668. Commission of the European Communities, Luxembourg.

HECHT, C.A. 1992. The Variscan evolution of the Culm Basin, south-west England. *Proceedings of the Ussher Society,* **8,** 33–38.

HEIN, U.F., BUCZKO, Z. & BEHR, H.J. 1995. The siderite vein mineralisation of North Devon/West Somerset (SW England): genetic implications from fluid inclusions, chlorite thermometry, stable isotopes and REE fractionation (extended abstract). *Conference abstract volume, SGA Meeting, Prague, August 1995.*

HENDERSON, C.M.B., MARTIN, J.S. & MASON, R.A. 1989. Compositional relations in Li-micas from S.W. England and France: an ion- and electron-microprobe study. *Mineralogical Magazine*, **53**, 427–449.

HENDRIKS, E.M.L. 1931. The stratigraphy of south Cornwall. *Report of the British Association for 1930,* 332.

HENDRIKS, E.M.L. 1937. Rock succession and structure in south Cornwall: a revision. With notes on the Central European facies and Variscan folding there present. *Quarterly Journal of the Geological Society of London,* **93,** 322–367.

HENDRIKS, E.M.L. 1939. The Start–Dodman–Lizard boundary zone in relation to the Alpine structure of Cornwall. *Geological Magazine*, **76**, 385–402.

HENDRIKS, E.M.L. 1959. A summary of present views on the structure of Devon and Cornwall. *Geological Magazine,* **96,** 253–257.

HENDRIKS, E.M.L. 1971. Facies variation in relation to tectonic evolution in Cornwall. *Transactions of the Royal Geological Society of Cornwall,* **20,** 114–151.

HENDRIKS, E.M.L., HOUSE, M.R. & RHODES, F.H.T. 1971. Evidence bearing on the stratigraphical succession in south Cornwall. *Proceedings of the Ussher Society,* **2,** 270–275.

HENLEY, S. 1971. Hedenbergite and sphalerite in the Perran Iron Lode. *Proceedings of the Ussher Society*, **2**, 329–334.

HENLEY, S. 1974, Geochemistry and petrogenesis of elvan dykes in the Perranporth area, Cornwall. *Proceedings of the Ussher Society*, **3**, 136–145.

HENWOOD, W.J. 1843. On the metalliferous deposits of Cornwall and Devon. *Transactions of the Royal Geological Society of Cornwall*, **5**, 1–386.

HEYWORTH, A. & KIDSON, C. 1982. Sea-level changes in southwest England and Wales. *Proceedings of the Geologists' Association*, **93**, 91–111.

HIGGS, R. 1983. The possible influence of storms in the deposition of the Bude Formation (Westphalian), north Cornwall and north Devon. *Proceedings of the Ussher Society,* **5**, 477–478.

HIGGS, R. 1984. Possible wave-influenced sedimentary structures in the Bude Formation (Lower Westphalian, south-west England), and their environmental implications. *Proceedings of the Ussher Society,* **6**, 88–94.

HIGGS, R. 1986. 'Lake Bude' (early Westphalian, S W England): storm-dominated siliclastic shelf sedimentation in an equatorial lake. *Proceedings of the Ussher Society,* **6**, 417–418.

HIGGS, R. 1987. The fan that never was?—Discussion of 'Upper Carboniferous fine–grained turbiditic sandstones from Southwest England: a model for growth in an ancient, delta-fed subsea fan'. *Journal of Sedimentary Petrology,* **57**, 378–382.

HIGGS, R. 1988. Fish trails in the Upper Carboniferous of south-west England. *Palaeontology,* **31**, 255–272.

HIGGS, R. 1991. The Bude Formation (Lower Westphalian), SW England: siliclastic shelf sedimentation in a large equatorial lake. *Sedimentology*, **38**, 445–469.

HILL, J.B. & MACALISTER, D.A. 1906. The geology of Falmouth and Truro and the mining district of Camborne and Redruth. *Memoir of the Geological Survey of Great Britain.*

HILL, P.I. & MANNING, D.A.C. 1987. Multiple intrusions and pervasive hydrothermal circulation in the St Austell Granite, Cornwall. *Proceedings of the Ussher Society*, **6**, 447–453.

HILLIS, R.R. 1988. *The geology and tectonic evolution of the Western Approaches Trough.* PhD Thesis, University of Edinburgh.

HILLIS, R.R. & CHAPMAN, T.J. 1992. Variscan structure and its influence on post-Carboniferous basin development, Western Approaches Basin, SW UK Continental Shelf. *Journal of the Geological Society of London,* **149,** 413–417.

HOBSON, D.M. 1976. The structure of the Dartmouth Antiform. *Proceedings of the Ussher Society,* **3,** 320–332.

HOBSON, D.M. & SANDERSON, D.J. 1983. Variscan deformation in Southwest England. *In*: HANCOCK, P.L. (ed) *The Variscan Fold Belt in the British Isles*, Hilger, Bristol, 108–129.

HOLDER, M.T. & LEVERIDGE, B.E. 1986a. A model for the tectonic evolution of south Cornwall. *Journal of the Geological Society of London*, **143**, 125–134.

HOLDER, M.T. & LEVERIDGE, B.E. 1986b. Correlation of the Rhenohercynian Variscides. *Journal of the Geological Society of London,* **143,** 141–147.

HOLDSWORTH, R.E. 1989. The Start–Perranporth Line: a Devonian terrane boundary in the Variscan orogen of SW England. *Journal of the Geological Society of London,* **146,** 419–421.

HOPKINSON, L. & ROBERTS, S. (1995). Ridge-axis deformation and coeval melt migration within layer 3 gabbros: evidence from the Lizard Complex, U.K. *Contributions to Mineralogy and Petrology*, **121**, 126–138.

HOSKING, K.F.G. 1966. Permo-Carboniferous and later mineralisation of Cornwall and south-west Devon. *In:* HOSKING, K.G.F. & SHRIMPTON G.H. (eds) *Present Views of some Aspects of the Geology of Cornwall and Devon.* 150th Anniversary Volume (for 1964). Royal Geological Society of Cornwall, Penzance, 201–245.

HOSKING, K.F.G. & SHRIMPTON, G.J. 1966. *Present views of some Aspects of the Geology of Cornwall and Devon.* 150th Anniversary Volume (for 1964). Royal Geological Society of Cornwall, Penzance.

HOUSE, M.R. 1963. Devonian ammonoid successions and facies in Devon and Cornwall. *Quarterly Journal of the Geological Society of London*, **70**, 315–321.

HOUSE, M.R. 1985. Correlation of mid-Palaeozoic ammonoid evolutionary events with global sedimentary perturbations. *Nature*, **313**, 17–22.

HUMPHREYS, B. & SMITH, S.A. 1989. The distribution and significance of sedimentary apatite in Lower to Middle Devonian sediments east of Plymouth Sound. *Proceedings of the Ussher Society,* **7,** 118–124.

INSTITUTE OF GEOLOGICAL SCIENCES, 1977. *Hydrogeological map of England and Wales.* Institute of Geological Sciences, London.

ISAAC, K.P. 1983. Tertiary lateritic weathering in Devon, England, and the Palaeogene continental environment of South West England. *Proceedings of the Geologists' Association,* **94**, 105–114.

ISAAC, K.P. 1985. Thrust and nappe tectonics in West Devon. *Proceedings of the Geologists' Association,* **96**, 109–128.

ISAAC, K.P., CHANDLER, P.J, WHITELEY, M.J. & TURNER P.J. 1983. An excursion guide to the geology of central SW England: report on the field meeting to west Devon and east Cornwall, 28–31 May, 1982. *Proceedings of the Geologists' Association*, **94**, 357–76.

ISAAC, K.P., TURNER, P.J., & STEWART, I.J. 1982.The evolution of the Hercynides of central SW England. *Journal of the Geological Society of London*, **139**, 521–531.

JACKSON, N.J. 1974. Grylls Bunny, a 'tin floor' at Botallack. *Proceedings of the Ussher Society*, **3**, 186–188.

JACKSON, N.J. 1975. The Levant Mine Carbona. *Proceedings of the Ussher Society*, **3**, 220–225.

JACKSON, N.J., HALLIDAY, A.N., SHEPPARD, S.M.F. & MITCHELL, J.G. 1982. Hydrothermal activity in the St Just mining district, Cornwall, England. *In:* EVANS, A.M. (ed). *Metallization Associated with Acid Magmatism*, Wiley: Chichester, 137–179.

JACKSON, N.J., WILLIS-RICHARDS, J., MANNING, D.A.C., & SAMS, M. 1989. Evolution of the Cornubian ore field, southwest England: Part II. Mineral deposits and ore-forming processes. *Economic Geology*, **84**, 1101–1133.

JACKSON, R.R. 1991. Vein-arrays and their relationship to transpression during fold development in the Culm Basin, central south-west England. *Proceedings of the Ussher Society*, **7**, 356–362.

JAMES, H.C.L. 1975a. A Pleistocene section at Gunwalloe Fishing Cove, Lizard Peninsula. *Proceedings of the Ussher Society*, **3**, 294–298.

JAMES, H.C.L. 1975b. An examination of recently exposed Pleistocene sections at Godrevy. *Proceedings of the Ussher Society*, **3**, 299–301.

JAMES, H.C.L. 1981. Pleistocene sections at Gerrans Bay, south Cornwall. *Proceedings of the Ussher Society*, **4**, 239–240.

JEFFERIES, N.L. 1985. The distribution of rare-earth elements within the Carnmenellis pluton, Cornwall. *Mineralogical Magazine*, **49**, 495–504.

JENKIN, A.K.H. 1948. *The Cornish miner.* Allen & Unwin, London.

JENKINS, D.G. 1982. The age and palaeoecology of the St Erth beds, southern England, based on foramininera. *Geological Magazine*, **119**, 201–205.

JOHNSON, N., PAYTON, P. & SPALDING, A (eds). 1996. *The Conservation value of metalliferous mine sites in Cornwall.* Cornwall Archaeological Unit and Institute of Cornish Studies, Truro.

JONES, K.A. 1994. The Most Southerly Point Thrust—an example of ductile thrusting in the Lizard Complex, south-west Cornwall. *Proceedings of the Ussher Society,* **8,** 254–261.

JONES, K.A. 1997. Deformation and emplacement of the Lizard Ophiolite Complex, SW England, based on evidence from the Basal Unit. *Journal of the Geological Society of London*, **154**, 871–885.

JONES, R.H. 1991. A seismic reflection survey as part of the geophysical investigation of the Carnmenellis Granite. *Proceedings of the Ussher Society,* **7,** 418–420.

JOWSEY, N.L., PARKIN, D.L., SMITH, A.P.C. & WALSH, P.T. 1988. Recent investigations into the Bovey Formation at Beacon Cottage Farm, St Agnes, Cornwall (Abstract). *Proceedings of the Ussher Society,* **7**, 105.

KING, A.F. 1966. Structure and stratigraphy of the Upper Carboniferous Bude Sandstones, north Cornwall. *Proceedings of the Ussher Society,* **1**, 229–232.

KIRBY, G.A. 1978. Layered gabbros in the Eastern Lizard, Cornwall, and their significance. *Geological Magazine*, **115**, 199–204.

KIRBY, G.A. 1979. The Lizard Complex as an ophiolite. *Nature, London*, **282,** 58–61.

LAMBERT, J.L.M. 1965. A reinterpretation of the breccias in the Meneage crush zone of the Lizard boundary, south-west England. *Quarterly Journal of the Geological Society of London*, **121,** 339–357.

LAMING, D.J.C. 1966. Imbrication, palaeocurrents and other sedimentary features in the Lower New Red Sandstone, Devonshire, England. *Journal of Sedimentary Petrology*, **36**, 940–959.

LAMING, D.J.C. 1982. The New Red Sandstone. *In*: DURRANCE, E.M. & LAMING, D.J.C. (eds), *The Geology of Devon,* University of Exeter Press, Exeter, 148–178.

LANE, A.N. 1966. The structure and stratigraphy of the Lower Devonian rocks of the Looe Area, S.E. Cornwall (Abstract). *Proceedings of the Ussher Society,* **1,** 215–217.

LANE, A.N. 1970. Possible Tertiary deformation of Armorican structures in south east Cornwall. *Proceedings of the Ussher Society,* **2,** 197–204.

LANE, H.R., SANDBERG, C.A. & ZIEGLER, W. 1980 Taxonomy and phylogeny of some lower Carboniferous conodonts and preliminary standard post-*Siphonodella* zonation. *Geologica et Palaeontologica*, **14**, 117–164.

LANG, W.H. 1929. On fossil wood (*Dadoxylon hendriksi,* n.sp) and other plant remains from the clay slates of S. Cornwall. *Annals of Botany,* **43,** 663–683.

LEAKE R.C., STYLES M.T. & ROLLIN, K.E. 1992. Exploration for vanadiferous magnetite and ilmenite in the Lizard Complex, Cornwall. *British Geological Survey Technical Report WF/92/1 (BGS Mineral Reconnaissance Programme Report 117).*

LEEDER, M.R. & GAWTHORPE, R.L. 1987. Sedimentary models for extensional tilt-block/half-graben basins. *In*: COWARD, M.P., DEWEY, J.F. & HANCOCK, P.L. (eds) *Continental Extensional Tectonics.* Special Publication of the Geological Society of London, **28,** 139–152.

LE GALL, B. 1990. Evidence of an imbricate crustal thrust belt in the southern British Variscides: contributions of South-Western approaches travserse (SWAT) deep seismic reflection profiling recorded through the English Channel and the Celtic Sea. *Tectonics*, **9**, 283–302.

LE GALL B., LE HÉRISSÉ, A. & DEUNFF, J. 1985. New palynological data from the Gramscatho Group at the Lizard Front (Cornwall); palaeogeographical and geodynamical implications. *Proceedings of the Geologists' Association*, **96**, 237–253.

LEVERIDGE, B.E. & HOLDER, M.T. 1985. Olistostromic breccias at the Mylor/Gramscatho boundary, south Cornwall. *Proceedings of the Ussher Society,* **6,** 147–154.

LEVERIDGE, B.E., HOLDER, M.T. & DAY, G.A. 1984. Thrust nappe tectonics in the Devonian of south Cornwall and the western English Channel. *In:* HUTTON, D.H.W. & SANDERSON, D.J. (eds) *Variscan Tectonics of the North Atlantic Region.* Special Publication of the Geological Society of London, **14,** 103–112.

LEVERIDGE, B.E., HOLDER, M.T. & GOODE, A.J.J. 1990. Geology of the country around Falmouth. *Memoir of the British Geological Survey.*

LEWIS, G.R. 1908. *The stannaries: a study of the medieval tin miners of Cornwall and Devon.* Reprinted Bradford Barton, Truro.

LINTON, D.L. 1955. The problem of tors. *Geographical Journal*, **121**, 480–487.

LONDON, D. & MANNING, D.A.C. 1995. Chemical variation and significance of tourmaline from southwest England. *Economic Geology*, **90**, 495–519.

LOTT, G.K., KNOX, R.W.O'B., BIGG, P.J., DAVEY, R.J. & MORTON, A.C. 1980. Aptian-Cenomanian stratigraphy in boreholes from offshore south-west England. *Report of the Institute of Geological Sciences, 80/8.*

LUCAS, F.W.A.A. 1993. *The distribution of Uranium and Thorium in the Cornubian batholith.* PhD Thesis, University of Exeter.

MACFADYAN, W.A. 1970. *Geological highlights of the West Country.* Butterworths, London.

MACKINTOSH, D.M. 1964. The Sedimentation of the Crackington Measures. *Proceedings of the Ussher Society,* **1**, 88–89.

McKEOWN, M.C., EDMONDS, E.A, WILLIAMS, M., FRESHNEY, E.C. & MASSON SMITH, D.J. 1973. Geology of the Country around Boscastle and Holsworthy. *Memoir of the Geological Survey of Great Britain.*

MANNING, D.A.C. & EXLEY, C.S. 1984. The origins of late-stage rocks in the St Austell granite–a reinterpretation. *Journal of the Geological Society of London*, **141**, 581–591.

MANNING, D.A.C. & HILL, P.I. 1990. The petrogenetic and metallogenetic significance of topaz granite from the S.W. England orefield. *In*: STEIN, H.J. & HANNAH, J.L. (eds) *Ore-bearing granite systems: petrogenesis and mineralising processes*. Geological Society of America Special Paper, **246**, 51–69.

MANNING, D.A.C., HILL, P.I. & HOWE, J.H. 1996. Primary lithological variation in the kaolinised St Austell Granite, Cornwall, England. *Journal of the Geological Society of London*, **153**, 827–838.

MAPEO, R.J.M. & ANDREWS, J.R. 1991. Pre-folding tectonic contraction and extension of the Bude Formation, North Cornwall. *Proceedings of the Ussher Society*, **4**, 350–355.

MARSH, B.D., 1982. On the mechanics of igneous diapirism, stoping and zone melting. *American Journal of Science*, **282**, 808–855.

MASSON, D.G. & ROBERTS, D.G. 1981. Late Jurassic–early Cretaceous reef trends on the continental margin S.W. of the British Isles. *Journal of the Geological Society of London*, **138**, 437–443.

MATTHEWS, S.C. 1969. A Lower Carboniferous conodont fauna from east Cornwall. *Palaeontology*, **12**, 262–275.

MATTHEWS, S.C. 1970. A new cephalopod fauna from the lower Carboniferous of east Cornwall. *Palaeontology*, **13**, 112–131.

MECHIE, J. & BROOKS, M. 1984. A seismic study of the deep geological structure in the Bristol Channel area, SW Britain. *Geophysical Journal of the Royal Astronomical Society*, **78**, 661–689.

MELVIN, J. 1986. Upper Carboniferous fine-grained turbiditic sandstones from southwest England: a model for growth in an ancient, delta-fed subsea fan. *Journal of Sedimentary Petrology*, **56**, 19–34.

MELVIN, J. 1987. Upper Carboniferous fine-grained turbiditic sandstones from southwest England: a model for growth in an ancient, delta-fed subsea fan-reply. *Journal of Sedimentary Petrology*, **57**, 378–382.

MILLER, J.A. & GREEN, D.H. 1961. Age determinations of rocks in the Lizard (Cornwall) area. *Nature*, **192**, 1175–1176.

MILLSON, J. 1987. The Jurassic evolution of the Celtic Sea basins. *In*: BROOKS, J. & GLENNIE, K.W. (eds), *Petroleum Geology of North West Europe*, Graham & Trotman, London, 599–610.

MITCHELL, G.F. 1960. The Pleistocene history of the Irish Sea. *Advancement of Science*, **17**, 313–325.

MITCHELL, G.F. 1972. The Pleistocene history of the Irish Sea: second approximation. *Scientific Proceedings of the Royal Dublin Society*, **A 4**, 181–199.

MITCHELL, G.F., CATT, J.A., WEIR, A.H., McMILLAN, N.F., MARGEREL, J.P. & WHATLEY, J.C. 1973. The Late Pliocene marine formation at St Erth, Cornwall. *Philosophical Transactions of the Royal Society of London*, B **266**, 1–37.

MITCHELL, G.F. & ORME, A.R. 1967. The Pleistocene deposits of the Isles of Scilly. *Quarterly Journal of the Geological Society of London*, **123**, 59–92.

MITCHELL, P. & BARR, D. 1995. The nature and significance of public exposure to arsenic—a review of its relevance to South West England. *Environmental Geochemistry & Health* (in press).

MONTADERT, L., ROBERTS, D.G., DE CHARPAL, O. & GUENNOC, P. 1979. Rifting and subsidence of the northern continental margin of the Bay of Biscay. *In*: MONTADERT, L. *et al., Initial Reports of the Deep Sea Drilling Project,* **48,** United States Government Printing Office, Washington, 1025–1060.

MOORE, F. & MOORE, D.J. 1979. Fluid inclusion study of mineralization at St

Michael's Mount, Cornwall. *Transactions of the Institute of Mining and Metallurgy (Section B: Applied Earth Sciences)*. **88**, 57–60.
MOORE, J.M. & JACKSON, N.J. 1977. Structure and mineralization in the Cligga Granite stock, Cornwall. *Journal of the Geological Society of London*, **133**, 467–480.
MORRISON, T.A. 1983. *Cornwall's central mines.* 2 vols. Alison Hodge, Penzance.
NATIONAL RADIOLOGICAL PROTECTION BOARD 1990a. Human exposure to Rn in homes. Recommendation for the practical application of the Board's Statement. *Documents of the NRPB*, **1** (1), 17–32.
NATIONAL RADIOLOGICAL PROTECTION BOARD 1990b. Radon affected areas: Cornwall and Devon. *Documents of the NRPB*, **1** (4), 37–43.
NATIONAL RIVERS AUTHORITY, 1994. *Abandoned mines and the water environment*. Report of the National Rivers Authority, March 1994.
NAYLOR, D. & MOUNTENEY, S.N. 1975. *Geology of the North-West European Continental Shelf* (Volume 1), Graham Trotman Dudley, London.
NOALL, C. 1983. *Geevor.* Geevor Tin Mines Plc, Pendeen.
PALMER, J. & NEILSON, R.A. 1962. The origin of granite tors on Dartmoor, Devonshire. *Proceedings of the Yorkshire Geological Society*, **33**, 315–340.
PAMPLIN, C.F. 1990. A model for the tectono-thermal evolution of north Cornwall. *Proceedings of the Ussher Society*, **7**, 206–211.
PAMPLIN, C.F. & ANDREWS J.R. 1988. Timing and sense of shear in the Padstow Facing Confrontation. *Proceedings of the Ussher Society*, **7**, 73–76.
PENHALLURICK, R.D. 1986. *Tin in antiquity.* Institute of Metals.
PHILLIPS, F.C. 1928. Metamorphism in the Upper Devonian of north Cornwall. *Geological Magazine*, **65**, 541–556.
PHILLIPS, W.J. 1841. *Figures and descriptions of the Palaeozoic fossils of Devon, Cornwall and west Somerset*. Longman, Brown, Green and Longman, London.
PINET, B., MONTADERT, L., MASCLE, A., CAZES, M. and BOIS, C. 1987. New insights on the structure and the formation of sedimentary basins from deep seismic profiling in Western Europe. *In*:. BROOKS J. & GLENNIE K, (eds) *Petroleum Geology of North West Europe* Graham and Trotman, London, 11–31.
POUND, C.J. 1983. The sedimentology of the Lower–Middle Devonian Staddon Grits and Jennycliff Slates on the east side of Plymouth Sound, Devon. *Proceedings of the Ussher Society,* **5**, 465–472.
POWER, M.R., ALEXANDER, A.C., SHAIL, R.R. & SCOTT, P.W. 1996. A re-interpretation of the internal structure of the Lizard complex ophiolite, South Cornwall. *Proceedings of the Ussher Society*, **9**, 63–67.
PRIMMER, T.J. 1985. A transition from diagenesis to greenschist facies within a major Variscan fold/thrust complex in south-west England. *Mineralogical Magazine*, **49**, 365–374.
PRYCE, W. 1778. *Mineralogica Cornubiensis, a treatise on minerals mines and mining*. Phillips (printer), London (reprinted 1972, Bradford Barton, Truro).
RAMSBOTTOM, W.H.C., CALVER, M.A., EAGER, R.M.C., HODSON, F., HOLLIDAY, D.W., STUBBLEFIELD, C.J. & WILSON, R.B. 1978. *A correlation of Silesian rocks in the British Isles.* Special Report No 12, Geological Society of London.
RATTEY, P.R. & SANDERSON, D.J. 1982. Patterns of folding within nappes and thrust sheets: examples from the Variscan of south-west England. *Tectonophysics*, **88**, 246–287.
RATTEY, P.R. & SANDERSON D.J. 1984. The structure of SW Cornwall and its bearing on the emplacement of the Lizard Complex. *Journal of the Geological Society of London*, **141**, 87–95.

REID, C. 1890. Pliocene Deposits of Britain. *Memoir of the Geological Survey of the United Kingdom.*

REID, C. 1907. The geology of the country around Mevagissey. *Memoir of the Geological Survey of Great Britain.*

REID, C., BARROW, G. & DEWEY, H. 1910. The geology of the country around Padstow and Camelford. *Memoir of the Geological Survey of Great Britain.*

REID, C., BARROW, G., SHERLOCK, R.L., MACALISTER, D.A. & DEWEY, H. 1911. The geology of the country around Tavistock and Launceston. *Memoir of the Geological Survey of Great Britain.*

REID, C. & SCRIVENOR, J.B. 1906. Geology of the country near Newquay. *Memoir of the Geological Survey of Great Britain.*

RICHARDS, H.G., PARKER, R.H., GREEN, A.S.P., JONES, R.H., NICHOLLS, J.D.M., NICOL, D.A.C., RANDALL, M.M., RICHARDS, S., STEWART, R.C. & WILLIS-RICHARDS, J. 1994. The performance and characteristics of the experimental hot dry rock geothermal reservoir at Rosemanowes, Cornwall (1985–1988). *Geothermics*, **23**, 73–109.

RICHTER, D. 1967. Sedimentology and facies of the Meadfoot Beds (Lower Devonian) in South-east Devon (England). *Geologische Rundschau,* **56,** 543–561.

RIPLEY, M.J. 1965. Structural studies between Holywell Bay and Dinas Head, North Cornwall. *Proceedings of the Ussher Society,* **1,** 174–175.

ROBERTS, S., ANDREWS, J.R., BULL, J.M. & SANDERSON, D.J. 1993. Slow-spreading ridge-axis tectonics: evidence from the Lizard complex, UK. *Earth and Planetary Science Letters,* **116,** 101–112.

ROBERTS, D.G., HUNTER, P.M. & LAUGHTON, A.S. 1977. *Bathymetry of the Northeast Atlantic. Sheet 2, Continental Margin around the British Isles,* 1:240 000 (Taunton: Hydrographer of the Navy for Institute of Oceanographic Sciences).

ROBERTS, R.L. & SANDERSON, D.J. 1971. Polyphase development of slaty cleavage and the confrontation of facing directions in the Devonian rocks of North Cornwall. *Nature, London,* **230**, 87–89.

ROBINSON, D. & READ, D. 1981. Metamorphism and mineral chemistry of greenschists from Trebarwith Strand, Cornwall. *Proceedings of the Ussher Society*, **5**, 132–138.

ROBINSON, D., MAZZOLI, C. & PRIMMER, T.J. 1994. Metabasite parageneses in south-west England. *Proceedings of the Ussher Society*, **8**, 231–236.

ROLLIN, K.E. 1986. Geophysical surveys on the Lizard Complex, Cornwall. *Journal of the Geological Society of London,* **143**, 437–446.

ROWE, J. 1975. The declining year of Cornish mining. *In:* PORTER, J. (ed) *Education and labour in the South-west.* Exeter Papers in Economic History, University of Exeter, Exeter.

RUDWICK, M.J.S. 1979. The Devonian: a System born from conflict. *Special Papers in Palaeontology*, **23**, 9–21.

RUDWICK, M.J.S. 1985. *The Great Devonian Controversy*. University of Chicago Press, Chicago & London.

RUFFELL, A.H. & COWARD, M.P. 1992. Basement tectonics and their relationship to Mesozoic megasequences in the Celtic Seas and Bristol Channel area. *In*: PARNELL, J. (ed.), *Basins on the Atlantic Seaboard: Petroleum Geology, Sedimentology and Basin Evolution*, Geological Society Special Publication, **62,** 385–394.

SADLER, P.M. 1973. An interpretation of new stratigraphic evidence from south Cornwall. *Proceedings of the Ussher Society,* **2,** 535–550.

SADLER, P.M. 1974a. Trilobites from the Gorran Quartzites, Ordovician of south Cornwall. *Palaeontology,* **17,** 71–93.

SADLER, P.M. 1974b. An appraisal of the 'Lizard–Dodman–Start thrust' concept. *Proceedings of the Ussher Society*, **3**, 71–81.

SAMS, M.S. & THOMAS-BETTS, A. 1988. Models of convective fluid flow and mineralization in south-west England. *Journal of the Geological Society of London*, **145**, 809–817.

SANDEMAN, H.A., CLARK, A.H., STYLES, M.T., SCOTT, D.J., MALPAS, J.G. & FARRAR, E. 1997. Geochemistry and U–Pb and ^{40}Ar–^{39}Ar geochronology of the Man of War Gneiss, Lizard Complex, SW England: pre-Hercynian arc-type crust with a Sudeten-Iberian connection. *Journal of the Geological Society of London*, **154**, 403–411.

SANDERSON, D.J. 1971. Superposed folding at the northern margin of the Gramscatho and Mylor Beds, Perranporth, Cornwall. *Proceedings of the Ussher Society*, **2**, 266–269.

SANDERSON, D.J. 1973. Development of fold axes oblique to the regional trend. *Tectonophysics*, **16**, 55–70.

SANDERSON, D.J. 1979. The transition from upright to recumbent folding in the Variscan fold belt of southwest England: a model based on the kinematics of simple shear. *Journal of Structural Geology*, **7**, 171–180.

SANDERSON, D.J. 1984. Structural Variation Across the Northern Margin of the Variscides in NW Europe *In:* HUTTON, D.H.W. & SANDERSON, D.J. (eds) *Variscan Tectonics of the North Atlantic Region.* Special Publication of the Geological Society of London, **14**, 149–165.

SANDERSON, D.J. & DEARMAN, W.R. 1973. Structural zones of the Variscan fold belt in SW England: their location and development. *Journal of the Geological Society of London*, **129**, 527–533.

SCHMITZ, C.J. 1979. *World non-ferrous metal production and prices 1700–1976.* Frank Cass.

SCHNEIDER, F. 1993. Modelling of zone alteration associated with W/Sn ore deposits in Southwest England. *In*: PARNELL, J., RIFFELL, A.H. & MOLES, N.R. (eds) Extended abstracts of Geofluids/93 Conference, Torquay, England, May 4–7, 1993, 280–283.

SCOURSE, J.D. 1987. Periglacial sediments and landforms in the Isles of Scilly and west Cornwall. *In*: BOARDMAN, J. (ed.) *Periglacial processes and landforms in Britain and Ireland.* Cambridge University Press, Cambridge, 225–236.

SCOURSE, J.D. 1991. Late Pleistocene stratigraphy and palaeobotany of the Isles of Scilly. *Philosophical Transactions of the Royal Society of London*, B **334**, 405–448.

SCOURSE, J.D. 1996. Late Pleistocene Stratigraphy of North and West Cornwall. *Transactions of the Royal Geological Society of Cornwall*, **22**, 2–56.

SCRIVENER, R.C. 1982. *Tin and related mineralisation in the Dartmoor Granite.* PhD Thesis, University of Exeter.

SCRIVENER, R.C., LEAKE, R.C., LEVERIDGE, B.E., & SHEPHERD, T.J. 1989. Volcanic-exhalative mineralisation in the variscan province of SW England. *Terra Abstracts*, **1**, 125.

SCRIVENER, R.C., DARBYSHIRE, D.P.F. & SHEPHERD, T.J. 1994. Timing and significance of crosscourse mineralization in SW England. *Journal of the Geological Society, London,* **151**, 587–590.

SEAGO, R.D. & CHAPMAN, T.J. 1988. The confrontation of stuctural styles and the evolution of a foreland basin in central SW England. *Journal of the Geological Society of London*, **145**, 789–800.

SEDGWICK, A. & MURCHISON, R.I. 1839. Classification of the older stratified rocks of Devonshire and Cornwall. *London and Edinburgh Philosophical Magazine and Journal of Science,* **14,** 241–360.

SELWOOD, E.B. 1960. Ammonoids and trilobites from the Upper Devonian and Lower Carboniferous of the Launceston area of Cornwall. *Palaeontology*, **3,** 153–85.

SELWOOD, E.B. 1961. The Upper-Devonian and lower Carboniferous stratigraphy of Boscastle and Tintagel, Cornwall. *Geological Magazine*, **98**, 162–67.

SELWOOD, E.B. 1966. Derived fossils from the Upper Culm Measures south of Launceston, Cornwall. *Proceedings of the Ussher Society,* **1**, 234.

SELWOOD, E.B. 1990. A review of basin development in central south-west England. *Proceedings of the Ussher Society*, **7**, 199–205.

SELWOOD, E.B. & DURRANCE, E.M. 1982. The Devonian rocks. *In:* Durrance, E.M. & LAMING, D.J.C. (eds) *The Geology of Devon,* University of Exeter Press, Exeter, 15–41.

SELWOOD, E.B. & STEWART, I.J. 1985. Devonian-Carboniferous biostratigraphy and conodont localities of the Launceston and St Mellion outlier. *In*: AUSTIN, R.L. & ARMSTRONG, M.A. *Field Excursion A. Devonian and Dinantian conodont localities in South-West England, 20–25 July*. Fourth European Conodont Symposium Field Guide. Nottingham, 63–90.

SELWOOD, E.B.. STEWART, I.J. & THOMAS J.M. 1985. Upper Palaeozoic sediments and structure in north Cornwall—a reinterpretation. *Proceedings of the Geologists' Association*, **96**, 129–41.

SELWOOD, E.B. & THOMAS, J.M. 1986a. Variscan facies and structure in central SW England. *Journal of the Geological Society of London.* **143**, 199–207.

SELWOOD, E.B. & THOMAS J.M 1986b. Upper Palaeozoic successions and nappe structures in north Cornwall. *Journal of the Geological Society of London*, **143**, 75–82.

SELWOOD, E.B. & THOMAS, J.M. 1988. The Padstow Confrontation, north Cornwall: a reappraisal. *Journal of the Geological Society of London*, **145**, 801–807.

SELWOOD, E.B. & THOMAS, J.M. 1993. The Tredorn Nappe, north Cornwall: a review. *Proceedings of the Ussher Society,* **8**, 89–93.

SELWOOD, E.B., THOMAS, J.M., BORLEY, G.D., & DEAN, A. 1993. A revision of the Upper Palaeozoic stratigraphy of the Trevone basin, north Cornwall, and its regional significance. *Proceedings of the Geologists' Association*, **104**, 137–148.

SELWOOD, E.B., THOMAS, J.M., WILLIAMS, B.J., CLAYTON, R.E. DURNING, B., GOODE, A.J.J., LEVERIDGE, B.E., SMITH, O. & WARR, L.N. (1998). Geology of the Country around Padstow and Camelford. *Memoir of the British Geological Survey.*

SHACKLETON, R.M., RIES, A.C. & COWARD, M.P. 1982 An interpretation of the Variscan structures in SW England. *Journal of the Geological Society of London*, **139**, 533–541.

SHAIL, R.K. 1989. Gramscatho–Mylor facies relationships; Hayle, south Cornwall. *Proceedings of the Ussher Society,* **7,** 125–130.

SHAIL, R.K. & WILKINSON, J.J. 1994. Late- to post-Variscan extensional tectonics in south Cornwall. *Proceedings of the Ussher Society,* **8,** 262–270.

SHEPHERD, T.J., MILLER, M.F., SCRIVENER, R.C. & DARBYSHIRE, D.P.F. 1985. Hydrothermal fluid evolution in relation to mineralisation in southwest England, with special reference to the Dartmoor–Bodmin area. *In: High heat production granites, hydrothermal circulation and ore genesis.* Institution of Mine and Metallurgy, London, 345–364.

SHEPPARD, S.M.F. 1977. The Cornubian batholith, SW England: D/H and $^{18}O/^{16}O$ studies of kaolinite and other alteration minerals. *Journal of the Geological Society of London*, **133**, 573–591.

SMITH, A.G., HURLEY, A.M., & BRIEDEN, J.C. 1981. *Phanerozoic Palaeocontinental World Maps* . Cambridge University Press, Cambridge.

SMITH, S.A. & HUMPHREYS, B. 1989. Lakes and alluvial sandflat-playas in the Dartmouth Group, south-west England. *Proceedings of the Ussher Society,* **7,** 112–117.

SMITH, S.A. & HUMPHREYS, B. 1991. Sedimentology and depositional setting of the Dartmouth Group, Bigbury Bay, south Devon. *Journal of the Geological Society of London,* **148,** 235–244.

SORBY, H.C. 1858. On the microscopical structure of crystals indicating the origin of minerals and rocks. *Quarterly Journal of the Geological Society of London,* **14,** 453–500.

STEELE, S.A. 1994. *The Start–Perranporth Zone; transpressional reactivation across a major basement fault in the Variscan orogen of S.W. England.* PhD Thesis, University of Durham.

STEPHENS, N. 1961. Re-examination of some Pleistocene sections in Cornwall and Devon. *Abstracts & Proceedings of the 4th Conference on the Geology and Geomorphology of South West England, Royal Geological Society of Cornwall,* 21–23.

STEPHENS, N. 1970. The West Country and southern Ireland. *In*: LEWIS, C.A. (ed.) *The glaciations of Wales and adjoining regions.* Longman, London, 267–314.

STEPHENS, N. 1973. South-west England. *In*: MITCHELL, G.F., PENNY, L.F., SHOTTON, F.W. & WEST, R.G. (eds) *A correlation of Quaternary deposits in the British Isles.* Geological Society Special Report, **4**, London.

STEPHENS, N. & SYNGE, F.M. 1966. Pleistocene shorelines. *In*: DURY, G.H. (ed.) *Essays in Geomorphology*. Heinemann, London, 1–51.

STEWART, I.J. 1981a. Late Devonian and lower Carboniferous conodont fauna from north Cornwall, and their stratigraphical significance. *Proceedings of the Ussher Society*, **5**, 179–185.

STEWART, I.J. 1981b. The Trekellend Thrust. *Proceedings of the Ussher Society*, **5**, 163–67.

STONE, M. 1966. Fold structures in the Mylor Beds, near Porthleven, Cornwall. *Geological Magazine,* **103,** 440–460.

STONE, M. 1968. A study of the Praa Sands elvan and its bearing on the origin of elvans. *Proceedings of the Ussher Society*, **2**, 37–42.

STONE, M. 1975. Structure and petrology of the Tregonning-Godolphin granite, Cornwall. *Proceedings of the Geologists' Association*, **86**, 155–170.

STONE, M. 1984. Textural evolution of lithium mica granites in the Cornubian batholith. *Proceedings of the Geologists' Association*, **95**, 29–41.

STONE, M. 1992. The Tregonning Granite: petrogenesis of Li-mica granites in the Cornubian batholith. *Mineralogical Magazine*, **56**, 141–155.

STONE, M. & EXLEY, C.S. 1986. High heat production granites of south-west England and their associated mineralisation: a review. *Transactions of the Institution of Mining and Metallurgy* (*Section B*: Applied Earth Sciences), **95**, 25–36.

STYLES, M.T. & KIRBY, G.A. 1980. New investigations of the Lizard Complex, Cornwall, England and a discussion of an ophiolite model. *In*: PANAYIOTOU, A. (ed.) *Proceedings of the International Ophiolite Symposium, Cyprus 1979*, 517–526.

STYLES, M.T. & RUNDLE, C.C. 1984. The Rb-Sr isochron age of the Kennack Gneiss and its bearing on the age of the Lizard Complex, Cornwall. *Journal of the Geological Society of London,* **141**, 15–19.

TANNER, P.W.G. 1985. Structural history of the Devonian rocks south of Liskeard, East Cornwall. *Proceedings of the Ussher Society,* **6,** 155–164.

TAYLOR, R.T. & WILSON, A.C. 1975. Notes on some igneous rocks of west Cornwall. *Proceedings of the Ussher Society,* **3,** 255–262.

The Tin Crisis 1987. SECOND REPORT OF THE HOUSE OF COMMONS TRADE AND INDUSTRY COMMITTEE, SESSION 1985–86. HMSO, Cmd 305–I.

THOMAS, A.C. 1985. *Exploration of a Drowned Landscape.* Batsford, London.

THORNTON, I., ABRAHAMS, P.W., CULBARD, E., ROTHER, J.A.P., & OLSON, B.H. 1983. The interaction between geochemical and pollutant metal sources in the environment: implications for the community. *In*: THORNTON, I. & HOWARTH, R. (eds) *Applied Geochemistry in the 1980s.* Graham & Trotman, London, 270–308.

TROUNSON, J.H. 1984. The Cornish mining industry in the 19th and 20th centuries. *Journal of the Trevithick Society,* **11**, 6–17.

TROUNSON, J.H. (eds BURT, R. & WAITE, P.) 1989. *The Cornish mineral industry 1937–1951.* University of Exeter Press, Exeter.

TROUNSON, J.H. (eds BURT, R. & WAITE, P.) 1993. *Cornwall's future mines: areas in Cornwall of mineral potential.* University of Exeter Press, Exeter.

TUCKER, M.E., 1969. Crinoidal turbidites from the Devonian of Cornwall and their palaeogeographic importance. *Sedimentology*, **13**, 218–290.

TURNER, P.J. 1982. The Anatomy of a Thrust: a study of the Greystone Thrust Complex, East Cornwall. *Proceedings, of the Ussher Society*, **5**, 270–78.

TURNER, P.J. 1984. Hercynian high-angle fault zones between Dartmoor and Bodmin Moor. *Proceedings of the Ussher Society,* **6**, 60–67.

TURNER, P.J. 1985. Stratigraphic and structural variations in the Lifton–Marystow area, West Devon, England. *Proceedings of the Geologists' Association*, **96**, 323–335.

TURNER, R.E., TAYLOR, R.T., GOODE, A.J.J. & OWENS, B. 1979. Palynological evidence for the age of the Mylor Slates, Mount Wellington, Cornwall. *Proceedings of the Ussher Society,* **4,** 274–283.

TYLER, D.J. 1988. Evidence and significance of limulid instars from trackways in the Bude Formation (Westphalian), south-west England. *Proceedings of the Ussher Society,* **7**, 77–80.

United Kingdom Minerals Yearbook 1994. 1995. British Geological Survey, Keyworth.

USSHER, W.A.E. 1890. The Devonian rocks of south Devon. *Quarterly Journal of the Geological Society of London,* **46,** 487–517.

USSHER, W.A.E. 1907. The geology of the country around Plymouth and Liskeard. *Memoir of the Geological Survey of Great Britain.*

USSHER, W.A.E., BARROW, G. & MACALISTER, D.A. 1909. The Geology of the Country around Bodmin and St Austell. *Memoir of the Geological Survey of Great Britain.*

VAN HOORN, B. 1987. The South Celtic Sea/Bristol Channel Basin; origin, deformation and inversion history. *Tectonophysics*, **137,** 309–334.

WALSH, P.T., ATKINSON, K., BOULTER, M.C. & SHAKESBY, R.A. 1987. The Oligocene and Miocene outliers of West Cornwall and their bearing on the geomorphological evolution of Oldland Britain. *Philosophical Transactions of the Royal Society of London,* **A323**, 211–245.

WARR, L.N. 1989. The structural evolution of the Davidstow Anticline, and its relationship to the Southern Culm Overfold, north Cornwall. *Proceedings of the Ussher Society*, **7**, 136–140.

WARR, L.N. 1993. Basin development and foreland basin development in the Rhenohercynian of south-west England. *In*: GAYER, R.A., GREILING, R. & VOGEL, A. (eds), *The Rhenohercynian and Sub-Variscan Fold Belts*, Earth Evolution Series, Vieweg & Sohn, 197–224.

WARR, L.N. & DURNING, B. 1990. Discussion on a reappraisal of the facing confrontation in north Cornwall: fold- or thrust-dominated tectonics? *Journal of the Geological Society of London*, **147**, 408–510.

WARR, L., PRIMMER, T.J. & ROBINSON, D. 1991. Variscan very low-grade metamorphism in southwest England: a diastathermal and thrust-related origin. *Journal of Metamorphic Geology*, **9**, 751–764.

WARR, L. & ROBINSON, D. 1991. The application of the illite crystallinity technique to geological interpretation: a case study from north Cornwall. *Proceedings of the Ussher Society*, **7**, 223–227.

WEIDNER, J.R. & MARTIN, R.F. 1987. Phase equilibria of a fluorine-rich leucogranite from the St Austell pluton, Cornwall. *Geochimica et Cosmochimica Acta*, **51**, 1591–1597.

WHITE, E.I. 1956. Preliminary note on the range of Pteraspids in Western Europe. *Bulletin of the Royal Institution of Natural Science, Belgium*, **22**, 1–10.

WHITELEY, M.J. 1981. The faunas of the Viverdon Down area, south-east Cornwall. *Proceedings of the Ussher Society*, **5**, 186–93.

WHITELEY, M.J. 1983. *The geology of the St Mellion Outlier, Cornwall, and its regional setting.* PhD Thesis, University of Exeter.

WHITELEY, M.J. 1984. Dinantian sandstones in southwest England. *European Dinantian Environments*, 1st Meeting, Abstracts, Open University, 5–11.

WHITLEY, N. 1882. The evidence of glacial action in Cornwall and Devon. *Transactions of the Royal Geological Society of Cornwall*, **10**, 132–141.

WHITTAKER, A. 1985. *Atlas of onshore sedimentary basins in England and Wales: post-Carboniferous tectonics and stratigraphy.* Blackie, London.

WILKINSON, J.J. 1990. The role of metamorphic fluid in the evolution of the Cornubian orefield–fluid inclusion evidence from South Cornwall. *Mineralogical Magazine*, **54**, 219–230.

WILKINSON, J.J. & KNIGHT, R.R. 1989. Palynological evidence from the Porthleven area, south Cornwall: implications for the Devonian stratigraphy and Hercynian structural evolution. *Journal of the Geological Society,* **146,** 739–742.

WILLIS-RICHARDS, J. & JACKSON, N.J. 1989. Evolution of the Cornubian ore field, southwest England: Part I. Batholith modeling and ore distribution. *Economic Geology*, **84**, 1078–2100.

WILLIS-RICHARDS, J. 1990. *Thermotectonics of the Cornubian batholith and their economic significance.* PhD Thesis, Camborne School of Mines.

WILLIS-RICHARDS, J. 1993. Controls on lithospheric strength, fracture opening and fluid movement in S.W England. *In*: PARNELL, J., RIFFELL, A.H. & MOLES, N.R. (eds) Extended abstracts of Geofluids'93 Conference, Torquay, England, May 4–7, 1993.

WILLOCK, A.D. 1982. An introduction of the geology of the area between Buckfastleigh and Ivybridge. *Proceedings of the Ussher Society*, **5**, 289–295.

WILSON, G. 1951. The tectonics of the Tintagel area, North Cornwall. *Quarterly Journal of the Geological Society of London*, **106**, 393–432.

WINTLE, A.G. 1981. Thermoluminescence dating of Late Devensian loesses in southern England. *Nature*, **289**, 479–480.

WITTE, G. 1983. *The structure and stratigraphy of the Launceston area, east Cornwall.* MSc thesis, University of Exeter.

ZIEGLER, P.A. 1987. Celtic Sea–Western Approaches area: an overview. *Tectonophysics,* **137**, 285–289.

ZIEGLER, P.A. 1990. *Geological atlas of western and central Europe.* Shell International, The Hague.

Subject Index

Locality Index

CPSIA information can be obtained
at www.ICGtesting.com
Printed in the USA
JSHW011826100423
40159JS00005B/32